Underground Construction Technology International Conference & Exhibition 2016 (UCT 2016)

Atlanta, Georgia, USA
3 - 4 February 2016

ISBN: 978-1-5108-2336-5

Printed from e-media with permission by:

Curran Associates, Inc.
57 Morehouse Lane
Red Hook, NY 12571

Some format issues inherent in the e-media version may also appear in this print version.

Printed by Curran Associates, Inc. (2016)

For permission requests, please contact Oildom Publishing Company of Texas, Inc.
at the address below.

Oildom Publishing Company of Texas, Inc.
PO Box 941669
Houston, Texas 77094-8669
USA

Phone: (281) 558-6930
Fax: (281) 558-7029

www.oildom.com

Additional copies of this publication are available from:

Curran Associates, Inc.
57 Morehouse Lane
Red Hook, NY 12571 USA
Phone: 845-758-0400
Fax: 845-758-2633
Email: curran@proceedings.com
Web: www.proceedings.com

TABLE OF CONTENTS

Manhole Rehabilitation: Design Engineering, Installation and Quality Assurance Testing

Norman E. "Ed" Kampbell, P.E.

Soil-Structure Interaction Systems Specialist, Rehabilitation Resource Solutions LLC, 4862 Sarasota Ct., Hilliard, OH 43026-9096, ekampbell@sbcglobal.net

Abstract

In a previous paper entitled "The Practical Design Approach to Coatings and Liners for the Rehabilitation of Cylindrical Shaped Manhole Structures" the author presented useful information on how the soil-structure interaction system impacts the lateral loading on a cylindrically-shaped vertical shaft (or manhole) and how it deviates significantly from the longitudinal vertical loading of horizontally oriented buried pipes; and with that knowledge, what loads are likely to be borne by the coatings or lining systems used in their renewal. In this paper the author presents how the engineer receives delivery on the assumptions used in his/her design through the employment of proper installation and quality assurance practices which are to be conveyed in the project specifications.

While all coatings rely on an adhesive bond with the existing wall surface of the manhole structure, liners can be either bonded or un-bonded to the wall surface of the structure. The above referenced paper on design demonstrated that the load coming onto these coatings and liners is, for all practical purposes, that of the external hydrostatic pressure generated by the groundwater surrounding the structure. In the design case of a bonded liner or coating, it is imperative that the level of adhesion accounted for in the engineer's design process be attained by the contractor's installation practices in the field. Further, because of the challenges of working in these structures; it is incumbent upon the engineer to lay out a quality assurance testing methodology that confirms that a minimum level of adhesion was achieved throughout the structure.

Un-bonded liners are, by definition, relying upon their own inherent strength to attain the stated design life of the renewal. Long-term structural stability of these un-bonded cylindrical shells is a function of the flexural modulus of the cured lining system, the level of uniformity of the cylindrical shell's shape (i.e. variations in the radii along the vertical axis), and the type of end termination detailed for the proposed shell structure at the bottom of the manhole. Additionally, the shell must be able to maintain its water-tightness at the bottom of the manhole. Flexible un-bonded shell materials must be designed to account for the linearly increasing external loading acting on the liner and the deformation that will result from this loading configuration over the designed service life of this renewal system. Seals are thus quite essential for these shells.

The installation and quality assurance practices must reflect the needs of the lining system(s) chosen by the Engineer; thereby assuring the owner that the performance expectations of the work will be achieved for the necessary service life period.

Introduction

Manhole renewal and/or rehabilitation can be accomplished using a variety of coating and lining materials; however, the success of the specified improvements comes first from properly assessing the existing conditions at the time of the proposed work and designing the recommended improvements to perform in the environment found. Examples of these are; low pH on the wall surfaces, spalling cementitious or brick and masonry wall components, presence of H

Evaluating the Existing Structures

Evaluation of the existing structures must begin with the project engineer (or engineer of record). This is because the amount and types of information needed by the proposing contractors requires much more time than would typically be made available between the advertising of the request for bidders and the opening of the bids. Discussed below is the information that must be put together by the project engineer and incorporated into the contract documents (plans and specifications).

The amount of adhesion that can be developed by a coating or liner is typically limited by the strength of the existing manhole structure's wall composition. Examples of these are brick and mortar and concrete (cast in place or precast). As the acceptable performance parameter for bonded coatings and liners is that the substrate fails first (the coating or liner brings some of the host wall with it); it is the responsibility of the project engineer to survey the tensile strength of the substrate in the manholes proposed for renewal using the methodology in ASTM D 7234 (Standard Test Method for Pull-Off Adhesion Strength of Coatings on Concrete Using Portable Pull-Off Adhesion Testers) and communicate this information on the plans and in the project specifications. Figure 1 (taken from the D 7234 standard) shows how the substrate failure is to be classified for a coating or liner.

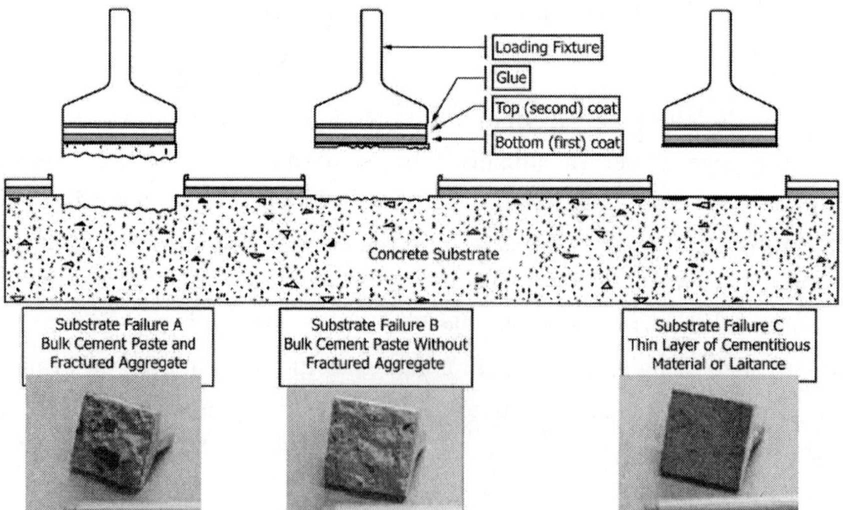

Figure 1. Substrate failure classification per ASTM D 7234.

The level of corrosion resistance needed of the coating or liner is based upon the pH values measured on the undisturbed wall surface and, if obtainable, the direct cause of those pH values. Using just the pH value alone without having an understanding of what mechanism(s) are creating them can lead to poor selection of the applicable renewal or rehabilitation products. Further, the cause of the pH values, especially low ones, could be the result of illegal discharges to the wastewater collection system that should be tracked down and dealt with accordingly.

The key design parameter for any structural renewal or rehabilitation is knowing where the phreatic surface (water table) is located relative to the flowline of the outlet pipe in the manhole. Arbitrarily assigning values when a water table is known to exist in the ground surrounding the structure is irresponsible on the part of the project engineer. When the stated design life of the proposed improvements is to be 50 years or more, it is reasonable to expect the project engineer or owner to expend the effort necessary to find out where the phreatic surface truly is located. Seasonal variations can be addressed with the factor of safety. It was quite common to install gauges in various manholes in conjunction with the determination of potential infiltration quantities.

Groundwater monitoring should always be conducted by the project engineer if no data are available. Two types of groundwater level measurement gauges are commonly used for sewer evaluation studies: the manhole gauge and the piezometer. The manhole gauge shown in Figure 2 is used to determine groundwater levels adjacent to manholes. These gauges are inexpensive and fairly easy to install. The downside to these gauges is they can become clogged from mineral deposits. This author also recommends augmenting the data from these monitoring points with information gleaned from the maintenance records (or personnel) who can give validation to the groundwater measurements.

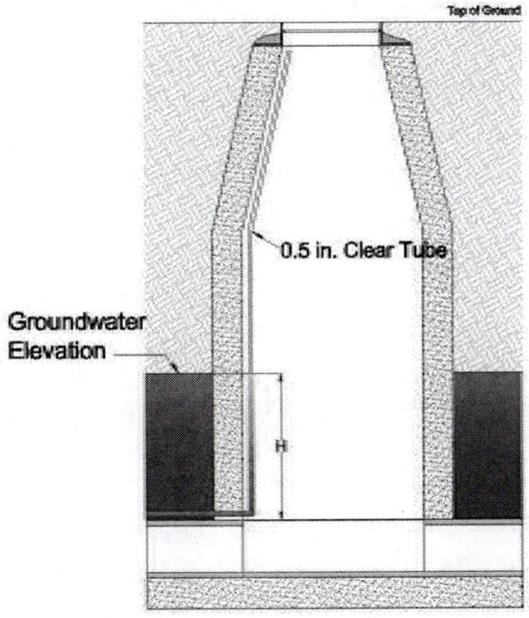

Figure 2. Manhole gauge placement showing groundwater elevation.

Design Engineering

In the earlier referenced paper on a pragmatic design approach to coating and liners installed in manhole structures, the author pointed out that the most significant load that will likely come onto the coating or liner after its installation is that of the external hydrostatic pressure caused by groundwater; specifically, the portion of the groundwater that is below the phreatic surface. In the case of adhesive-bonded coatings and liners, this load can be borne completely by the in situ adhesion with the existing manhole wall, borne partially by the adhesion with help from the lining's own structural resistance, or borne completely by the lining's structural resistance to this load (the un-bonded shell design scenario).

In order to determine the level of adhesive resistance that is available, the design engineer relies on the coating or lining manufacturer's tested and experience-established ability to bond to the various materials of the manhole wall's construction. When the stated level of adhesion of the coating or lining and the level of adhesion performance remaining in existing manhole wall structure are equal to or greater than two times the estimated groundwater pressure acting on the manhole wall externally; it can be concluded that the bonded condition is achievable. Otherwise, the lining will need to be designed as a partially bonded or un-bonded shell structure. For the readers' information, groundwater at a depth of 100 feet generates an external hydrostatic pressure of 43.3 psi; implying that an in situ adhesive resistance value in the existing manhole wall of 100 psi would be sufficient to develop a factor of safety of more than 2.0. Therefore, a sprayed-on, bonded liner's thickness will be driven by the project engineer's own experiences as to the likely size of the maximum area where no bond is achieved due to a lack of proper surface preparation or other faults by the lining installer which is discussed below. In the referenced paper on design the author demonstrated how to calculate the stress and strain on an un-bonded area estimated to be, at most, approximately 4 inches in diameter (i.e. a clamped circular plate under a uniformly distributed load). The height of the "blister", h, that would be expected to form was estimated using Equation 1 below. The maximum displacement occurs at the center of the blister where r = 0; and r_0 is equal to the radius of the un-bonded area. The maximum stress also occurs at the center of the blister; and can be estimated using Equation 2. As an example, assuming a 4-in. un-bonded circular area with 21.6 psi (50 feet of Head) external hydrostatic pressure applied continuously for 50 years, the maximum height of the blister calculates to be 0.01 inches; and the maximum stress in the liner is estimated to be approximately 674 psi (a factor of safety a little greater than 11).

$$h = \frac{pr_0^4}{64D}\left[1 - \left(\frac{r}{r_0}\right)^2\right]^2 \qquad \text{(EQN 1)}$$

$$\sigma_{max} = \frac{3(1+v)pr_0^2}{8h^2} \qquad \text{(EQN 2)}$$

$$D = \frac{Et^3}{12(1-v^2)} \qquad \text{(EQN 3)}$$

While the above solution was first presented in the context of cylindrically shaped manholes (or shafts), the reader is advised that rectangular structures needing lining where essentially the only real loading is groundwater pressure can also be designed using this maximum un-bonded area approach.

Unfortunately, the solution for a completely un-bonded, or standalone, shell is much more complicated in its analysis and is still under development by the author as of this writing. The design issues with standalone thermoplastic and thermoset resin systems is the magnitude of deformation that can occur over time under a sustained loading from the hydrostatic pressure and thus its impact on the long-term fit that is necessary to produce water tightness (if that is one of the end goals of the renewal process). That being said, water tightness is best achieved by continuing the lining onto the bench and invert thereby creating a monolithic structure within the old structure. Figure 3 below is an example of that technique.

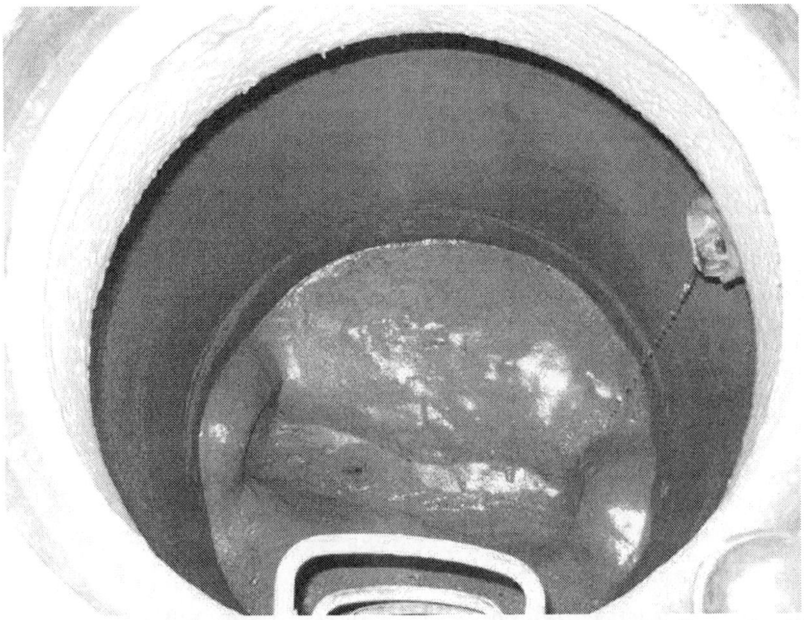

Figure 3. Example of a monolithic lining producing an essentially water-tight vessel.

Content of the Contract Documents

The contract documents must always convey the desired performance outcomes or expectations of the proposed renewal work. Far too often the efforts put into the creation of the project specifications focus only on the "legality" of the writing while completely ignoring the communication of the end goal of the owner that prompted doing the work in the first place. In fact, performance-based specifications are far better suited for renewal and/or rehabilitation projects versus the more traditionally used prescriptive-based specifications. This is due to the rate of innovation in this engineering segment and the difficulty most engineers find in keeping up to date by attending the various trade shows and technical conferences.

Employing performance-based specifications, however, still requires the project engineer to pre-qualify which alternatives he/she believe to be technically sound and appropriate for the owner's end goals when considering the site specific environmental conditions (both internal and external) present in the structure(s) to be renewed and/or rehabilitated. There are four categories to these materials; cementitious mortar lining, epoxy resin system lining, polyurethane resin system lining, and specialty lining systems. The specialty lining system category currently contains two systems; the multi-component stress skin panel liner system and the PVC bonded composite liner system.

At the heart of performance-based specifications are the performance work statement (PWS) submittal and the quality assurance testing that will be performed to assure the owner and the engineer that a proper installation was achieved. The PWS includes, among other things, information on the product or products that will be installed, confirmation of the required thicknesses of the product in each structure and how those thicknesses were arrived at, the key installation steps that will be followed by the installation personnel (i.e. surface preparation, humidity controls ...), and an outline of the defects that may occur and the manufacturer's recommended remediation for them.

Good Installation Practices

The use of spray-applied coatings and liners requires adhering to the system manufacturer's stated requirements with respect to the surface profile, the amount of material that can be applied per pass and the maximum time that can elapse between subsequent passes; the latter typically referred to as the recoat window. Ensuring that a uniform coating or liner develops a monolithic membrane or wall structure is a key requirement of the system's subsequent long-term performance.

Prior to the placement of any spray-applied coating or liner all active leaks must be stopped and the interior surfaces of the manhole structure must be cleaned and prepared. The steps that should be addressed in the PWS are as follows:

- Surfaces to be coated shall be cleaned sufficiently to provide a sound, clean, and pH neutralized surface suitable for the coating or lining material to be applied.
    - Excessive debris, sediment, root intrusion or other foreign materials which may impact the effectiveness of the surface preparation process shall be removed prior to the commencement thereof.
    - Offset structural components, lids, covers, frames, etc. shall be repaired, replaced, or reset prior to the commencement of surface preparation.
    - External soil voids shall be remediated and/or stabilized by replacement or injection of a stabilizing grout
    - Oils, grease, incompatible existing coatings, efflorescence, or other contaminants which may affect the performance and adhesion of the coating to the substrate shall be removed in accordance with SSPC-SP 1 Solvent Cleaning (published standard of the Society of Protective Coatings) or other suitable method.
    - Surface preparation method, or combination of methods, that may be used include high-pressure water cleaning, abrasive blasting, detergent water cleaning, steam or hot water cleaning, and others referenced in industry accepted standards such as:
        - NACE No. 6/SSPC SP-13 Surface Preparation of Concrete
        - ASTM D-4258 Standard Practice for Surface Cleaning Concrete for Coating and ASTM D-4529 Standard Practice for Abrading Concrete,
        - ICRI Technical Guideline No. 03732 Selecting and Specifying Concrete Surface Preparation for Sealers, Coatings, and Polymeric Overlays...
            - For application of cementitious materials; at least a CSP-2

- For application of thermoset materials (i.e. epoxy); CSP-4 to CSP-6 for thicknesses less than 125 mils and CSP-7 or greater for lining thicknesses of 125 mils or more.

  - All loose or defective brick, grout, ledges, steps, and protruding ledges shall be removed to provide an even surface prior to the application of the coating or lining material.

Engineered Cementitious Liners (ECL) are the new generation of cementitious lining materials that through their proprietary mix designs deliver application-specific formulations of fiber-reinforced cement-based mortar systems engineered for the environmental conditions normally found in sewers. The most successful of these type systems are applied to the prepared wall surface using a high velocity centrifugal casting in place process incorporating a bi-directional spinning head that literally is flinging the mortar system in place. Because this process uses centrifugal energy to propel the mortar and not air, the resulting finished liner is a densely packed cementitious mortar lining. Figure 4 below illustrates the spin-casting process.

Figure 4. Centrifugal casting a cementitious manhole lining.

The alternative to the spin-casting method is by spraying the cementitious mortar onto the wall surface and then using a trowel to densify and smooth the lining on the wall. In this lining methodology it is generally recommended to hold the nozzle no more than 8 inches from the wall with the nozzle man directing the movement manually to ensure that the maximum thickness per pass is not exceeded. That being said, any troweling must be minimized to avoid compromising the quality of the finished lining.

Polymeric coatings and linings consist of epoxy resin systems, fiber-reinforced epoxy resin systems, and polyurethane resin systems. The formulations of these resin systems must be compatible with the environmental conditions during their application and throughout their proposed service life. The two-part polymeric systems consist of a resin and a hardener which must be combined precisely to the ratios stated by the manufacturer. Alterations to this ratio in the field will affect the performance of the finished coating or lining significantly. This means that "cutting back" on the amount of hardener to be used because the resin system's components were stored at too high a temperature is not an allowable option for the installer.

Structures renewed using epoxy resin systems may be returned to full operational service after they have adequately cured, generally 4 to 6 hours in manholes; however, this time might need to be lengthened if the surface(s) to which they are being applied are at too cool a temperature (i.e. below 55°F). In these instances, the resin system manufacturer will have given the installer prior written instruction (which was to be included in the PWS) as to the length of time required at this lower temperature (i.e. set time is given as 2 hours at 77°F, but 8 hours at 40°F). Also, because epoxy resin systems are thermosetting materials (give off heat when curing), the curing time is inversely proportional to the thickness; the thicker the applied material thickness, the greater amount of heat that is generated, which in turn produces a shorter curing time.

Polyurethane resin systems typically cure in just a few minutes which can allow the structure to be returned to service much more quickly. The downside to this rapid curing inside buried structures is the very short time given to the resin to wet out the surface to which it has been applied. It is for this reason that field adhesion test results are usually much lower than those achieved with epoxy resin systems; 200 – 250 psi versus 600+. However, as one can see from the design discussion earlier this shouldn't be considered a problem as the external hydrostatic design loading is far less than this value. That being said, the leading supplier of polyurethane lining materials has chosen to design their resultant liners as if no bonding was achieved throughout the whole structure (the un-bonded standalone shell design condition).

Performance Verification

Long-term performance of a coating or liner must be verified through in situ testing to assure the owner that the installation followed the correct steps in its application. This quality assurance testing is a must to verify the thickness of liner that was installed, confirm that the level of adhesion achieved exceeds the minimum required by the design, that the adhesion between coats achieved the needed monolithic performance and that there are no holidays in the coating or lining that would compromise the corrosion barrier performance of the coating or liner.

For polymeric materials the coating thickness and various adhesion performance parameters may be obtained by performing adhesion pull-off testing per the methodology given in ASTM D-7234 (discussed earlier). Ideally each manhole lined would be tested in at least three locations to verify the installed thickness and the adhesion performance achieved. Figure 5 below graphically illustrates the additional failure modes that can be gleaned from the adhesion pull tests (Figure 1 and Figure 5 go together for the total coding).

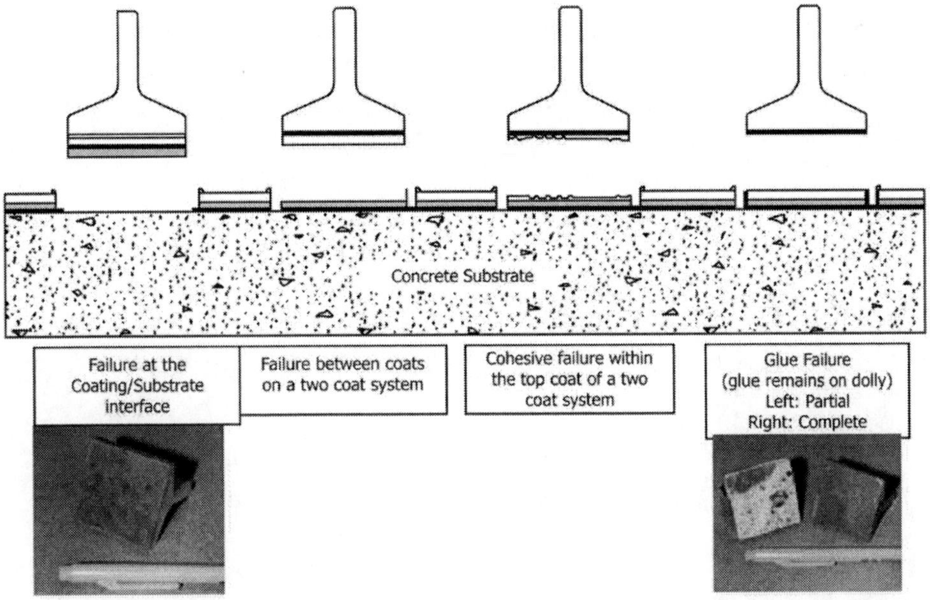

Figure 5. Pictures of the various levels of the surface profile required for adhesion

Conclusions / Recommendations

There is no doubt that the caliber of the innovation in cementitious and polymeric coating and lining materials provides the world's infrastructure with some very good, cost-effective tools to renew the pipe access structures such as shafts, manholes, and junction box structures. While a wall thickness design for each and every such structure is not warranted for the majority of the renewal and rehabilitation projects taken on by the owners of these stormwater and wastewater systems, there are design steps that the project engineer should be following in order to make sure that the proper renewal systems are being chosen for the site specific conditions found on the proposed project. Awareness of what they are finding beyond the obvious as the conduct their survey for putting together the plans and specifications is the hallmark of the engineering profession; such as, when the pH is abnormally low they should be taking the time to ascertain the cause of this condition. Also, they should consider the impact that this finding carries over into the lining of the interconnecting pipes to and from these structures. A standard polyester resin system CIPP is not suitable in a system where the pH is 1.0!

Modest thicknesses of coatings and linings installed in keeping with the advice given herein. Presented below is a table from one of the author's earlier papers (labeled Table 2 herein) on this subject which demonstrates the required minimum wall thicknesses for bonded coatings and liners in manholes.

Table 2. Required minimum wall thicknesses for bonded coatings and linings in manholes

	pH > 7.0	3 < pH ≤ 7	2 < pH < 3	pH ≤ 2	0<H<5'	5'<H<15'	15'<H<30'
CML – Type I	Yes	Yes			0.5"	0.75"	1.0"
CACML	Yes	Yes	Yes		0.5"	0.75"	1.0"
CACML + Epoxy	Yes	Yes	Yes	Yes	0.500/0.100	0.750/0.100	1.000/0.100
Epoxy only	Yes	Yes	Yes	Yes	0.100-0.125	0.100-0.125	0.125-0.150
Polyurethane	Yes	Yes	Yes	Yes	0.100-0.125	0.100-0.125	0.125-0.150

And finally, careful consideration should be given before mixing different material systems together to make sure the details of their connection with one another are sufficient enough to ensure long-term performance.

NASSCO's Inspector Training & Certification Program For CIPP Update

Lynn E. Osborn, P.E.

NASSCO Technical Director

February 3, 2016

ITCP History

- ITCP for CIPP developed in 2007
- First training session 2008
- Over 1600 inspectors trained to date
- 1.5 day course
- Certification requires passing an open book exam

ITCP CIPP Manual

NASSCO's Inspector Training & Certification Program

Introduction

- The purpose is to ensure that a high quality CIPP product is installed
- The intent is provide a comprehensive understanding of CIPP
- Also, how to implement the QA/QC program outlined in the specifications
- Key: Performance specifications and educated inspectors
- Periodic updates have been made to course materials
- However, after 9 years, time for a major overhaul
- Numerous modifications were made to the text, photos and so on
- Several significant changes were made as follows

NASSCO's Inspector Training & Certification Program

Terminology

- Tube – The dry felt or reinforced tubular product prior to the introduction of resin. Includes any thermoplastic coatings.
- Liner – The wet out tubular product prior to curing.
- CIPP – The final product once cured.

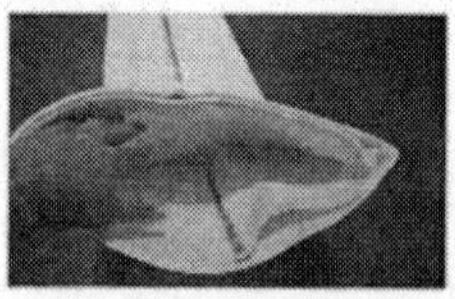

Tube

Liner

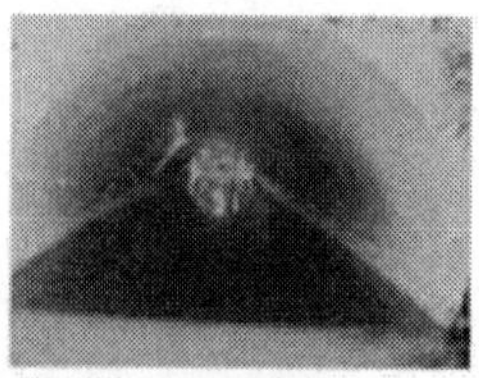

CIPP

NASSCO's Inspector Training & Certification Program

Air Inversion/Steam Cure

- Air inversion/steam cure (AISC) has become installation process of choice for small sewers for many CIPP installers
- Not widely used in 2007
- AISC is more technically challenging than water invert, water cure
- More information added to the ITCP CIPP manual

Air Inversion Unit

Steam Cure

NASSCO's Inspector Training & Certification Program

UV Cure Process

- Liner is pulled-in and inflated with air
- Light train is pulled through the liner, curing the resin
- 10 years ago first UV cure process introduced in the US
- Usage steadily increasing
- ITCP CIPP class attendees are seeking more information
- More descriptive information added to the manual

UV Cure Light Train

Light Cure Liner

Continuous Temperature Monitoring

- Curing the resin is one of the most important processes in CIPP installation
- Cured resin is the structure of the CIPP
- Time tested method for monitoring cure – Thermocouple sensors
- New technology – Fiber optic cable for essentially continuous monitoring

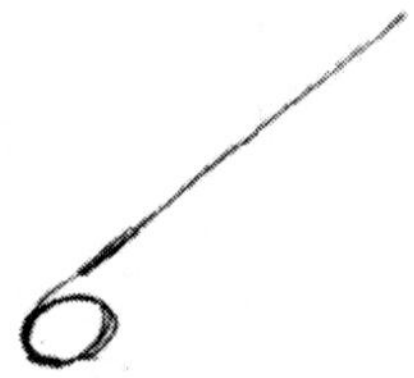

Thermocouple Sensor

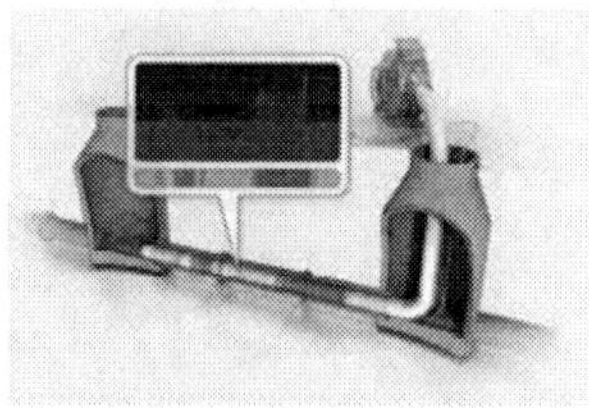

Continuous Temperature Graph

Styrene Rulings

- Styrene is a major component of many resins for CIPP
- It functions well and has a relatively low cost
- However, it also has a very pungent odor
- Evaluated for decades for regarding its effects on humans
- Studies in North America and Europe have found either:
  - No carcinogenic potential, or
  - So low that it is unlikely to present any carcinogenic hazard
- In 2011, the 12th Report on Carcinogens published by HHS listed styrene as a "reasonably anticipated carcinogen"
- Report strongly opposed.
  - Legal actions
  - Follow-up review by National Academy of Sciences
- The ruling was upheld
- So far impact on CIPP industry has been minor
  - Labeling requirements
  - Updating Safety Data Sheets (SDS)

NASSCO's Inspector Training & Certification Program

Performance Specifications

- NASSCO recommends that CIPP specifications be structured as performance specifications
- Appendix B of the ITCP CIPP Manual contains the NASSCO "Performance Specification Guideline for the Installation of CIPP"
- This Specification Guideline is updated periodically
- The most current version will be a part of the updated ITCP CIPP Program.

PERFORMANCE SPECIFICATION GUIDELINE

FOR THE INSTALLATION OF

CURED-IN-PLACE PIPE (CIPP)

NASSCO's Inspector Training & Certification Program

ASTM Standard References

- Three ASTM Standard Practices cover the installation of CIPP
  - ASTM F 1216 – Inversion Process
  - ASTM F 1743 – Pull-in Process
  - ASTM F 2019 – Pull-in Glass Reinforced Process
- Appendix F of the ITCP CIPP Manual contains a chart that compares these 3 Standards on several subjects:
  - Resin impregnation
  - Installation method
  - Curing techniques
  - Etc.
- This comparison table was reviewed and changes were made to reflect any modifications to the three ASTM Standards

Descriptive Concepts

- Chapter 2 of the ITCP CIPP Manual is an overview of the CIPP technology
- While teaching this chapter, instructors discuss different CIPP concepts with the class
  - Inspector's role
  - Goals of pipeline rehabilitation
  - Troubleshooting
  - Etc.
- To make the revised manual more useful as a reference document, several of these concepts have been added.

NASSCO's Inspector Training & Certification Program

Wet Out Checklist

- Course attendees are encouraged to visit a wet out facility
- During a CIPP project, the inspector may visit the wet out facility to perform several quality checks
  - Method of tube shipment
  - Method of tube storage
  - Condition of the dry tubes
  - Type of resin used
  - Amount of resin used
- A Wet Out Checklist has been added to the manual to help with the organization of the inspector's wet out visit

Factory Wet Out Process

NASSCO's Inspector Training & Certification Program

Other Pipe Renewal Technologies

- Chapter 5 of the manual describes several other commonly used technologies
- This chapter was reviewed, and technologies that emerged since 2007 were added

NASSCO's Inspector Training & Certification Program

Defects

- Inspectors should be knowledgeable in recognizing different defects
- Defect categories:
  - Minor defects that can be left in place; no action required
  - Defects that need to be repaired to be acceptable
  - Defects that cannot be repaired. Remove & replace.
- Which category defects fall into should be spelled out in the specifications
- The defects in the original ITCP CIPP manual were reviewed and compared to current practices

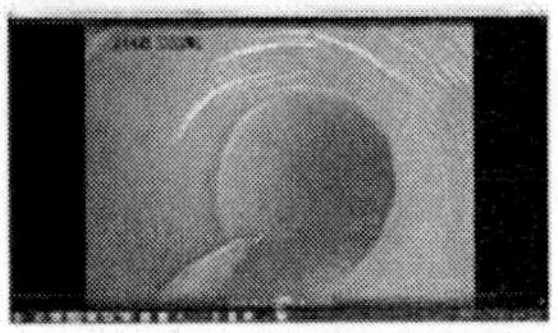

Fin at 7:00 O'clock

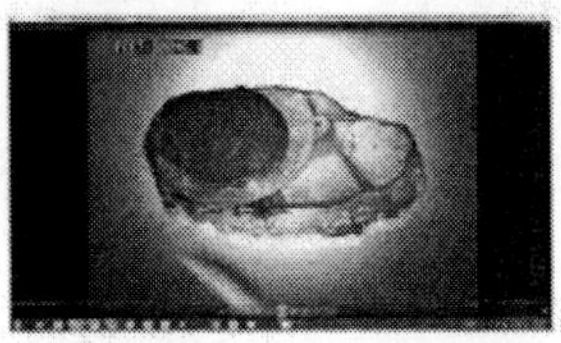

Overcut Lateral Connection

NASSCO's Inspector Training & Certification Program

CIPP for Pressure Pipe

- Inspecting CIPP for pressure pipe involves added responsibilities
- Achieving a high quality CIPP is critical in pressure pipe projects
- Several pressure pipe rehabilitation concepts are discussed in the manual:
  - Host pipe cleaning
  - Shipping and storing tubes
  - Tube wet out
  - Hydrostatic pressure test
  - Physical testing
- Most of the same inspection concepts that apply to gravity sewers also apply to pressure pipes
- However, when installing CIPP for pressure pipes, there are several key inspection areas that should be closely monitored

NASSCO's Inspector Training & Certification Program

Other Actions

- Updated course presentation materials
- Updated course certification test
- For those inspectors already certified for CIPP:
  - Online recertification module is being developed
  - Recertification will be required for prior certifications 3 years or older
  - Recertification will be available for any prior certification

NASSCO's Inspector Training & Certification Program

Questions?

THE TOTAL SOURCE

educational sessions

PennDOT Interstate 84 Culvert Rehabs

Dingmans Ferry, Pennsylvania

Existing 72" dia, 5% ellipsed CSP – Bituminous Coated with Paved Invert built in 1968

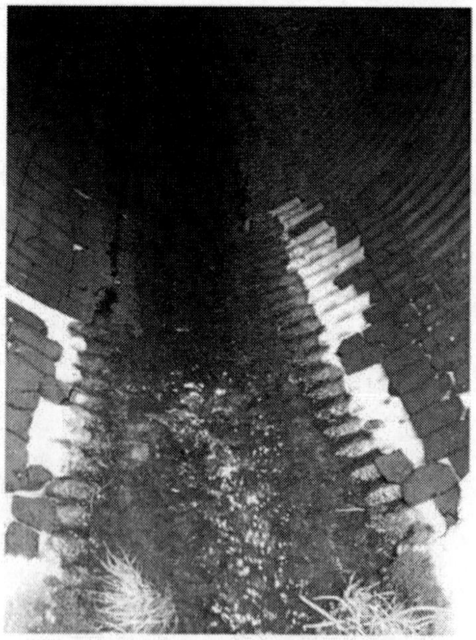

Existing 72" dia, 5% ellipsed CSP – Bituminous Coated with Paved Invert

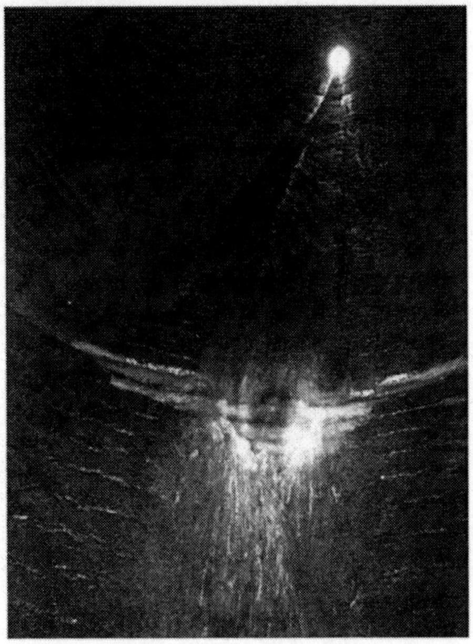

Existing 72" dia, 5% ellipsed CSP – Bituminous Coated with Paved Invert

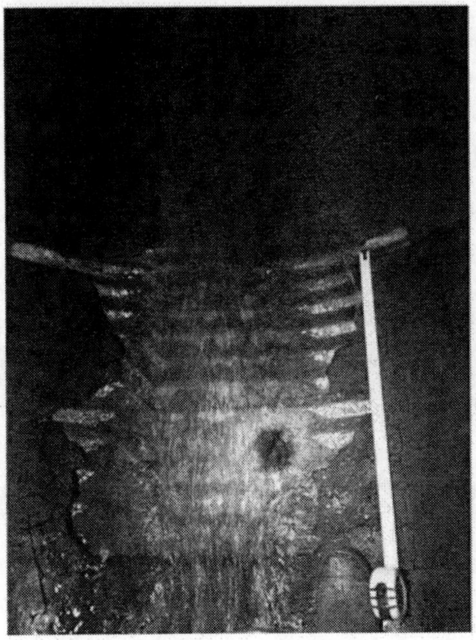

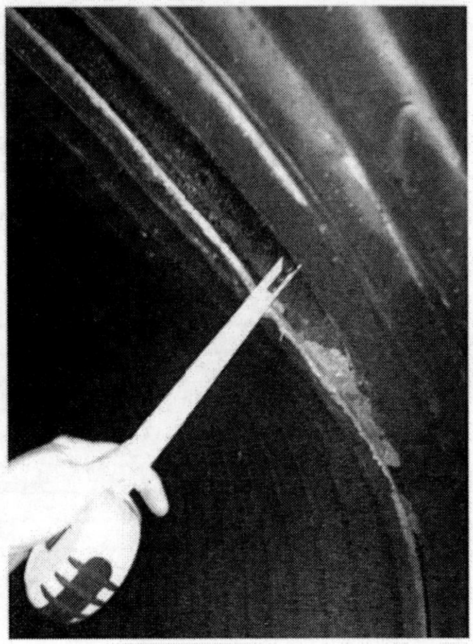

Rehab the 45 year old culverts now with federal money

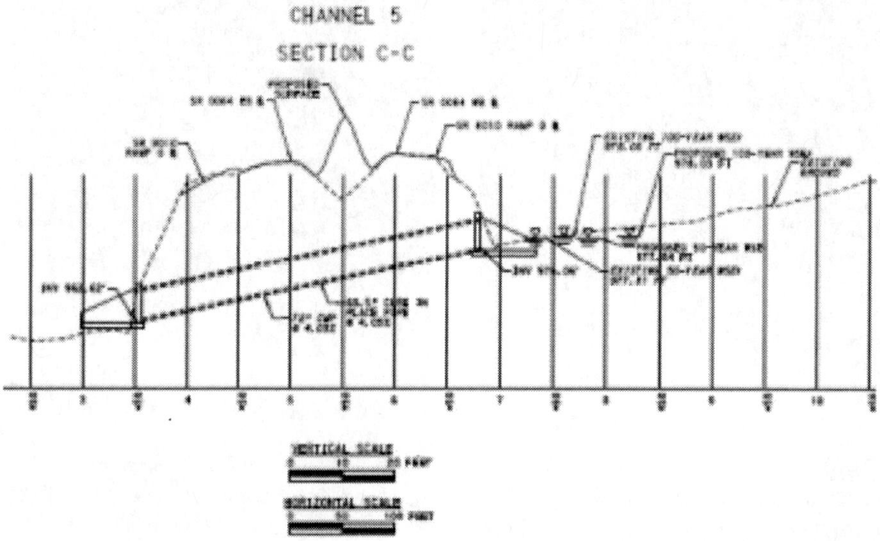

PennDot SR 84 Culvert Rehabs

HY-8 Analysis Results

Culvert Summary Table - Channel 5

Culvert Crossing: Sta 1458+25 Existing 100-Year

Total Discharge (cfs)	Culvert Discharge (cfs)	Headwater Elevation (ft)	Inlet Control Depth(ft)	Outlet Control Depth(ft)	Flow Type	Normal Depth (ft)	Critical Depth (ft)	Outlet Depth (ft)	Tailwater Depth (ft)	Outlet Velocity (ft/s)	Tailwater Velocity (ft/s)
0.00	0.00	975.91	0.00	0.0*	0-NF	0.00	0.00	0.00	0.00	0.00	0.00
30.00	30.00	976.63	0.72	0.0*	1-S2n	1.00	1.42	1.01	0.53	9.37	4.05
60.00	60.00	977.22	1.31	0.0*	1-S2n	1.43	2.04	1.44	0.74	11.36	4.88
90.00	90.00	977.87	1.96	0.0*	1-S2n	1.79	2.54	1.79	0.89	12.67	5.42
98.92	98.92	978.06	2.15	0.0*	1-S2n	1.88	2.66	1.88	0.92	13.03	5.56
150.00	150.00	979.19	3.28	0.0*	1-S2n	2.35	3.31	2.35	1.11	14.57	6.19
180.00	180.00	979.80	3.89	0.0*	1-S2n	2.59	3.65	2.59	1.21	15.36	6.48
210.00	210.00	980.37	4.46	0.0*	1-S2n	2.83	3.95	2.84	1.29	15.92	6.74
240.00	240.00	980.92	5.01	0.0*	1-S2n	3.07	4.24	3.07	1.37	16.48	6.98
270.00	270.00	981.46	5.55	0.0*	1-S2n	3.29	4.48	3.30	1.44	16.95	7.19
300.00	300.00	982.01	6.10	0.0*	5-S2n	3.52	4.72	3.52	1.51	17.38	7.38

HY-8 Analysis Results

Culvert Summary Table - Channel 5

Culvert Crossing: Sta 1458+25 Proposed 100-Year

Total Discharge (cfs)	Culvert Discharge (cfs)	Headwater Elevation (ft)	Inlet Control Depth(ft)	Outlet Control Depth(ft)	Flow Type	Normal Depth (ft)	Critical Depth (ft)	Outlet Depth (ft)	Tailwater Depth (ft)	Outlet Velocity (ft/s)	Tailwater Velocity (ft/s)
0.00	0.00	976.06	0.00	0.0*	0-NF	0.00	0.00	0.00	0.00	0.00	0.00
30.00	30.00	976.78	0.72	0.0*	1-S2n	0.63	1.44	0.64	0.53	18.77	4.05
60.00	60.00	977.35	1.29	0.0*	1-S2n	0.87	2.08	0.93	0.74	21.41	4.88
90.00	90.00	977.88	1.82	0.0*	1-S2n	1.11	2.57	1.19	0.89	23.39	5.42
98.92	98.92	978.03	1.97	0.0*	1-S2n	1.16	2.71	1.26	0.92	23.79	5.56
150.00	150.00	979.05	2.99	0.0*	1-S2n	1.41	3.37	1.59	1.11	25.63	6.19
180.00	180.00	979.67	3.61	0.0*	1-S2n	1.56	3.70	1.78	1.21	26.43	6.48
210.00	210.00	980.28	4.22	0.0*	1-S2n	1.71	4.02	1.95	1.29	27.10	6.74
240.00	240.00	980.88	4.82	0.0*	1-S2n	1.82	4.28	2.12	1.37	27.71	6.98
270.00	270.00	981.49	5.43	0.0*	1-S2n	1.93	4.54	2.28	1.44	28.23	7.19
300.00	300.00	982.11	6.05	0.0*	5-S2n	2.04	4.73	2.44	1.51	28.73	7.38

CHANNEL 2 AND 5

Existing and Proposed Conditions

In existing conditions, Channel 5 is conveyed to Channel 2 and Wetland B through a 72" Corrugated Metal Pipe (CMP). Channels 2 and 5 are Unnamed Tributaries to Sloat Brook, which is classified as an Exceptional Value (EV) and Migratory Fishes (MF) stream in Chapter 93. It is anticipated that the 72" CMP will be slip-lined as part of the reconstruction project. The slip-lining will be a cured-in-place lining that has a Manning's n coefficient of 0.009. The inside diameter of the lining will be 68.5". Capacity will be maintained by the lower Manning's n coefficient of the slip-liner. Use of the CIPP lining will maintain the existing alignment, elevations, and grades of the existing crossings and thereby would be able to accommodate the existing opportunity for aquatic movements through the crossings.

This pipe was installed with the original construction of the interstate alignment, it has aged and a slip-lining will lengthen the service life of the pipe. The increased velocity and the slight increase in the headwater for the 50-year event are negligible and will not adversely impact the surrounding area.

PennDot SR 84 Culvert Rehabs

The project specification required CIPP but allowed any reline material that met FHWA Pub CFL/TD-05-003

16. Abstract

Trenchless technology can be used with a wide range of methods, materials, and equipment for rehabilitation or replacement of damaged or deteriorated pipe culverts. To date, several trenchless lining techniques have been used for the rehabilitation of existing culverts. Colorado State University was contracted to establish guidelines and specifications for the use of culvert liners. Task 1 involved a thorough literature review to collect information describing the state-of-the-practice in culvert pipe lining. Task 2 involved the development of a decision-making methodology for choosing appropriate culvert liners based on various factors. Task 3 required CSU to prepare and submit a final report whereby providing guidance to both the Design and Construction Branches of the FHWA and to State DOTs.

PennDot SR 84 Culvert Rehabs

Issues in the general permit were discovered including:

- misinterpretation of the HY-8 hydraulic analysis results regarding flow characteristics (inlet control vs outlet control)
- statements by the DEP biologist were in error
- the full impact of high outlet velocities was overlooked
- use of CIPP would likely not provide a 100 year service life in this environment (abrasion level 4, freeze/thaw zone)

PennDot SR 84 Culvert Rehabs

It was challenging to find a CIPP subcontractor who could line 72" CMP under an Interstate in NE Pennsylvania.

For both 72" culverts, Contech proposed 66" round and ellipsed Aluminized Steel Type 2 CMP that met PennDOT's own Design Manual requirements for 100 year service life.

PennDot SR 84 Culvert Rehabs

Aluminized Steel Type 2 CMP typically lasts 3+ times longer than galvanized steel CMP in the same environment.

In ground prototype installations date to 1952 and many DOT's assign it a service life of 75 years or more, including PennDOT at 100 years for 12 gage AL-T2.

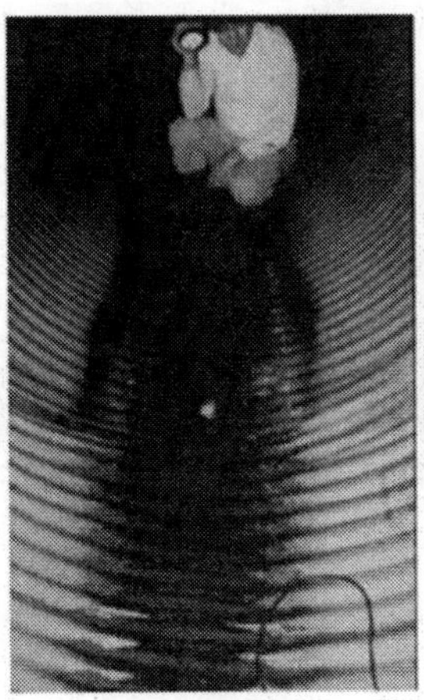

Appendix B – Fill Height Tables for Pipes and Pipe Arches

Publication 13M (DM-2)
Change #1 – 2/06

TABLE 10.B.6 (ENGLISH)
ALLOWABLE FILL HEIGHTS FOR
CORRUGATED ALUMINIZED STEEL PIPES
(2 2/3 in × 1/2 in CORRUGATIONS)

DESIGN LIFE - 100 YR								
PIPE DIAMETER (in)	HEIGHTS OF FILL (ft)							
	16 Gage		14 Gage		12 Gage		10 Gage	
	MIN	MAX	MIN	MAX	MIN	MAX	MIN	MAX
21	—	—	—	—	1.5	23	—	—
24	—	—	—	—	1.5	20	—	—
30	—	—	—	—	1.5	16	—	—
36	—	—	—	—	2.0	14	1.5	58
42	—	—	—	—	2.0	12	2.0	50
48	—	—	—	—	2.5	10	2.0	44
54	—	—	—	—	3.0	9	2.0	39
60	—	—	—	—	3.5	8	2.0	35
66	—	—	—	—	—	—	2.5	32
72	—	—	—	—	—	—	2.5	29

HY-8 Analysis Results

Culvert Summary Table - Culvert 1

Culvert Crossing: CH5 100 year -proposed

Total Discharge (cfs)	Culvert Discharge (cfs)	Headwater Elevation (ft)	Inlet Control Depth(ft)	Outlet Control Depth(ft)	Flow Type	Normal Depth (ft)	Critical Depth (ft)	Outlet Depth (ft)	Tailwater Depth (ft)	Outlet Velocity (ft/s)	Tailwater Velocity (ft/s)
0.00	0.00	976.14	0.00	0.0*	0-NF	0.00	0.00	0.00	0.00	0.00	0.00
30.00	30.00	977.07	0.93	0.0*	1-S2n	0.97	1.46	0.98	0.48	10.33	4.70
60.00	60.00	977.62	1.48	0.0*	1-S2n	1.38	2.11	1.40	0.67	12.51	5.66
90.00	90.00	978.12	1.98	0.0*	1-S2n	1.72	2.61	1.73	0.81	14.09	6.30
98.92	98.92	978.32	2.18	0.0*	1-S2n	1.81	2.75	1.81	0.84	14.50	6.45
150.00	150.00	979.36	3.22	0.0*	1-S2n	2.27	3.41	2.28	1.02	16.13	7.19
180.00	180.00	979.92	3.78	0.0*	1-S2n	2.51	3.75	2.52	1.10	16.98	7.53
210.00	210.00	980.44	4.30	0.0*	1-S2n	2.75	4.04	2.76	1.18	17.63	7.84
240.00	240.00	980.95	4.81	0.0*	1-S2n	2.98	4.32	2.99	1.25	18.21	8.11
270.00	270.00	981.44	5.30	0.0*	1-S2n	3.21	4.54	3.22	1.32	18.72	8.36
300.00	300.00	981.92	5.78	0.0*	5-S2n	3.45	4.73	3.45	1.38	19.12	8.59

PennDot SR 84 Culvert Rehabs

Field measurements were taken every 10'

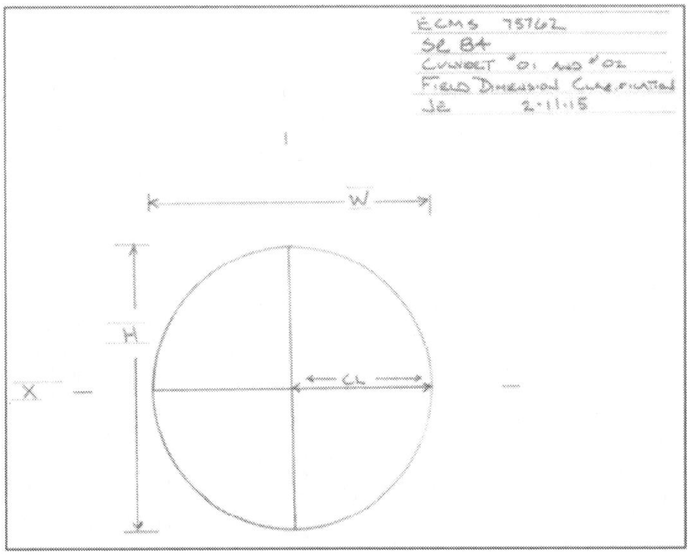

PennDot SR 84 Culvert Rehabs

Pipe dimensions were verified, sizes adjusted.

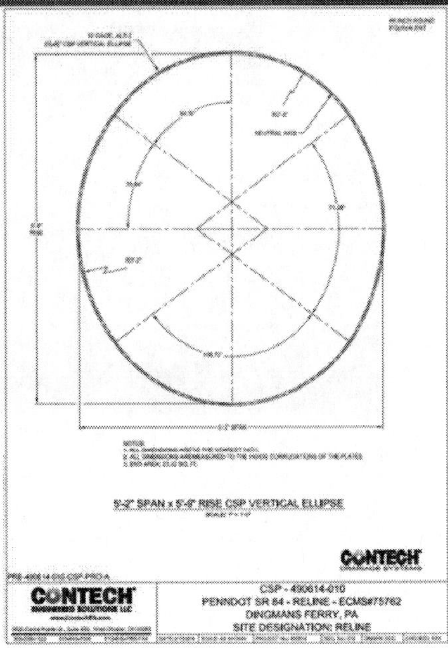

PennDot SR 84 Culvert Rehabs

Shop drawings were submitted and approved, then manufacturing commenced.

.

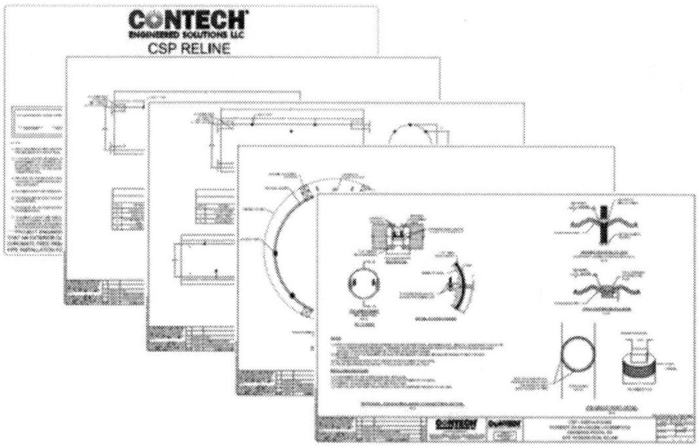

PennDot SR 84 Culvert Rehabs

PennDot SR 84 Culvert Rehabs

Culvert #2

PennDot SR 84 Culvert Rehabs

Sliplining sequence

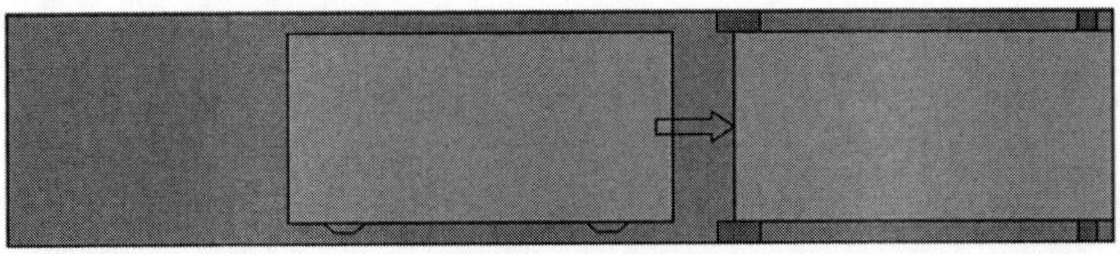

Profile View

PennDot SR 84 Culvert Rehabs

Installation sequence

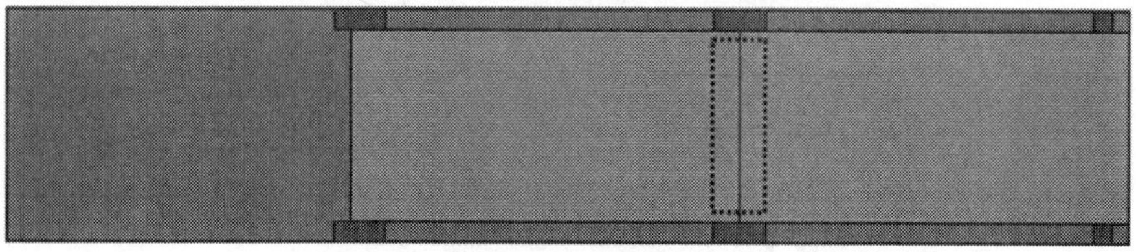

Profile View

PennDot SR 84 Culvert Rehabs

Installation sequence

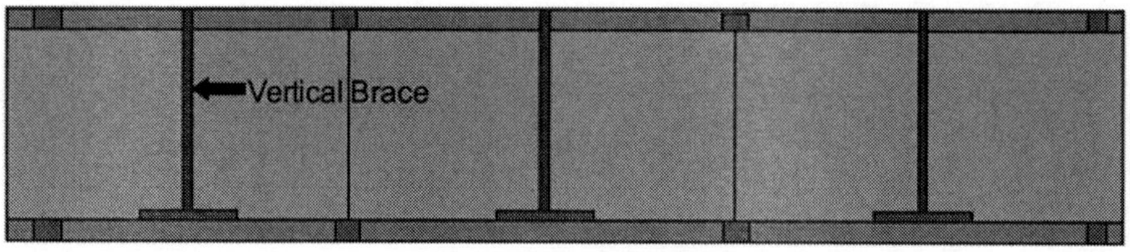

Profile View

PennDot SR 84 Culvert Rehabs

Grouting strategy was fine tuned with Mar-Allen, Contech and JDM

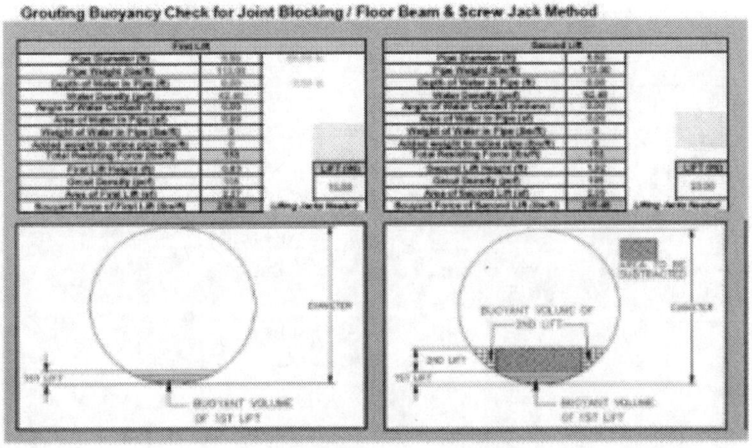

PennDot SR 84 Culvert Rehabs

Culvert #2 – Post Grouting before End Treatment

PennDot SR 84 Culvert Rehabs

Culvert #2 – Headwall/Wingwall Units were Precast

PennDot SR 84 Culvert Rehabs

Culvert #2 – Completed Structure

PennDot SR 84 Culvert Rehabs

Questions?

THE TOTAL SOURCE

educational sessions

A sink hole, a 24" sewer and its 90 feet in the ground

The Kingston Tunnel UV Cured Fiberglass CIPP installation

The Tannery Brook – the law of the unattended consequences – the Cause of the Sink Hole

The Tannery Brook from Above

The Sink Hole Spring 2012

The Washington St Tunnel from Lucas to Elizabeth St – 4,500 LF only 3 manholes

The Slip Line Plan

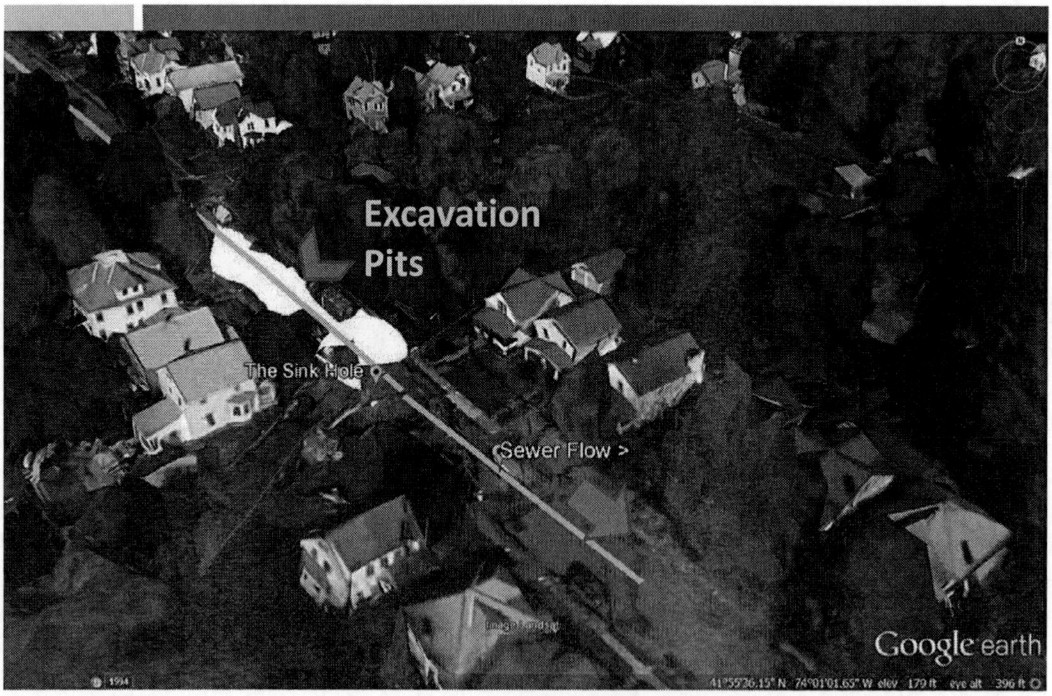

How the debris was entering system – Broken VCP 187 LF of it

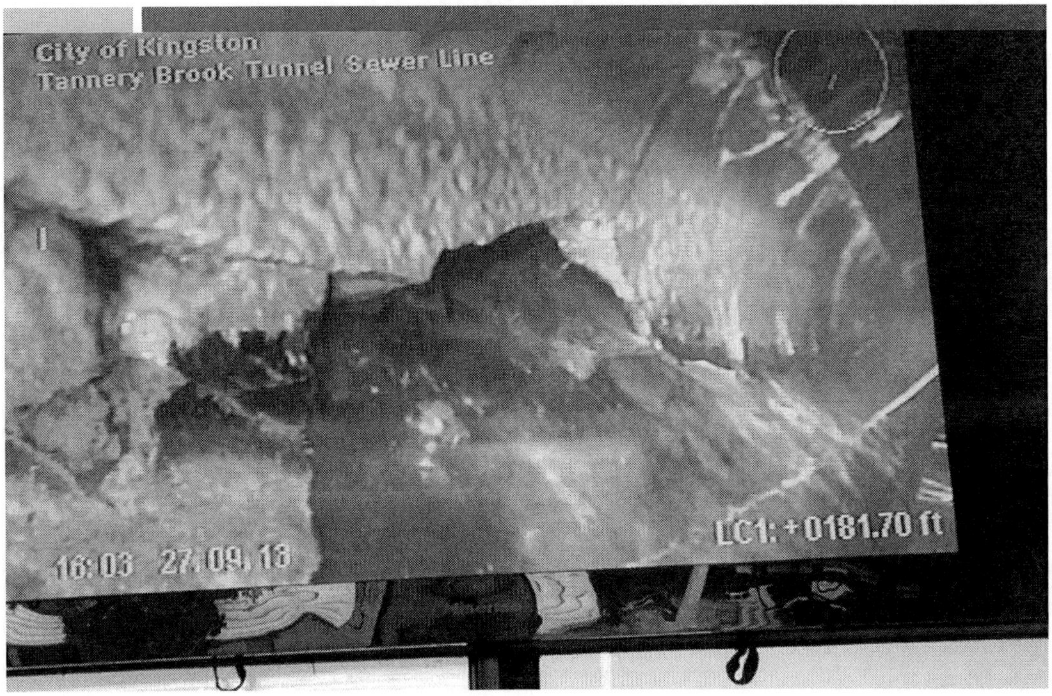

One way in – basically only 1 way out – 90 LF straight down - All equipment and men through this one man hole

Your Choices once you reached bottom

24" pipe to the Left

48" link pipe to the right nearly always with 2 feet of water

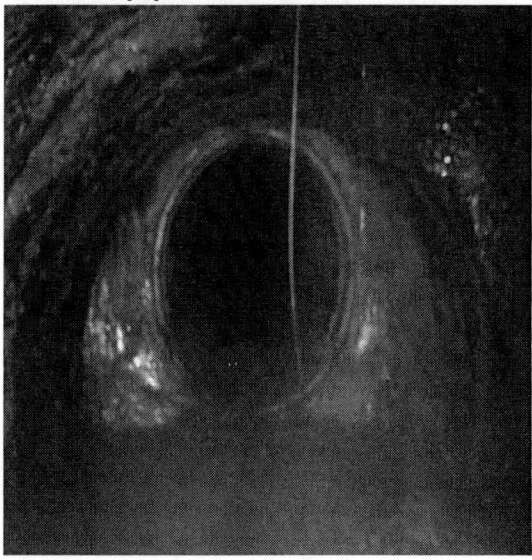

Custom Equipment – all designed and made onsite and MODULAR

The Upstream End and the Head Wall

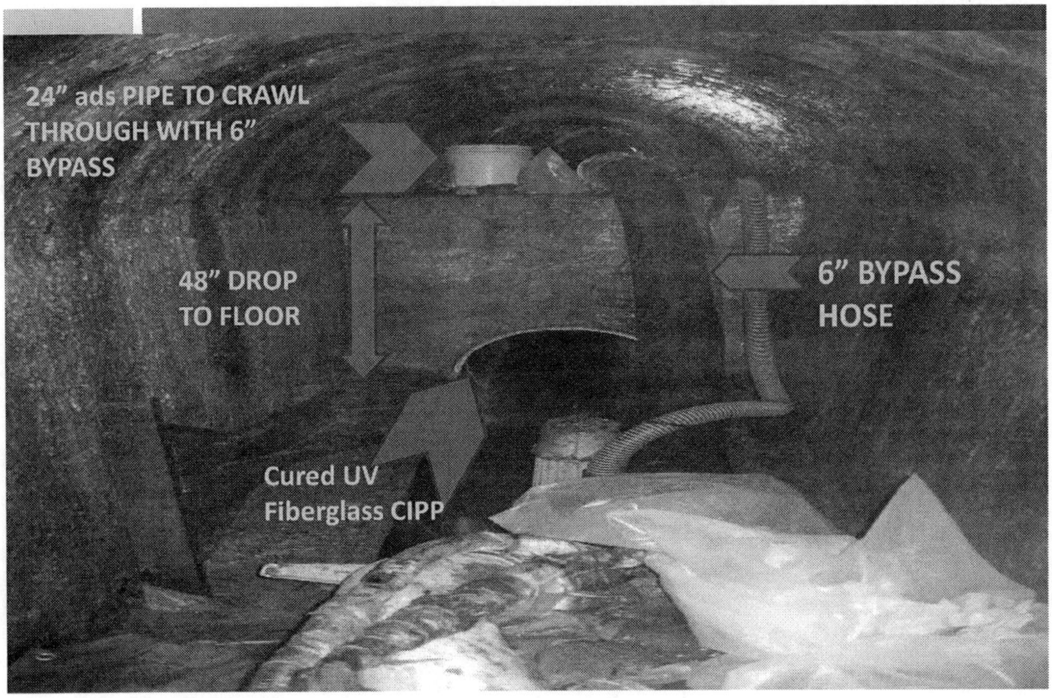

Not a spare inch to work the access to the sewer is 27" long, 15 " wide and 17" deep in concrete to the crown of the 24" sewer

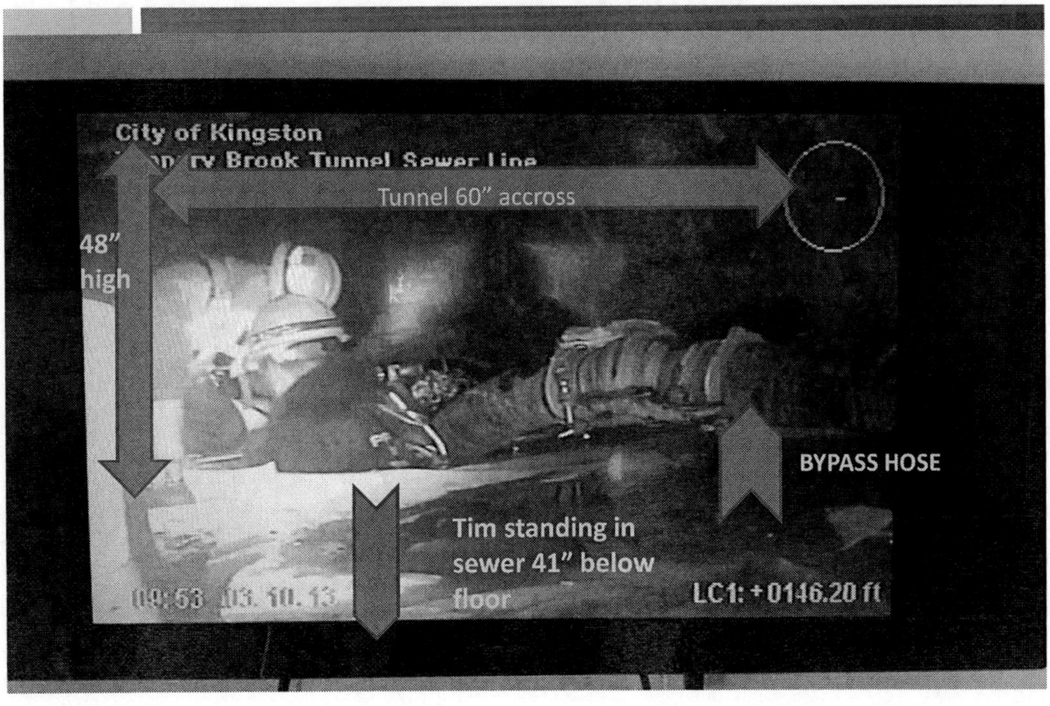

Communication – The largest problem –CCTV / Hand Signals and Dry Erase Boards

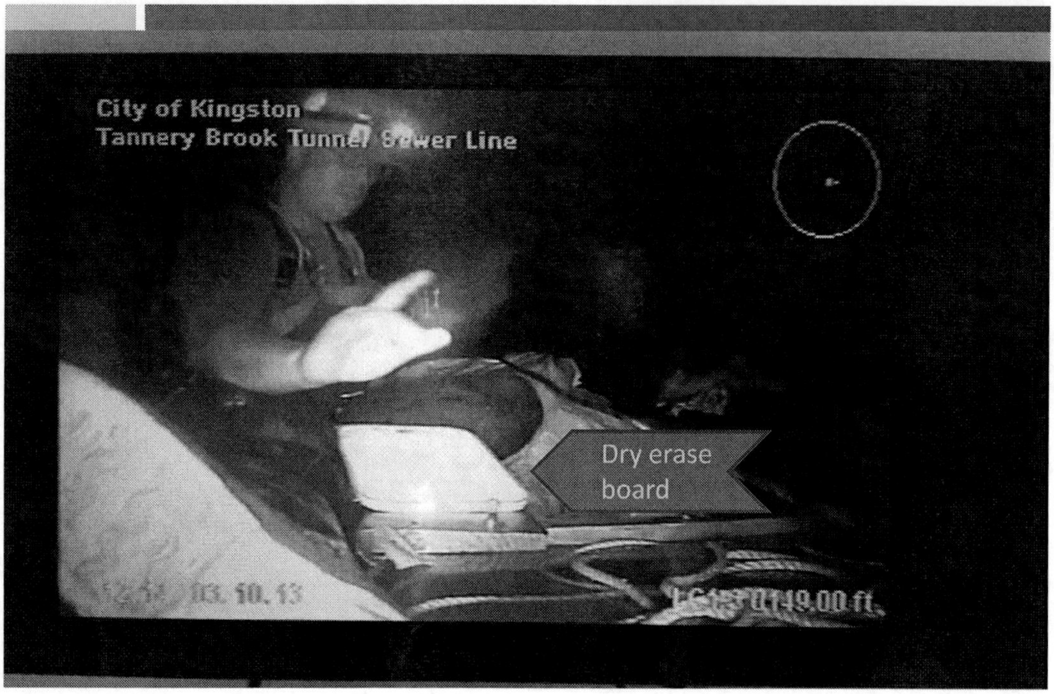

How many far did we move the liner?
a total of 1,300 LF from crate to install!

From the Crate lower down 90 LF feet

And then pull it 400 LF down the shaft and then pt it back into the sewer – BACK and FORTH

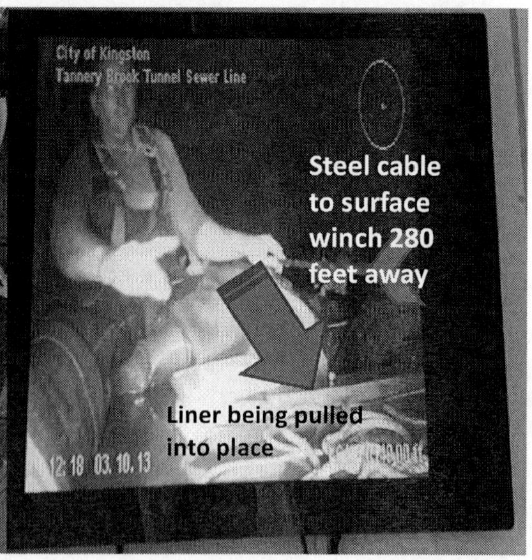

Nearing Completion of Cure 10 hours later

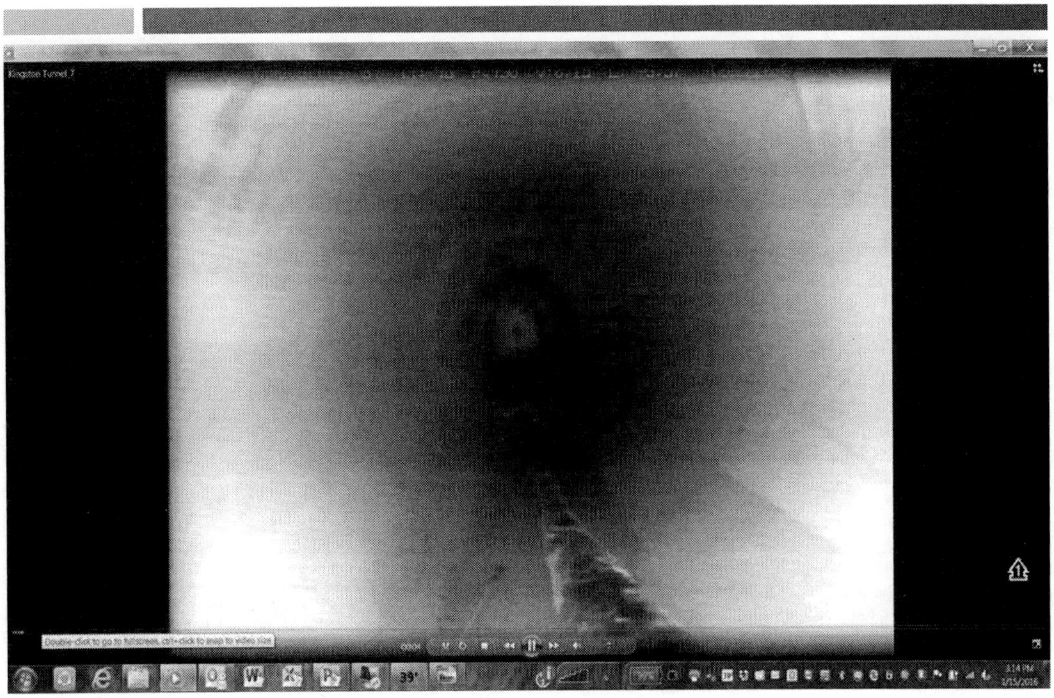

Real Time Data and Visual Feeds – See the Cure and Inspect

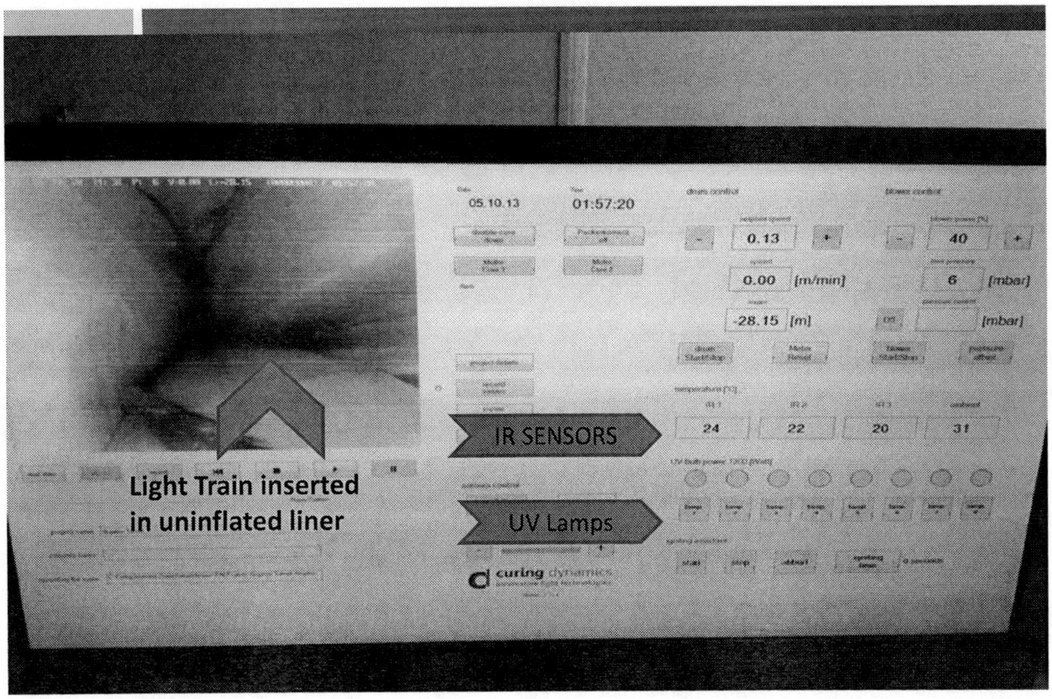

Project Completed

Thank You for your time

PRECISION

Precision Trenchless LLC

Mike App
Senior Project Manager
Precision Trenchless LLC
mapp@pim-inc.com

Please visit us at BOOTH 801

Lining solutions for a future life

SAERTEX multiCom

3D LIDAR for 3D Mapping

What is 2D Profiling?

- Profiling systems – 2D
- Non-stable Prone to error

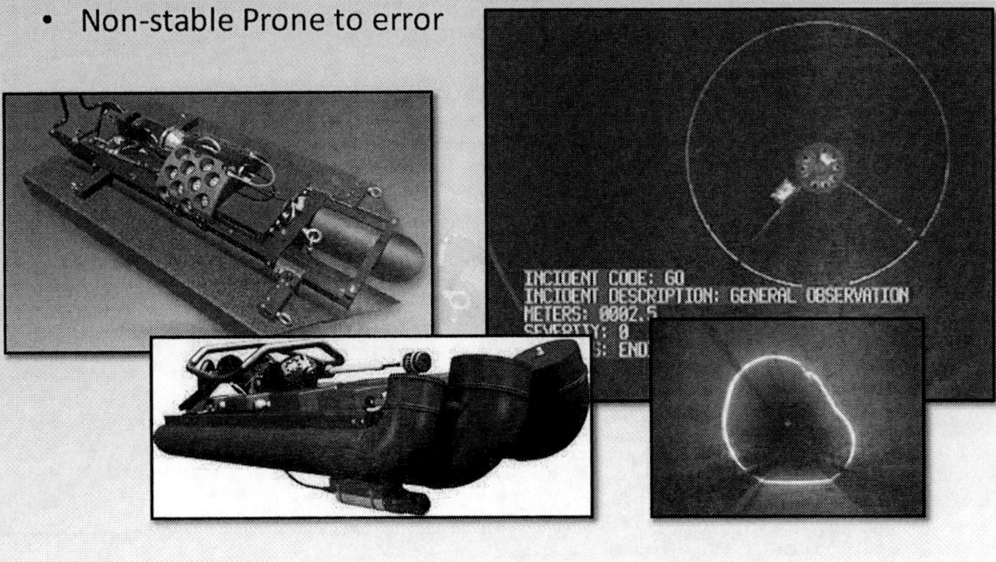

What is 3D LIDAR?

- Forward Looking – 3D
- Stable – accurate x-section

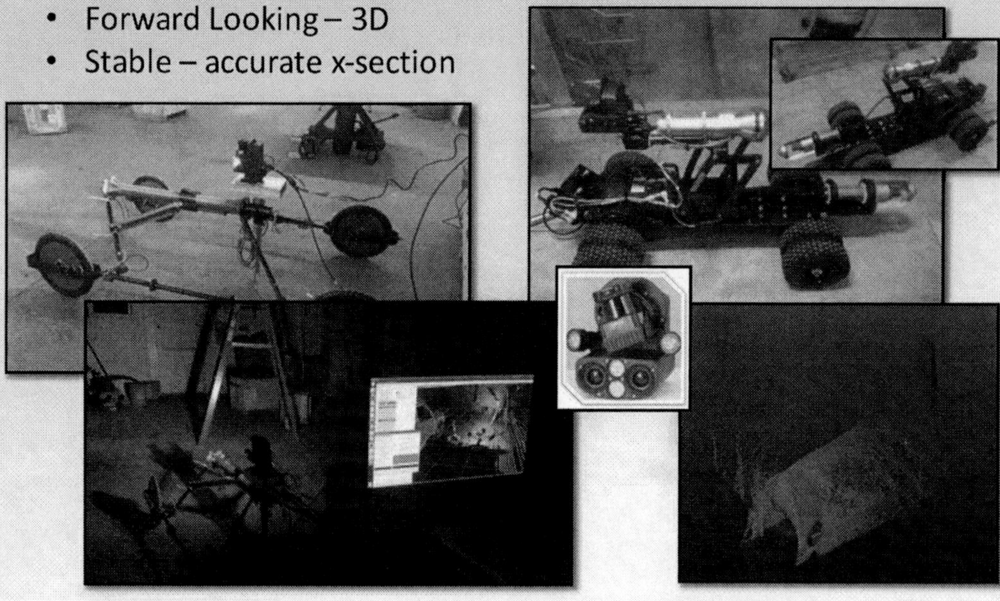

Application – Drill Location & Rehab

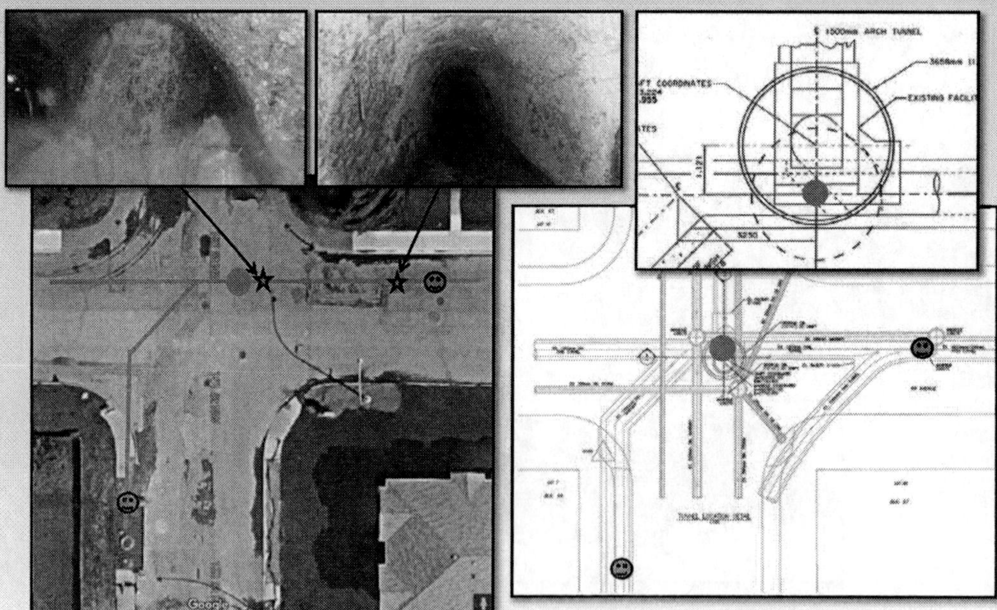

Deployment – MH Data

- Begin model construction

Offset MH – Plumb Line

- GPS Coordinate Transfer

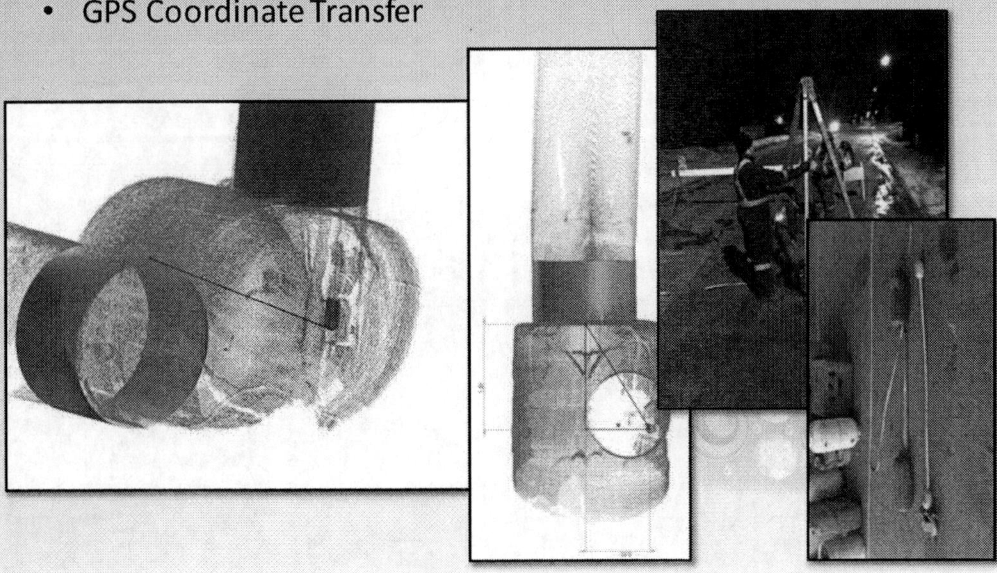

N-S By-Pass Pipe - Bend Analysis

- Import point clouds into CAD
- Define pipe wall and CL
- Measure angles
- Update 3D model

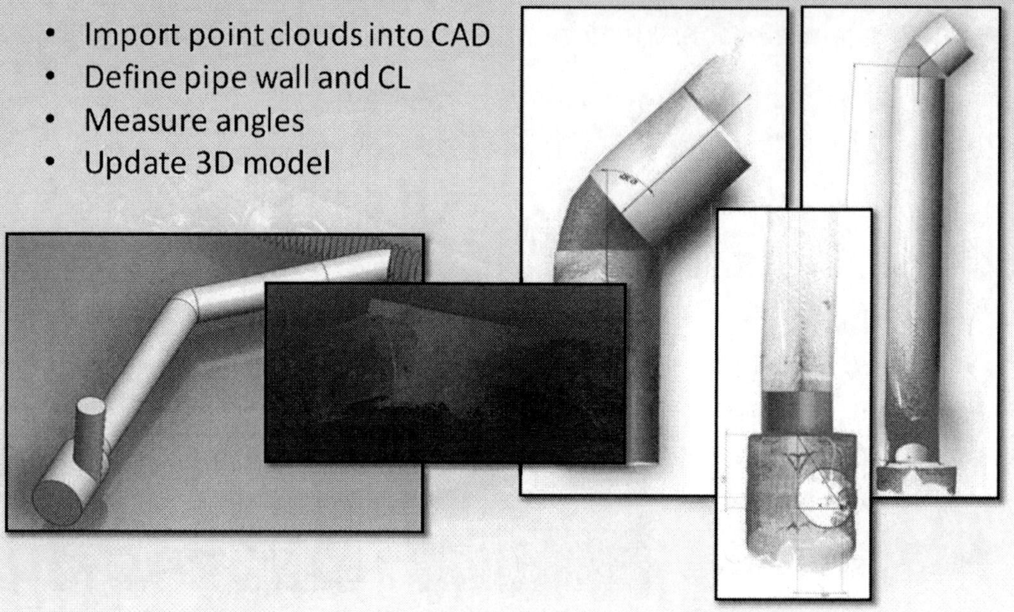

E-W Tunnel Cross-Sections

- Point clouds are aligned and merged
- Super-cloud is sliced into 1 ft. slices
- Each cross section is drawn and dimensioned

Tie the Pipe and Tunnel Together

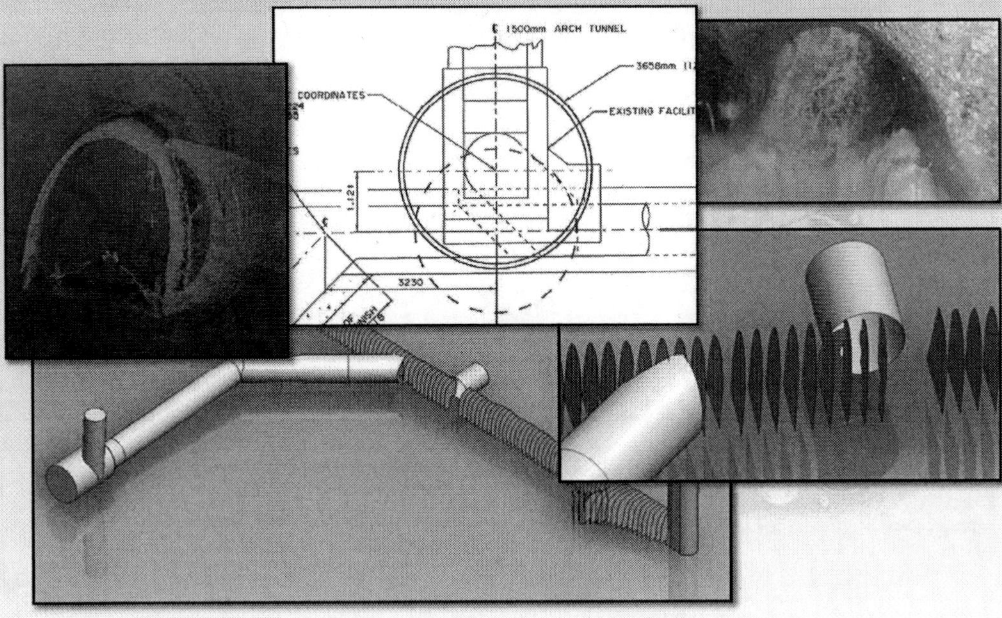

Deliverable - Cross Sections

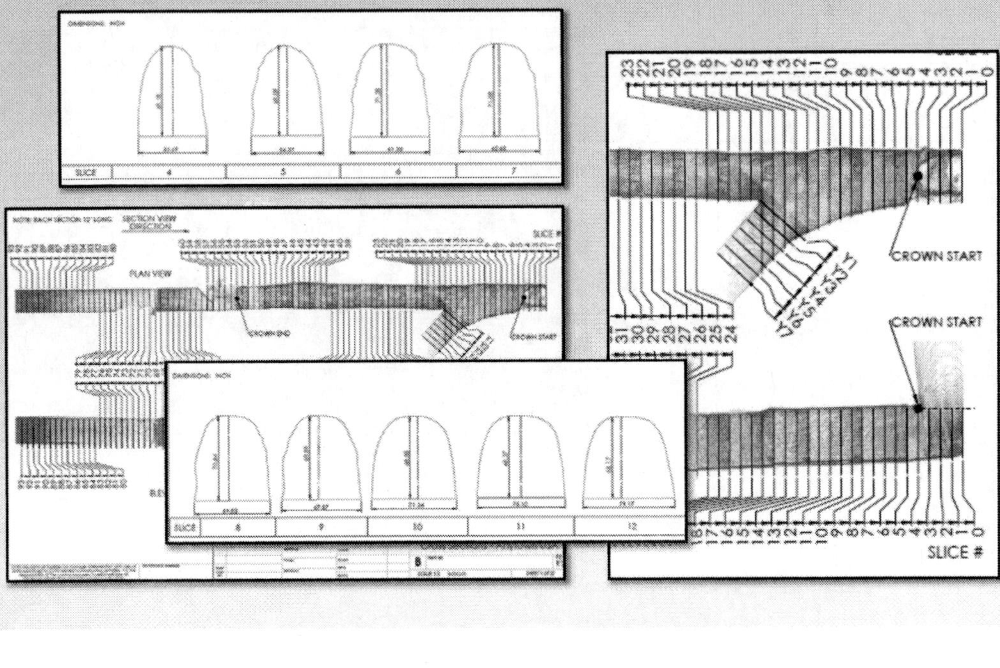

Deliverable – Access Shaft Dig Coords.

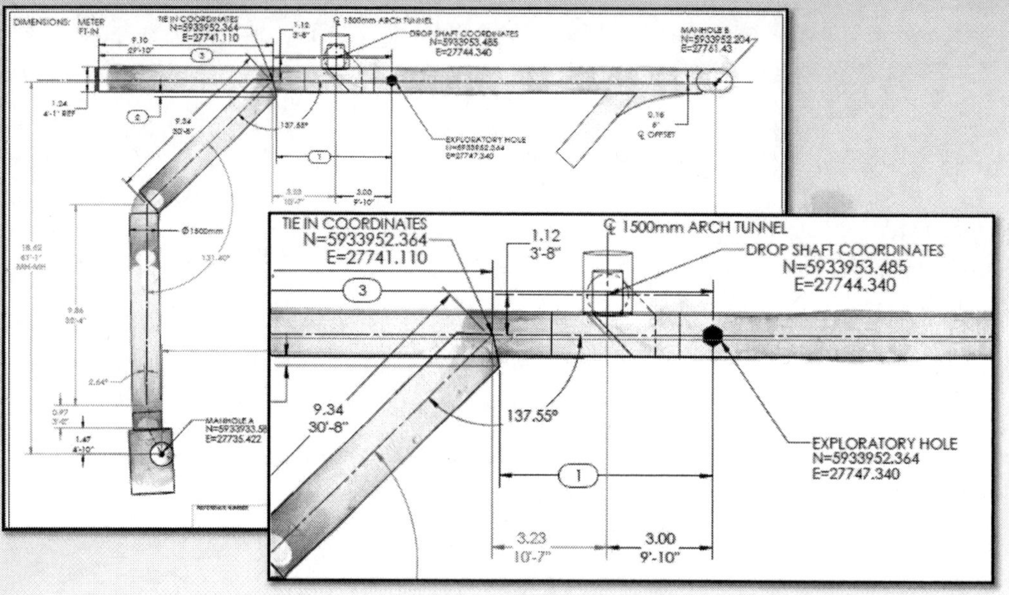

Results - Y 'nose' not there

- As built of Y
- Pipe jacking concerns

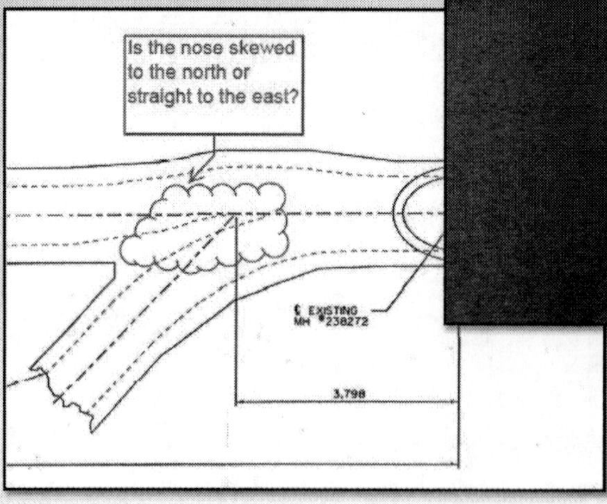

Results – Exploratory Hole for Shaft

- String line to South wall – 500mm
- String line to North wall – 490 mm

Gyroscopic 3D Mapping

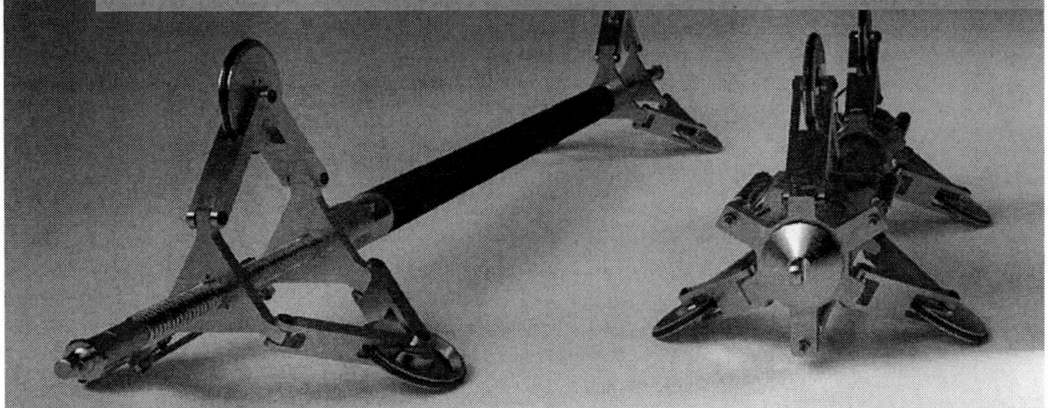

Gyroscopic 3D Mapping

- Collects 100 positional measurements/sec in all 3 axes
- Simultaneously collects linear distance & velocity measurements

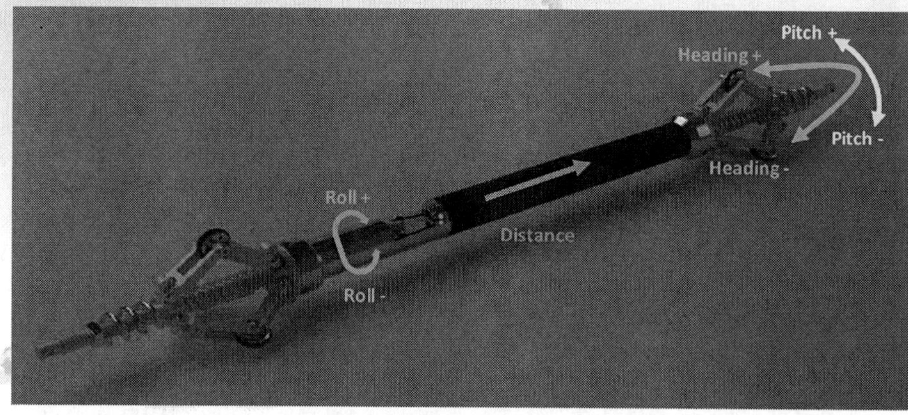

Projects and Applications – Cherry Hill

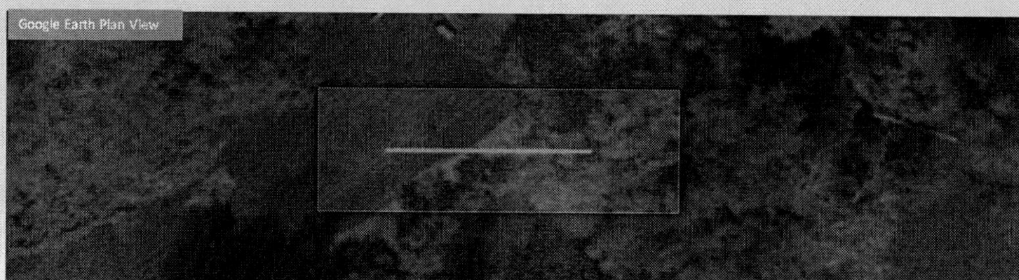

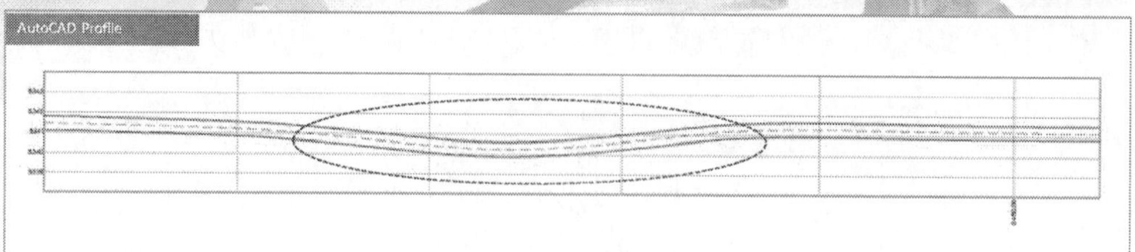

Projects and Applications – Cherry Hill

Inclination Report

Point Number	X	Y	Z	Station	Horiz Distance Between Points (ft)	Elevation Change Between Points (ft)	Segment Slope	Segment Grade (%)
10	3148602.100	1661374.550	5341.230	9.0026	0.9940	-0.0200	-0.0201	-2.01
11	3148601.180	1661374.160	5341.190	10.0018	0.9992	-0.0400	-0.0400	-4.00
12	3148600.260	1661373.760	5341.150	11.0050	1.0032	-0.0400	-0.0399	-3.99
13	3148599.340	1661373.370	5341.090	12.0043	0.9992	-0.0600	-0.0600	-6.00
14	3148598.420	1661372.990	5341.020	12.9997	0.9954	-0.0700	-0.0703	-7.03
15	3148597.490	1661372.630	5340.950	13.9969	0.9972	-0.0700	-0.0702	-7.02
16	3148596.560	1661372.270	5340.870	14.9941	0.9972	-0.0800	-0.0802	-8.02
17	3148595.620	1661371.930	5340.780	15.9937	0.9996	-0.0900	-0.0900	-9.00
18	3148594.690	1661371.570	5340.680	16.9910	0.9972	-0.1000	-0.1003	-10.03
19	3148593.760	1661371.210	5340.590	17.9882	0.9972	-0.0900	-0.0902	-9.02
20	3148592.840	1661370.860	5340.500	18.9726	0.9843	-0.0900	-0.0914	-9.14
21	3148591.910	1661370.500	5340.420	19.9698	0.9972	-0.0800	-0.0802	-8.02
22	3148590.980	1661370.130	5340.340	20.9707	1.0009	-0.0800	-0.0799	-7.99
23	3148590.050	1661369.760	5340.280	21.9716	1.0009	-0.0600	-0.0599	-5.99
24	3148589.120	1661369.390	5340.240	22.9725	1.0009	-0.0400	-0.0400	-4.00
25	3148588.200	1661369.020	5340.230	23.9641	0.9916	-0.0100	-0.0101	-1.01
26	3148587.270	1661368.640	5340.240	24.9688	1.0046	0.0100	0.0100	1.00
27	3148586.340	1661368.260	5340.280	25.9734	1.0046	0.0400	0.0398	3.98
28	3148585.430	1661367.870	5340.330	26.9635	0.9901	0.0500	0.0505	5.05
29	3148584.510	1661367.480	5340.410	27.9627	0.9992	0.0800	0.0801	8.01
30	3148583.600	1661367.090	5340.490	28.9528	0.9901	0.0800	0.0808	8.08
31	3148582.680	1661366.690	5340.600	29.9560	1.0032	0.1100	0.1096	10.96
32	3148581.770	1661366.290	5340.700	30.9500	0.9940	0.1000	0.1006	10.06
33	3148580.860	1661365.890	5340.810	31.9440	0.9940	0.1100	0.1107	11.07
34	3148579.950	1661365.490	5340.930	32.9380	0.9940	0.1200	0.1207	12.07
35	3148579.040	1661365.080	5341.030	33.9361	0.9981	0.1000	0.1002	10.02
36	3148578.140	1661364.680	5341.130	34.9210	0.9849	0.1000	0.1015	10.15
37	3148577.220	1661364.280	5341.190	35.9242	1.0032	0.0600	0.0598	5.98
38	3148576.300	1661363.890	5341.230	36.9235	0.9992	0.0400	0.0400	4.00

Projects and Applications – Oak Island

Google Earth Plan View

AutoCAD Profile

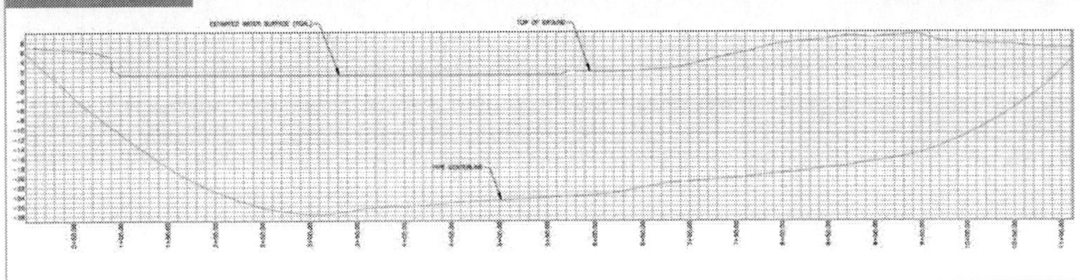

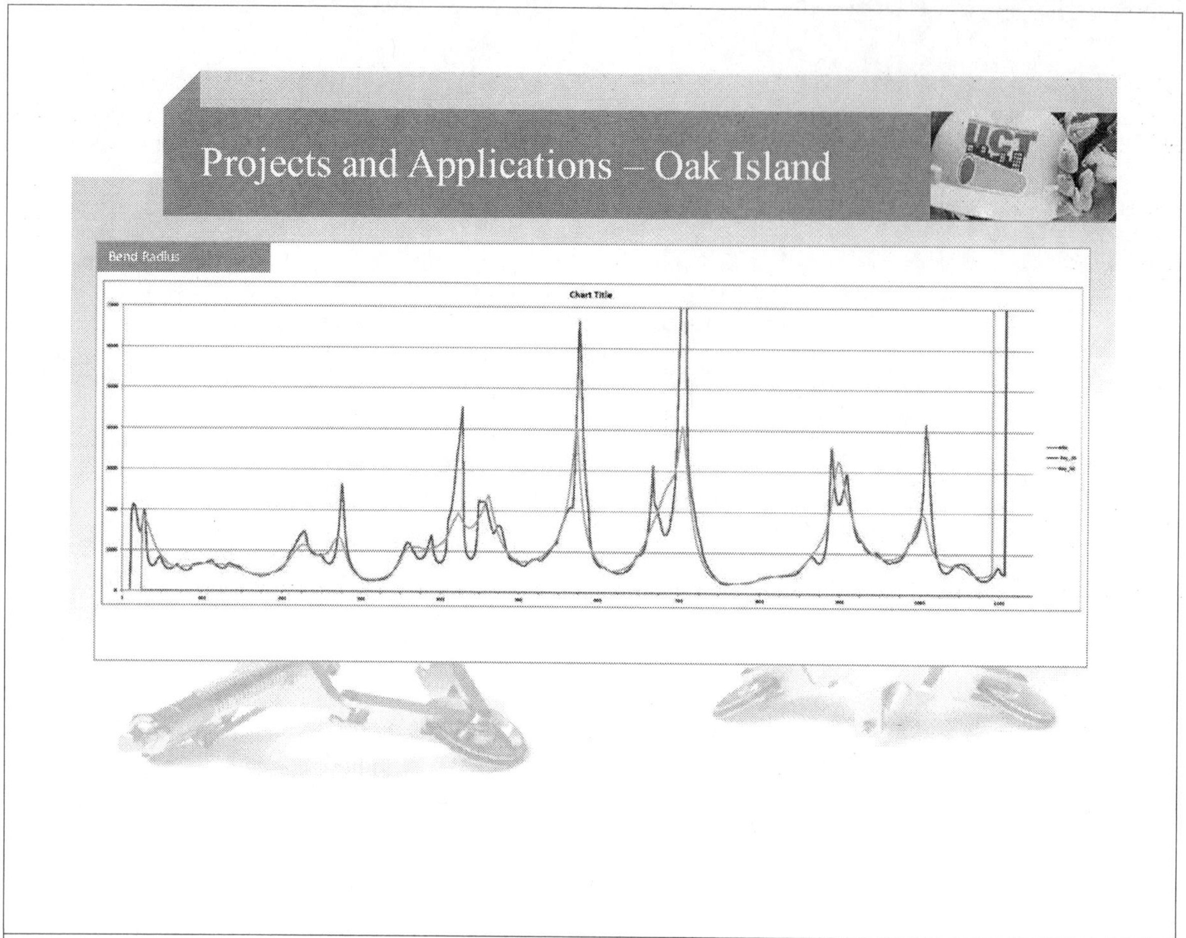
Projects and Applications – Oak Island
Bend Radius
Chart Title

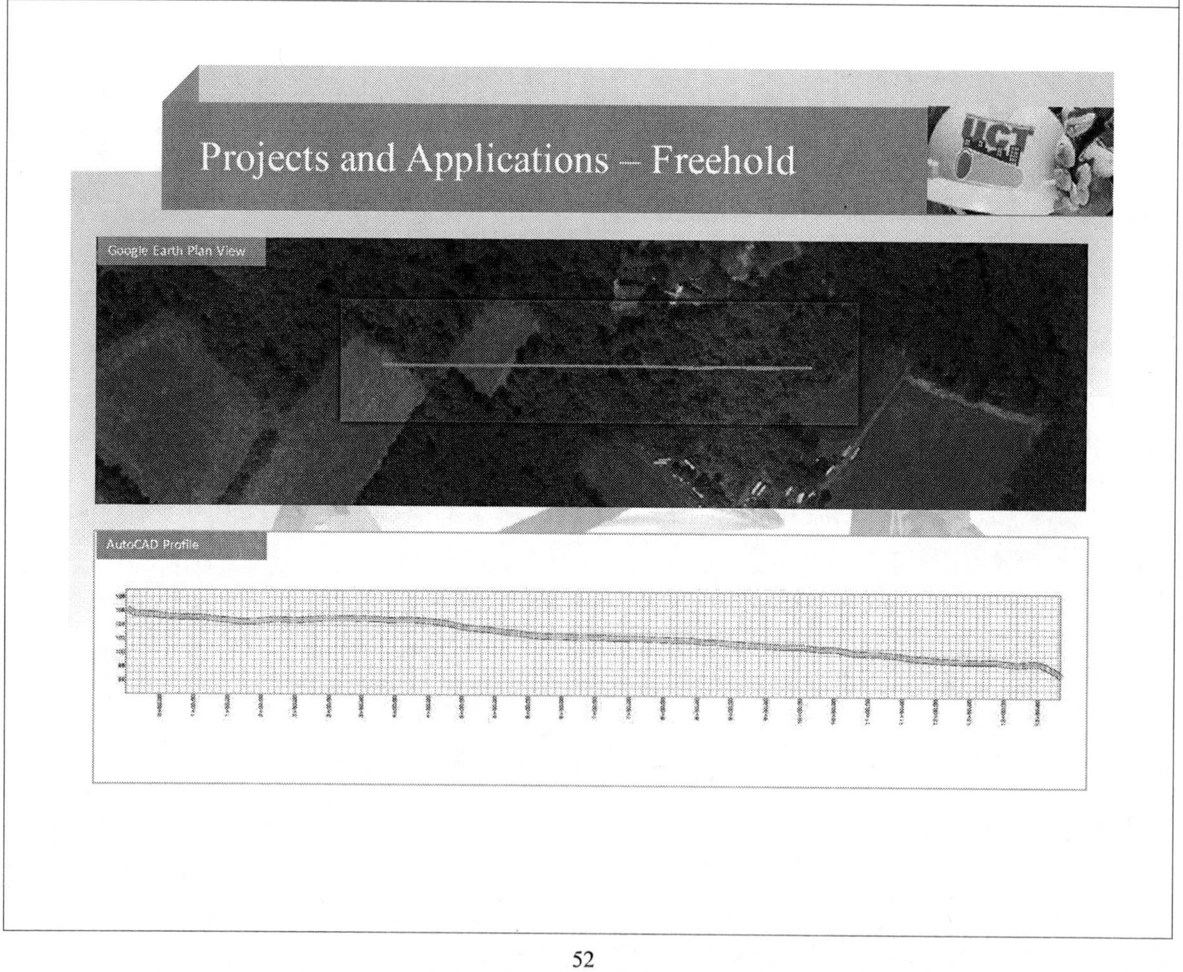
Projects and Applications – Freehold
Google Earth Plan View
AutoCAD Profile

Projects and Applications – Freehold

Inclination Report

Point Number	X	Y	Z	Station	Horiz Distance Between Points (ft)	Elevation Change Between Points (ft)	Segment Slope	Segment Grade (%)
1	553374.840	494631.580	106.000	0.0000	0.0000	0.0000	0.0000	0.00
2	553376.120	494632.780	105.890	1.7545	1.7545	-0.1100	-0.0627	-6.27
3	553376.300	494632.940	105.870	1.9954	0.2408	-0.0200	-0.0830	-8.30
4	553377.020	494633.630	105.760	2.9926	0.9972	-0.1100	-0.1103	-11.03
5	553377.740	494634.310	105.680	3.9830	0.9904	-0.0800	-0.0808	-8.08
6	553378.460	494635.000	105.610	4.9802	0.9972	-0.0700	-0.0702	-7.02
7	553379.190	494635.690	105.540	5.9847	1.0045	-0.0700	-0.0697	-6.97
8	553379.920	494636.370	105.480	6.9824	0.9976	-0.0600	-0.0601	-6.01
9	553380.640	494637.060	105.430	7.9796	0.9972	-0.0500	-0.0501	-5.01
10	553381.360	494637.750	105.390	8.9768	0.9972	-0.0400	-0.0401	-4.01
11	553382.090	494638.440	105.360	9.9813	1.0045	-0.0300	-0.0299	-2.99
12	553382.810	494639.130	105.330	10.9786	0.9972	-0.0300	-0.0301	-3.01
13	553383.540	494639.820	105.310	11.9831	1.0045	-0.0200	-0.0199	-1.99
14	553384.260	494640.510	105.290	12.9803	0.9972	-0.0200	-0.0201	-2.01
15	553384.980	494641.200	105.280	13.9776	0.9972	-0.0100	-0.0100	-1.00
16	553385.700	494641.890	105.270	14.9748	0.9972	-0.0100	-0.0100	-1.00
17	553386.420	494642.580	105.260	15.9721	0.9972	-0.0100	-0.0100	-1.00
18	553387.150	494643.270	105.250	16.9765	1.0045	-0.0100	-0.0100	-1.00
19	553387.870	494643.970	105.240	17.9807	1.0042	-0.0100	-0.0100	-1.00
20	553388.600	494644.660	105.240	18.9852	1.0045	0.0000	0.0000	0.00
21	553389.320	494645.350	105.230	19.9825	0.9972	-0.0100	-0.0100	-1.00
22	553390.040	494646.040	105.230	20.9797	0.9972	0.0000	0.0000	0.00
23	553390.760	494646.730	105.230	21.9770	0.9972	0.0000	0.0000	0.00
24	553391.490	494647.420	105.220	22.9815	1.0045	-0.0100	-0.0100	-1.00
25	553392.210	494648.110	105.220	23.9787	0.9972	0.0000	0.0000	0.00
26	553392.930	494648.800	105.220	24.9759	0.9972	0.0000	0.0000	0.00
27	553393.660	494649.490	105.210	25.9804	1.0045	-0.0100	-0.0100	-1.00
28	553394.380	494650.180	105.210	26.9777	0.9972	0.0000	0.0000	0.00
29	553395.110	494650.870	105.200	27.9822	1.0045	-0.0100	-0.0100	-1.00
30	553395.830	494651.550	105.200	28.9725	0.9904	0.0000	0.0000	0.00

Advantages of Gyroscopic Mapping

- Accurate information for use in As-Built drawings
- Gyro data easily uploads into Auto CAD and GIS Platforms
- Map utilities not available to conventional survey (e.g. river crossings, deep utilities, HDD lines)
- Integration of exact positional location with CCTV- identified anomalies allows for accurate, cost-effective spot repairs
- Not subject to soil type, adjacent utilities, equipment depth limitations
- Locates and quantifies defects such as pipe sags, misaligned joints, horizontal and vertical design affects
- Provides pipeline center-line 3D geometry:
  - Bending radius
  - Line and Arc Segmentation
- Able to map under water

THE TOTAL SOURCE
educational sessions

A Hitchhiker's Guide to the CIPP Galaxy

Author: Gary Heusser, P.E.

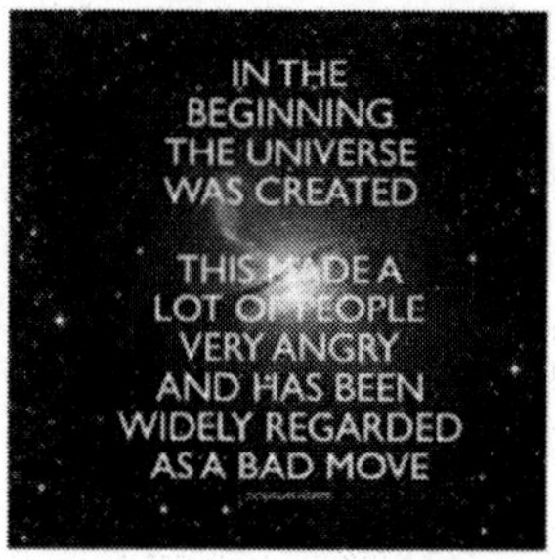

Littlejohn
An S&ME Company

A Hitchhiker's Guide to the CIPP Galaxy

What is Cured-In-Place Pipe (CIPP)??

- **Trenchless Rehabilitation to repair existing pipelines by forming a new pipe within an existing pipe**
- **A resin impregnated flexible tube (typically felt) coated on outside with impermeable plastic (once inverted, will be the inside layer of finished pipe) which is installed into the existing pipeline by inversion from a either a hydrostatic head, air pressure or pull-in method**

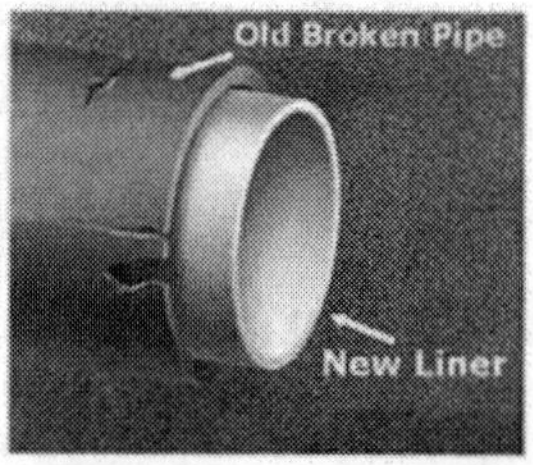

- **Once inverted, the resin curing process is accomplished by circulating hot water or the introduction of controlled steam throughout the entire length to cure the resin into a hard, impermeable pipe**
- **The CIPP extends the full length of the original pipe and provides a structurally sound, joint-less, close fitting and corrosion resistant "pipe within a pipe"**

A Hitchhiker's Guide to the CIPP Galaxy

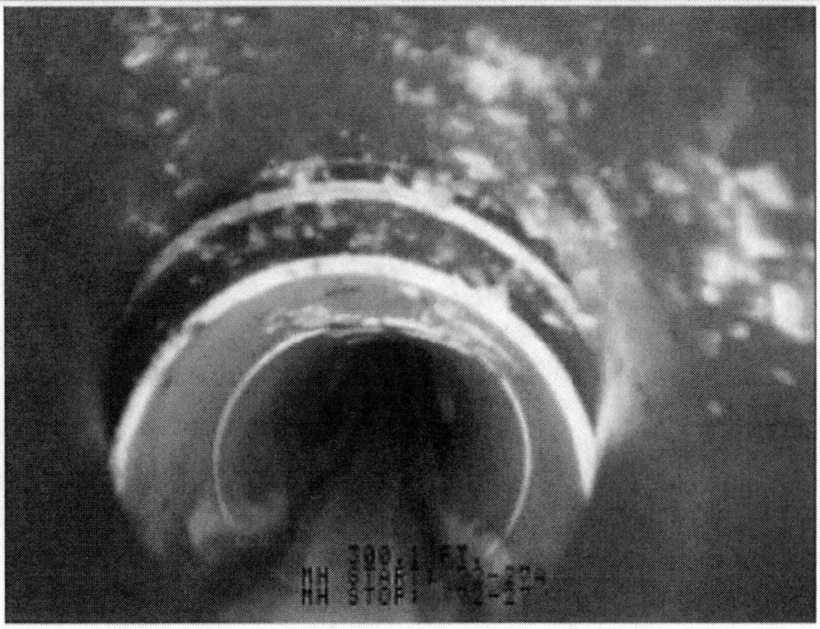

A Hitchhiker's Guide to the CIPP Galaxy

A Hitchhiker's Guide to the CIPP Galaxy

https://www.youtube.com/watch?v=I9CUcF-r6Us

The Overall CIPP Construction Process:

- Cleaning, CCTV Inspection & Lateral Verification
- Wet-Out Facility – Impregnating Tube with Resin
- Prep-work – Laying out hoses, Plugging, Bypassing, Notifying homeowners etc…
- Install Hydrophilic End Seals
- Inversion of Tube
- Curing Resin
- Cool-Down
- Lateral Reinstatement
- Trimming CIPP to MH Connection
- Site Cleanup

A Hitchhiker's Guide to the CIPP Galaxy

The Engineer's CIPP Process:

- Specification Writing
- Submittal Review
- Resident Project Representation

General Specifications:

- Proper Reference Standards – ASTM, ANSI, ISO etc...
- Quality Assurance (requirements will vary depending upon scope of work):
  - CIPP System Manufacturer Performance Record – Installed XX LF
  - Contractor/Installer Experience – Installed XX LF and XX Diameter and years with specific CIPP product
  - On Site Field Superintendent Experience – Installed XX LF and XX Diameter
  - Resin Class – Wet-Out Facilities and Resin Sample Analysis

A Hitchhiker's Guide to the CIPP Galaxy

Product Specification Writing:

- Resin Type and Composition
  - Acceptable Resins – Polyester and Vinyl Ester Resins
  - Not allowing resins containing filler, additives or enhancement agents
- Catalyst Systems
  - Acceptable Primary and Secondary Catalysts
  - Not allowing quick cure or accelerated resin systems
- Liner Tube
  - Two or more layers of non-woven felt
  - Limiting the number of splices per XX LF
  - Homogeneous across entire wall thickness and uniform thickness
  - Manufacturer's marking for identification per XX LF
  - Length allowance for samples (typically 18-inches)
  - Addition of dye (blue) to indicate full wet out of liner

A Hitchhiker's Guide to the CIPP Galaxy

Product Specification Writing:

- CIPP Design

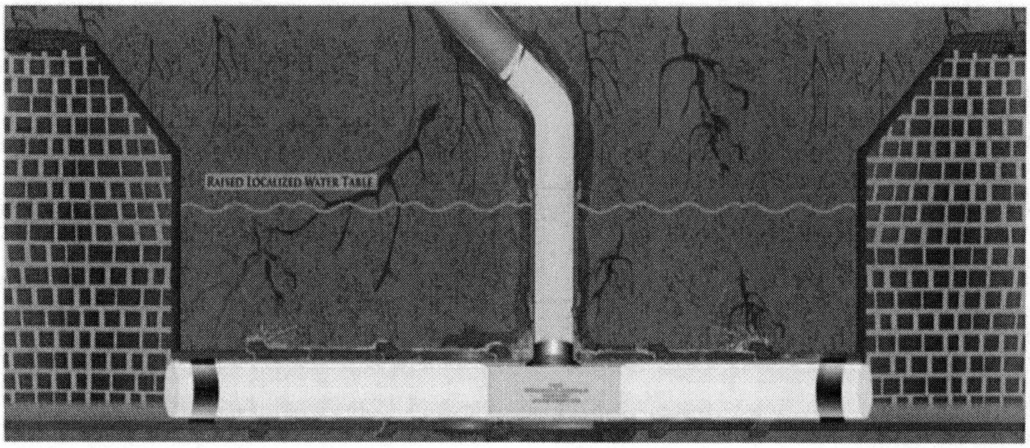

A Hitchhiker's Guide to the CIPP Galaxy

Execution Specification Writing:

- Follow CIPP manufacturer's curing schedule - warm-up, exotherm and cool down temperatures and durations
- Resin Impregnation (Wet-Out)
    - Ensuring proper refrigeration to prevent premature curing
    - Resin shall be sufficient to fill all voids in the tube
    - Excess resin (7-10%) adjustment due to polymerization and migration into defects of host pipe

A Hitchhiker's Guide to the CIPP Galaxy

Submittals:

- Proposed Curing Schedule from Manufacturer including step curing procedures, cooking times and cool down procedures
- Post-cure hold times and temperature tables
- Catalyst system and resin/catalyst ratios
- Quality Assurance requirements met
- Up to date Certifications

A Hitchhiker's Guide to the CIPP Galaxy

Testing:

- Allowance within Bid for potential CIPP testing
- Air Leakage Test before lateral reinstatement
- Sample Testing:
  - Required Thickness
  - Flexural Strength and Flexural Modulus

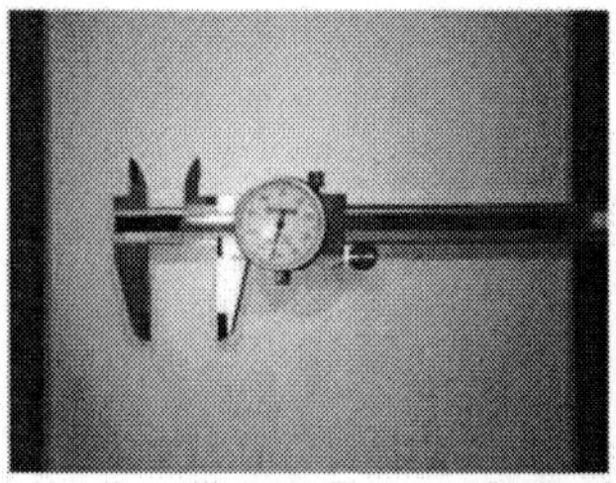

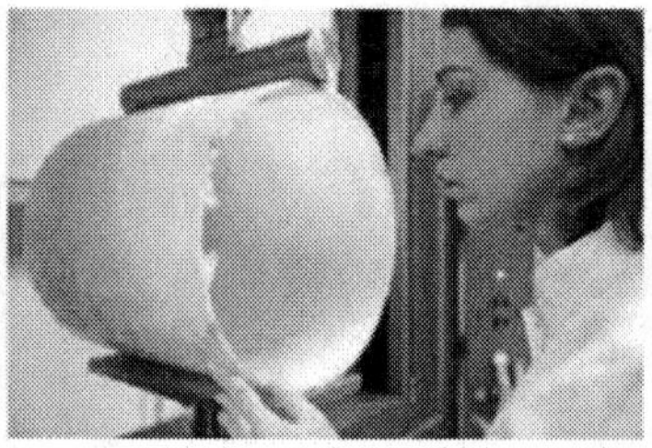

A Hitchhiker's Guide to the CIPP Galaxy

Post CCTV Inspection:

- Importance that camera stops and pans closely at potential defects

- One-year warranty included, so closely inspection final CIPP

- What to look for?
  - Wrinkles
  - Blistering
  - Lifts
  - Pinholes
  - Over/Undercut lateral reinstatements
  - Resin slugs
  - Manhole connections not cut properly
  - Overcooked CIPP

A Hitchhiker's Guide to the CIPP Galaxy

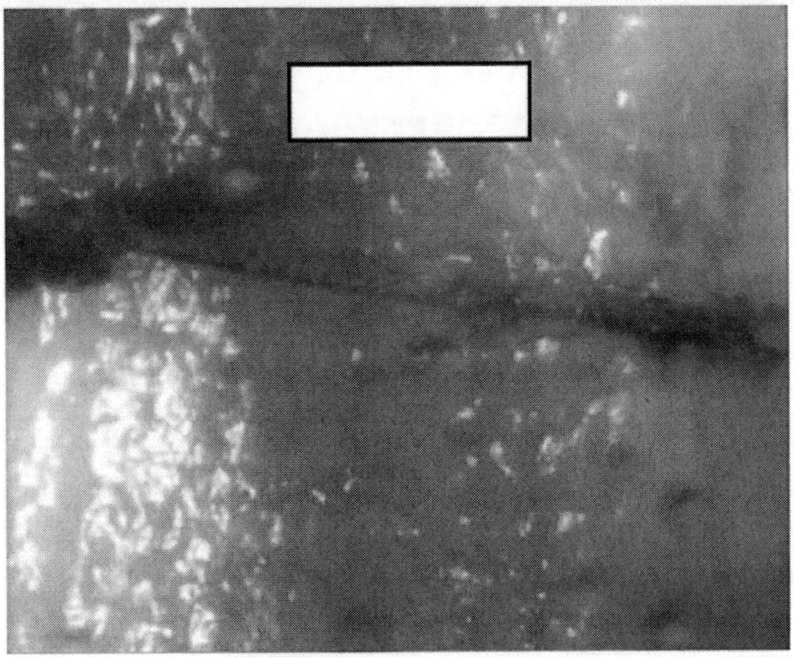

A Hitchhiker's Guide to the CIPP Galaxy

Resident Project Representative's Role:

- **Monitor cure schedule followed according to approved manufacturer's recommendations**
- **Reading independent thermocouples along with contractor**
- **Inspecting bag upon arrival**
- **Monitoring inversion process**
- **Watch live lateral reinstatements**
- **Collect CIPP sample for each pipe segment**
- **Witness air leakage pressure test**
- **Proper non-leak bypass pumping and downstream screening installed correctly**

A Hitchhiker's Guide to the CIPP Galaxy

Conflict Resolution:

- **Representing Owner's best interests**
- **Thin CIPP samples**
- **Split CIPP**
- **Failed Leakage Test**

A Hitchhiker's Guide to the CIPP Galaxy

Lessons Learned:

- **Pay close attention to resin/catalyst composition from what is specified to what is submitted**
- **Having independent field temperature readings to verify accordance with curing schedule**
- **Keeping a punchlist with observations and recommendations for each CIPP segment**
- **Testing any CIPP sample that are "questionable" in the Engineer's mind**
- **Downstream screening is vital**
- **Resident Project Representation is essential to the Owner and Engineer**

A Hitchhiker's Guide to the CIPP Galaxy

Author: Gary Heusser, P.E.
gheusser@leainc.com
210 Hamm Rd Suite B, Chattanooga, TN 37405

Littlejohn
An S&ME Company

THE TOTAL SOURCE

educational sessions

The Challenges of Constructing a Siphon Underneath the San Antonio River

Presented by

Joseph Cotton, PE, ENV SP

rjngroup

40 years of collection system solutions

Agenda

- Project Overview
- Project Challenges
- Challenge Resolution
- Lessons Learned
- Questions?

Project Overview

Construction of

- Bore Launching and Receiving Pits
- Jack and Bore Installation of 54" and 60" Steel Casings (approximately 210 linear feet each)
- Triple Barrel Inverted Siphon
  - 2 - 24" Fusible PVC
  - 1 - 16" Fusible PVC
- 12" Air Jumper
- 2 Siphon Boxes
- Replacement of 36" Double Wall Brick Gravity Main with 36" PVC Gravity Main
- Temporary By-Pass Gravity Mains

Challenge – Depth and Size of Bore Pits

- 15 Feet Wide
- 45 Feet Long
- 50 Feet Deep

Resolution – Depth and Size of Bore Pits

- 55 Foot Deep Shafts Drilled for I-Beam Installation
- I-Beams Driven into Ground for Support of Steel Sheeting

Challenges – Depth and Size of Bore Pits

- Steel Sheeting Driven into the Ground between the I-Beams
- Soil Excavated in Lifts
- Steel Sheeting and I-Beams Welded Together by Hand
- Steel I-Beams Installed for Cross Support of Bore Pit

Challenges – Previous and Existing Siphons

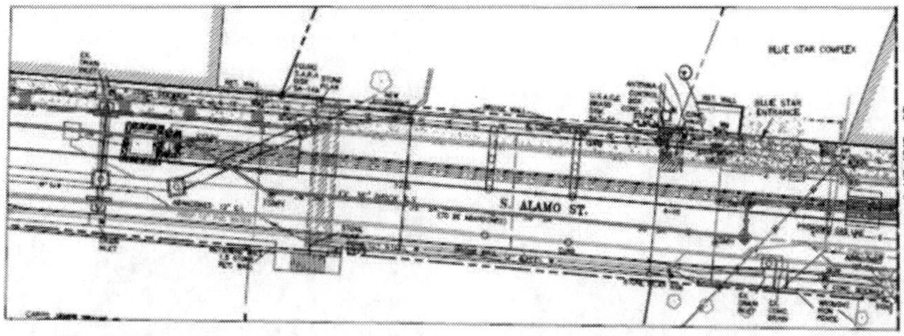

- Original Double Barrel Inverted Siphon Constructed in 1920s – 10” and 12” Barrels
- Existing Inverted Siphon Constructed In The 1970s – Single 36” Barrel

Resolution – Previous and Existing Siphons

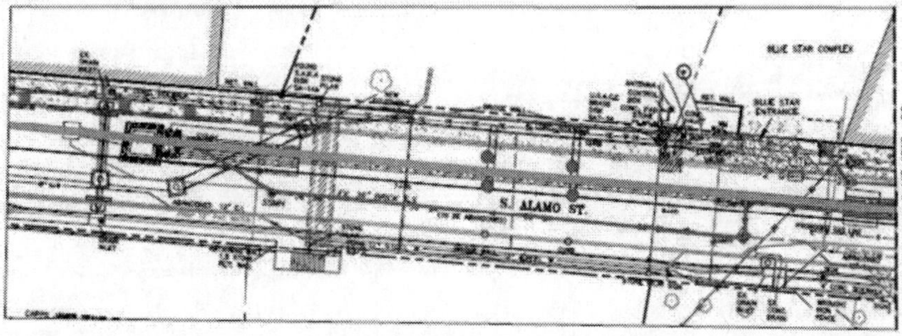

- Two Steel Casing Stacked Vertically – 54” and 60” Barrels
- Located Between the Existing Bridge Piers

Resolution – Previous and Existing Siphons

- Two Steel Casings Stacked Vertically – 54" and 60" Barrels

Challenge – Previous Unknown Flood Gate Piers

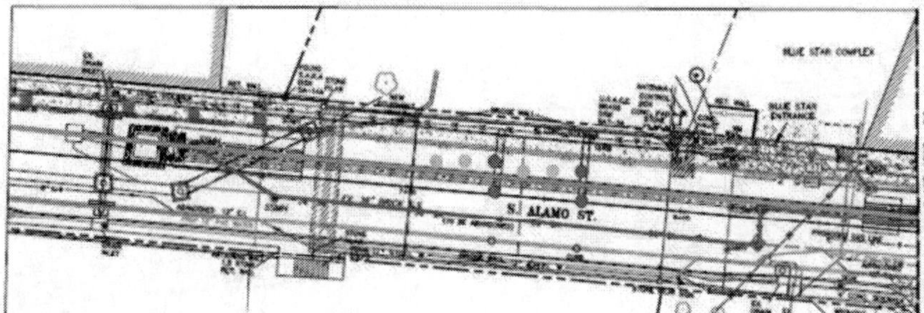

- Local River Authority Informed the Contractor and Owner of Potential Conflict with Flood Gate Piers During Construction

Resolution – Previous Unknown Flood Gate Piers

- Adjustment of Alignment to Account for Gate Piers
- Visual Confirmation of Gate Pier Clearance

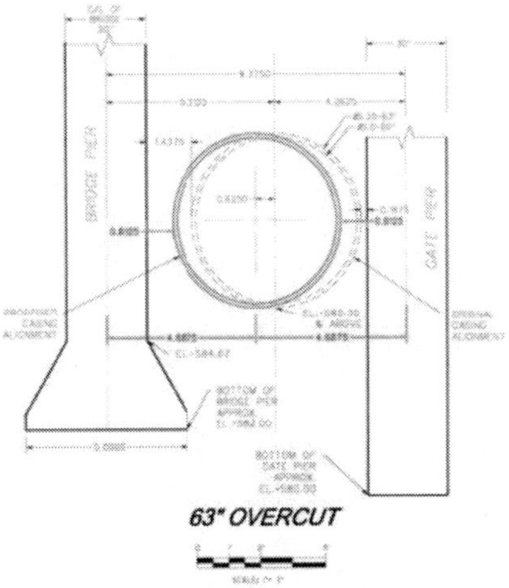

Challenge / Resolution

- Environmental Concerns
  - Contaminated Soil
    - Gas Station
    - Dry Cleaner
- Environmental Mitigation Plan and Exhibits Included with Bid Documents

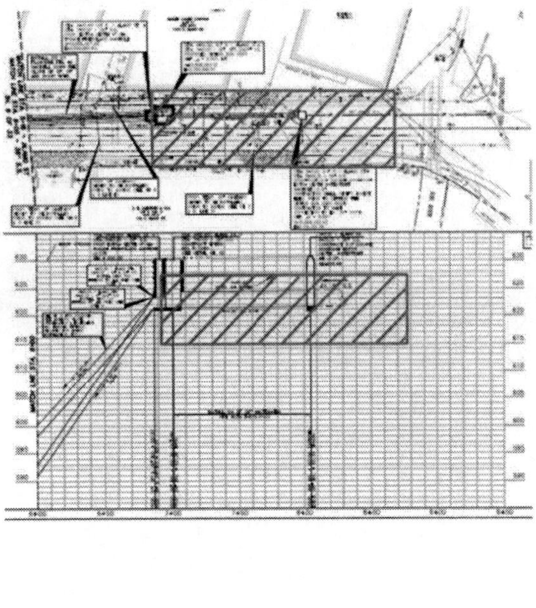

Challenge – Public Access to Surrounding Businesses Across from San Antonio River

- First Friday Art Walk
- Historic Business District Revitalization
- Historic Landmarks (Gunther House)

Resolution – Public Access to Surrounding Businesses Across from San Antonio River

- Fencing
- Bury By-Pass Pumping Pipes at Driveways

Challenge – Groundwater Intrusion

- Groundwater Encountered at 12' Below Existing Ground
- Inflow Rate from 50 to 250 Gallons per Minute

Resolution – Groundwater Intrusion

- Installation of 450 GPM Sump Pumps at Corners of Bore Pits

Resolution – Groundwater Intrusion

- Sump Pumps Transported Water into a Sedimentation Basin (Rolloff Dumpster)
- Water from Sedimentation Basin Gravity Flowed into Municipal Sanitary Sewer System

Groundwater Intrusion Setback

- Sump Pump Malfunction during a Large Rain Storm Caused Bore Pit to Fill with Storm Water
- Took 24 hours to Pump the Water Within the Bore Pit to an Acceptable Level

Constructions Challenge – Backfill Compaction

- Bottom of the Bore Pit was Consistently Saturated Due to Groundwater

Constructions Challenge – Backfill Compaction

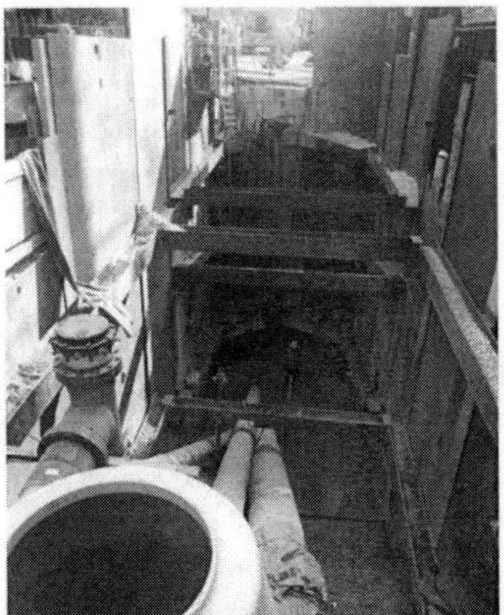

- Bore Pits Backfilled with Flowable Fill in 4' Vertical Lifts
- Flowable Fill used to Completely Fill Bore Pits

Construction Completion

- Complete Reconstruction of Roadway by Local Municipality
- Restoration of Pedestrian and Vehicle Traffic to Better Than Pre-Construction Conditions

Lessons Learned

- Wet Construction Seasons Lead to Increased Groundwater Quantities
- Pre-Construction Coordination with Other Agencies Does Not Always Prevent Changes During Construction
- Even the "Best" Plan Changes

Questions???

rjngroup

40 years of collection system solutions

THE TOTAL SOURCE

educational sessions

Three Dimensional Scanning for Repair Planning

Harry R. Price, P.E.
NTH Consultants, Ltd.
hprice@nthconsultants.com

Authors

- Harry Price, P.E. Senior Principal Engineer
  NTH Consultants, Ltd.
- Jason Edberg, P.E. Vice President
  NTH Consultants, Ltd.
- Jason Mayer, P.E. Senior Project Manager
  Giffels Webster Engineers
- Fritz Klingler, P.E. President
  F.K. Engineering Associates
- Sid Lockhart, P.E. Chief Engineer
  Office of the Oakland County Water Resources Commissioner
- Jason Matteo, P.E. Chief Engineer
  Office of the Macomb County Public Works Commissioner

Area Map

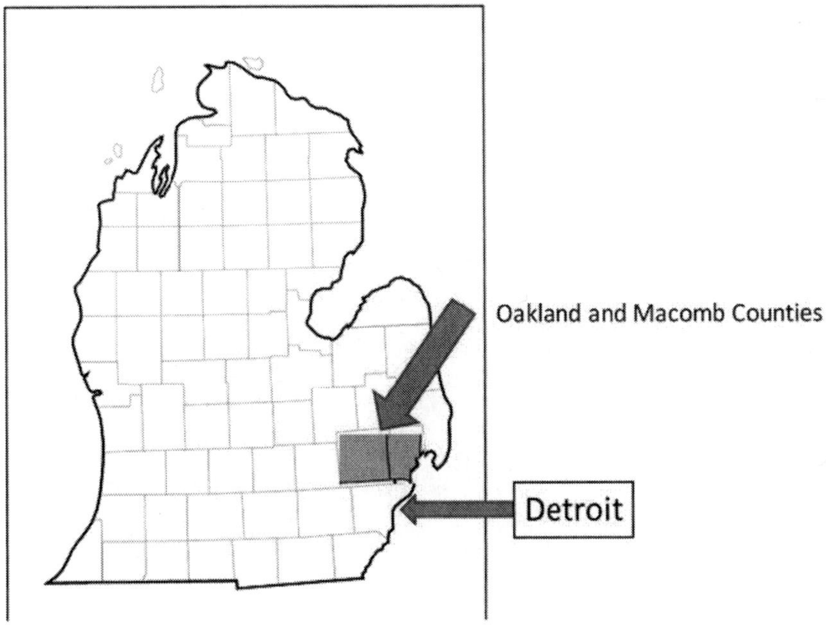

Base Map

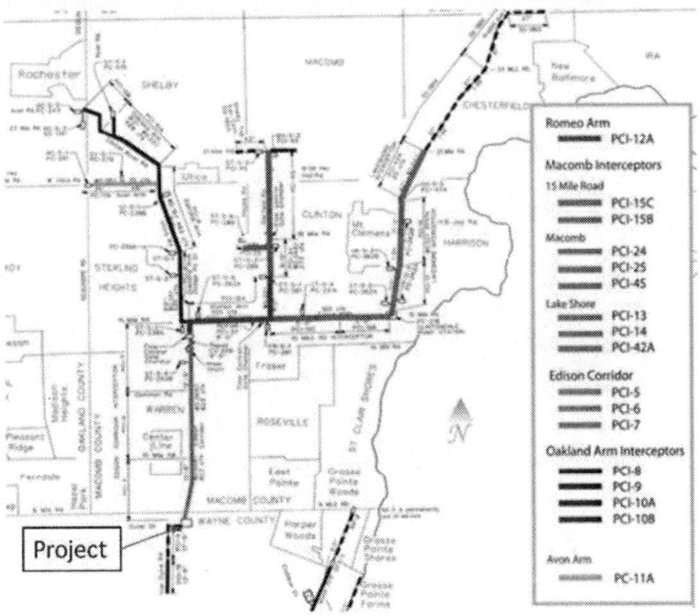

PLAN VIEW

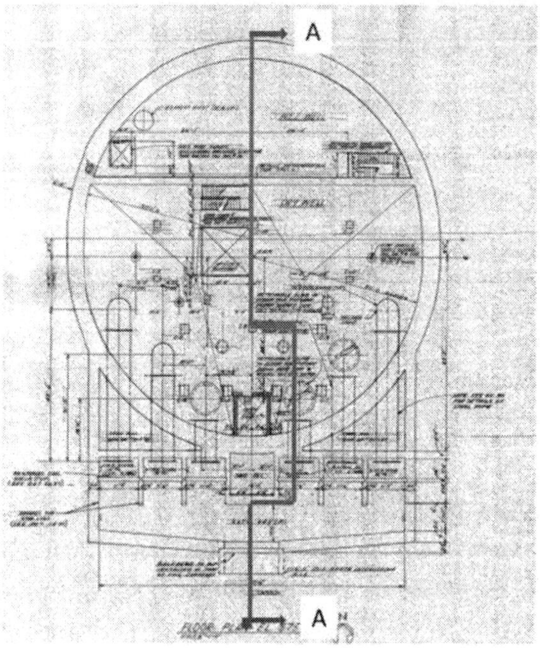

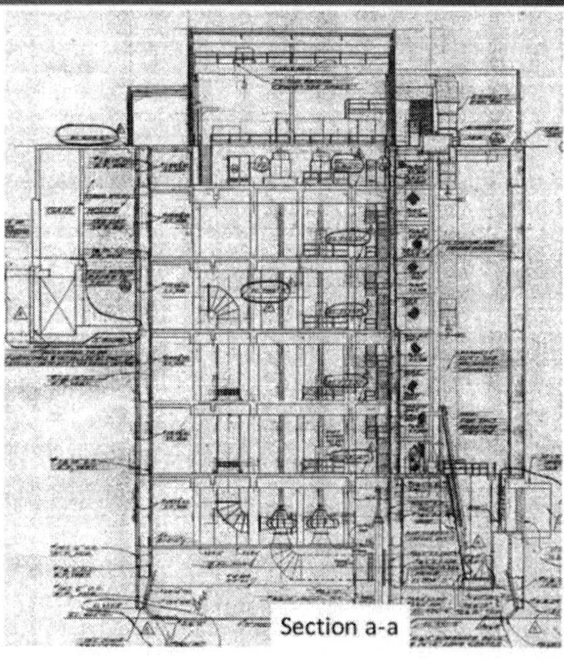

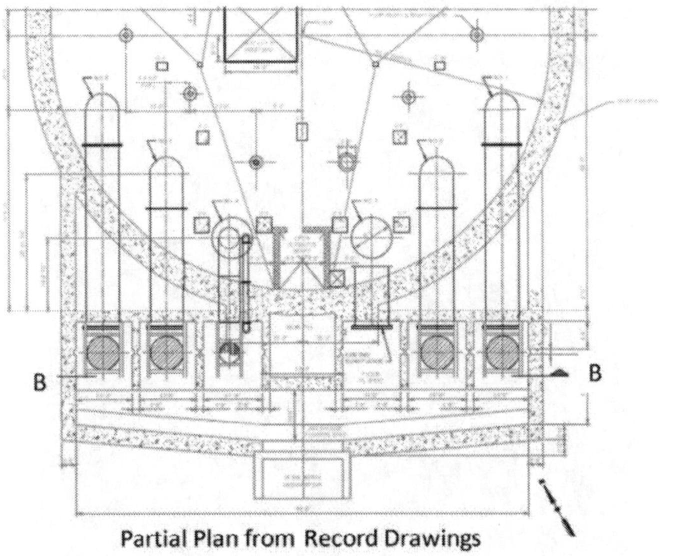

Partial Plan from Record Drawings

Section Showing Pump Discharges as Designed

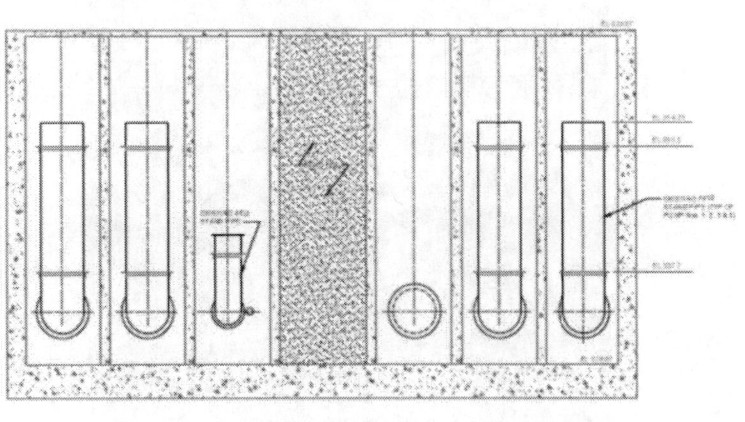

Section B-B

Discharge Pump Number 6

Walls Near Pump Number 5

Walls Between Pumps 4 and 5

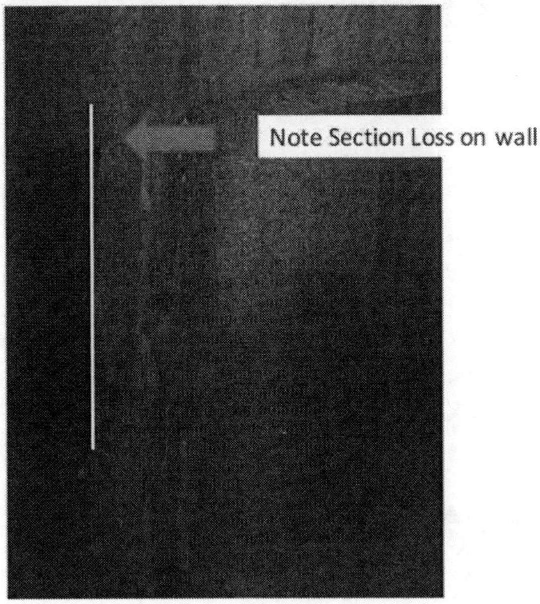

Reinforcing Steel Exposed above outlet

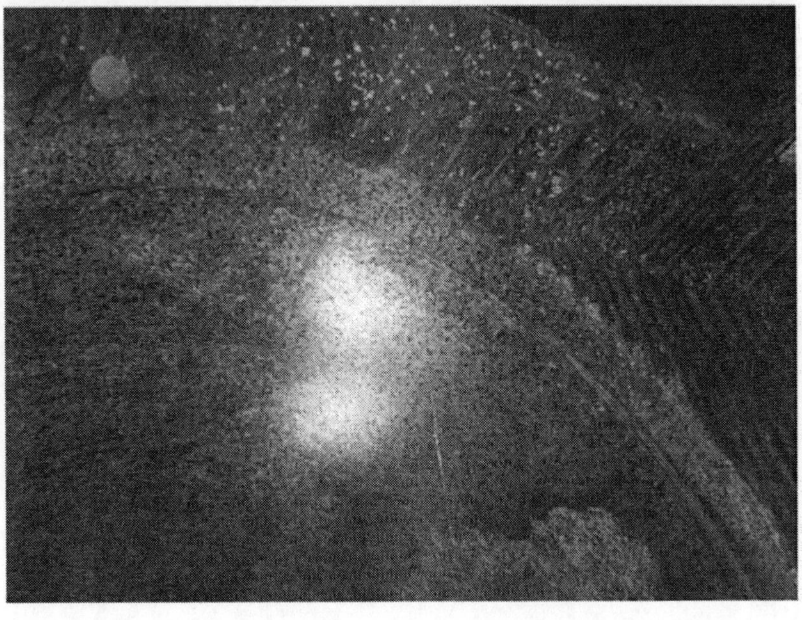

Alingnment of Discharge of Discharge Pipe As Scanned

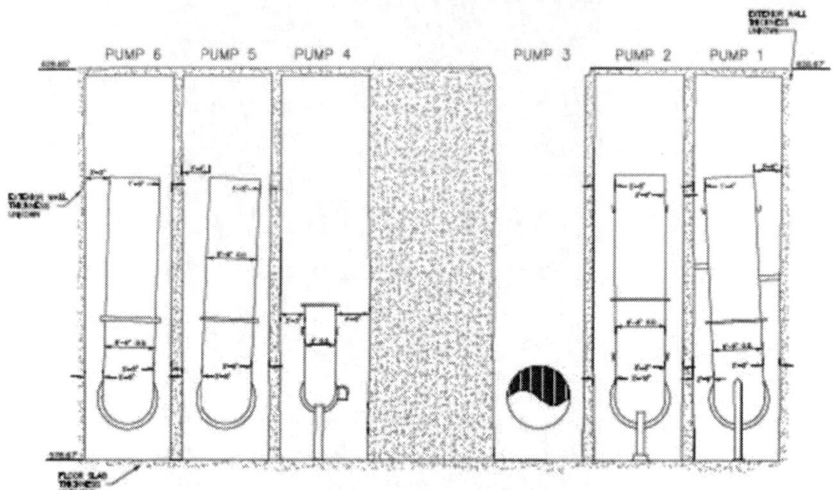

Section B from 3D Scanning

Section through Discharge Channel facing Discharge Pipe

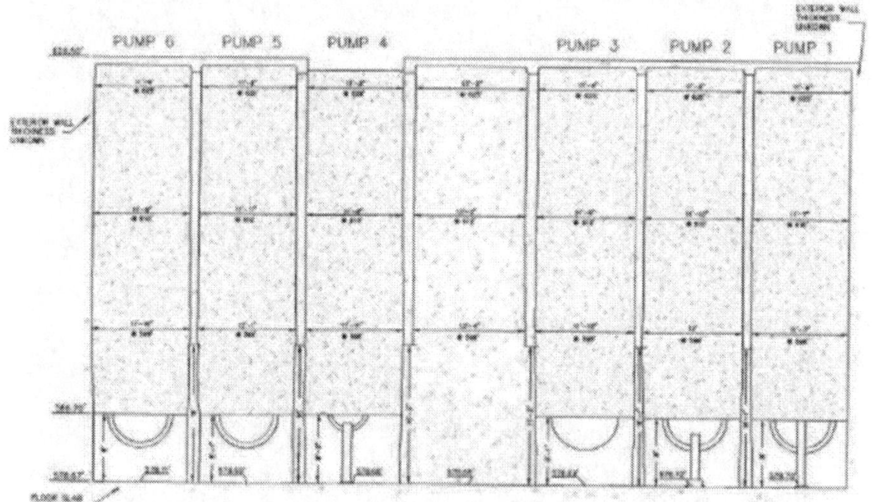

Plan View 37 Feet Above Base

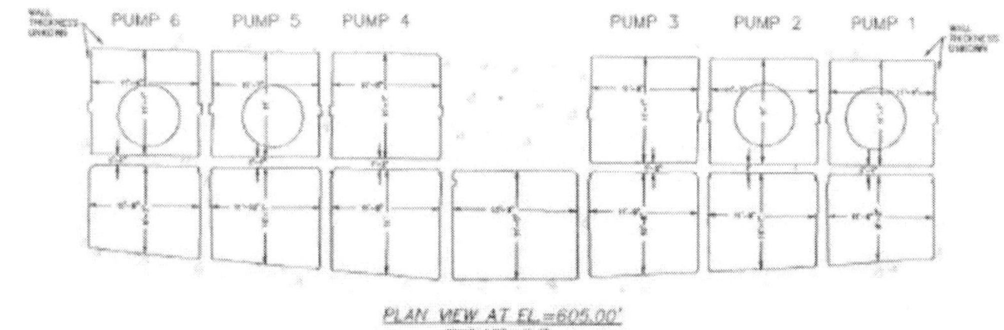

Plan View at Base

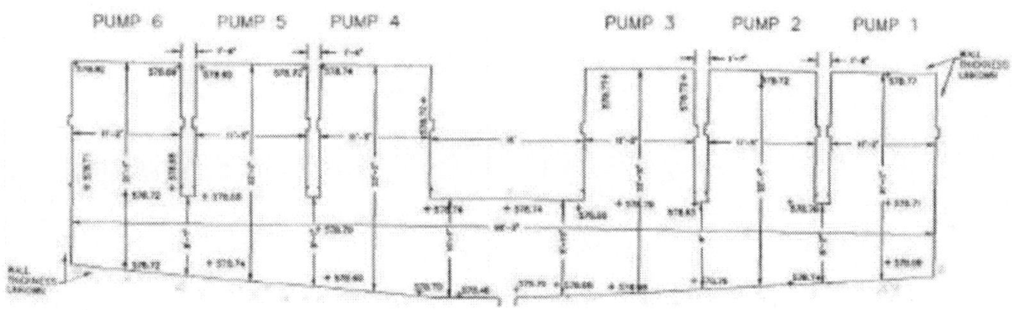

Section Facing Discharge

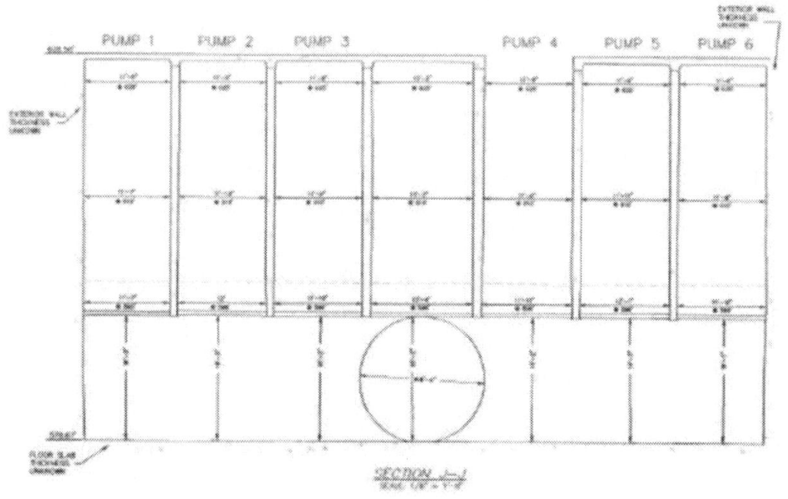

Plan View at Mid-Point of Discharge

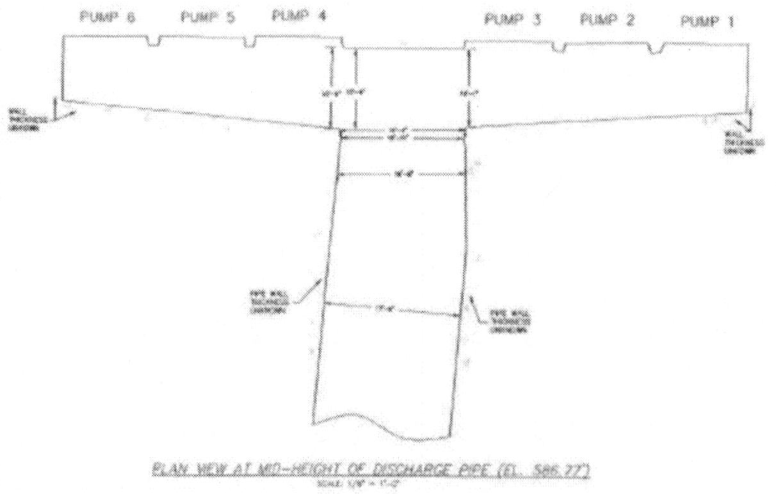

Section through Modified Discharge Chamber

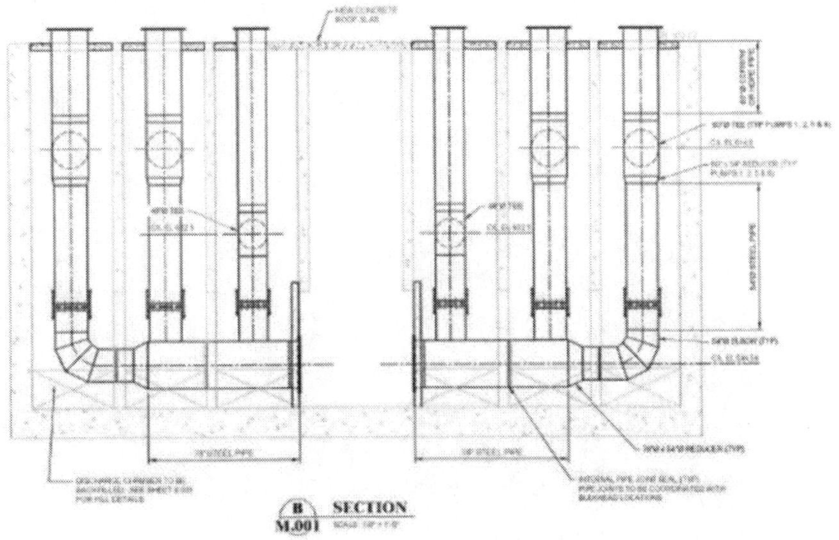

Section Perpendicular to Discharge Chamber

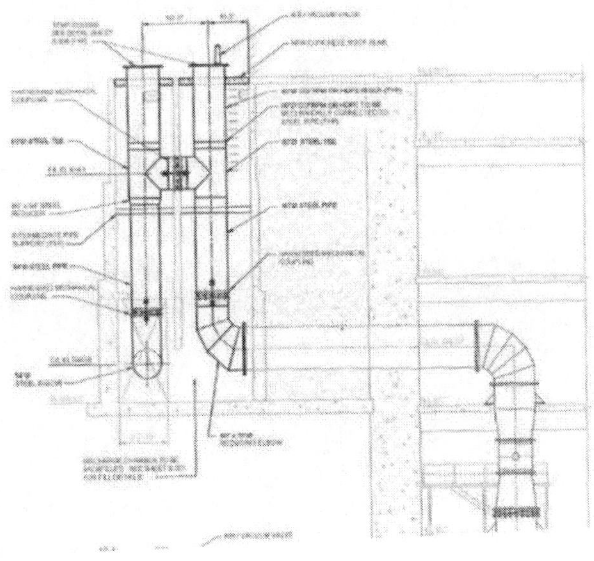

From Pump 3 Discharge Looking at Manifold

Looking Down at Manifold at Pump 2

Looking West From Pump 1

Conclusions

- 3D Scanning is useful in confined areas with limited work time
- With a structure attacked by MIC it is very useful for document existing conditions
- In confined structures it may be necessary to construct additional access

Questions

Harry R. Price, P.E.
NTH Consultants, Ltd.
hprice@nthconsultants.com

THE TOTAL SOURCE

educational sessions

Sewer Rehabilitation Strategy Workshop

Developing a Performance-Based Strategy

Patrick Stevens, P.E.
ADS Environmental Services
(256) 508-0333 | PStevens2@idexcorp.com

George Kurz, P.E., DEE
Consulting Engineer
(615) 714-6120 | george.kurz@comcast.net

Seminar Objectives

- Identify key elements of a performance-based strategy
- Make the best use of flow monitoring
- Provide a logical framework for applying various products and tools
- Development of principles used in the "System Approach"
- Cost-Effectiveness & Cost-Benefit

Seminar Program

- **A 10 Step Plan for Sewer Rehabilitation – Kurz**
- ------ 5 minute break ------
- **"If I had this to do over . . . " A 12 Step Program to Successfully Measure Sewer Rehabilitation – Stevens**
- ------ 5 minute break ------
- **Nashville – Removing 3.6 Billion Gallons of I/I- Kurz**
- ------ 5 minute break ------
- **Service Laterals as a Critical Part of the Rehab Program - Kurz**

THE TOTAL SOURCE

educational sessions

A Road Map for Sewer Rehab *

A ten-step strategic plan

PRESENTED BY

George Kurz, P.E., DEE

George.kurz@comcast.net

** A significant portion of this work was conducted with CTE-AECOM as part of the Nashville Overflow Abatement Program 1991-2005*

? ? ? ?

Do we *REALLY* have an I/I Problem?

IS SEWER REHABILITATION EFFECTIVE ?

WHAT LEVEL OF REDUCTION DO WE REASONABLY EXPECT ?

?

HOW DO WE ACHIEVE SUCCESS ? ?

? ? ?

2

I/I INDICATOR – BOD INFLUENT

Domestic Sewage "Strength":

Weak	100 – 150 mg/l
Medium	150 – 200 mg/l
Strong	200 - 250 mg/l

Strength of Domestic Sewage: ~ 350 mg/l

3

Magnitude of the I/I Problem

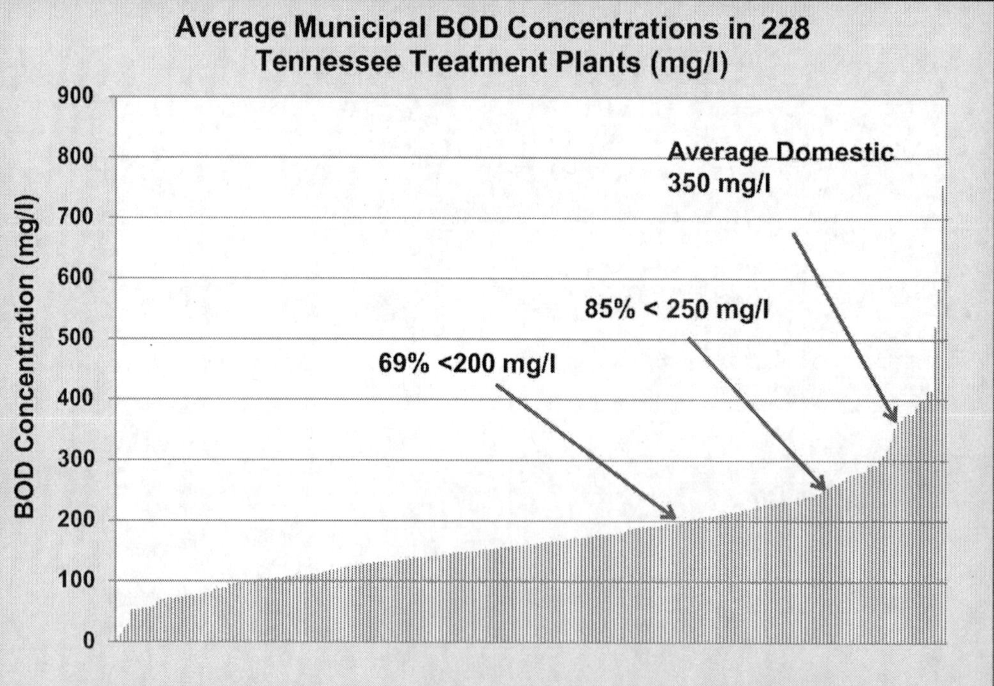

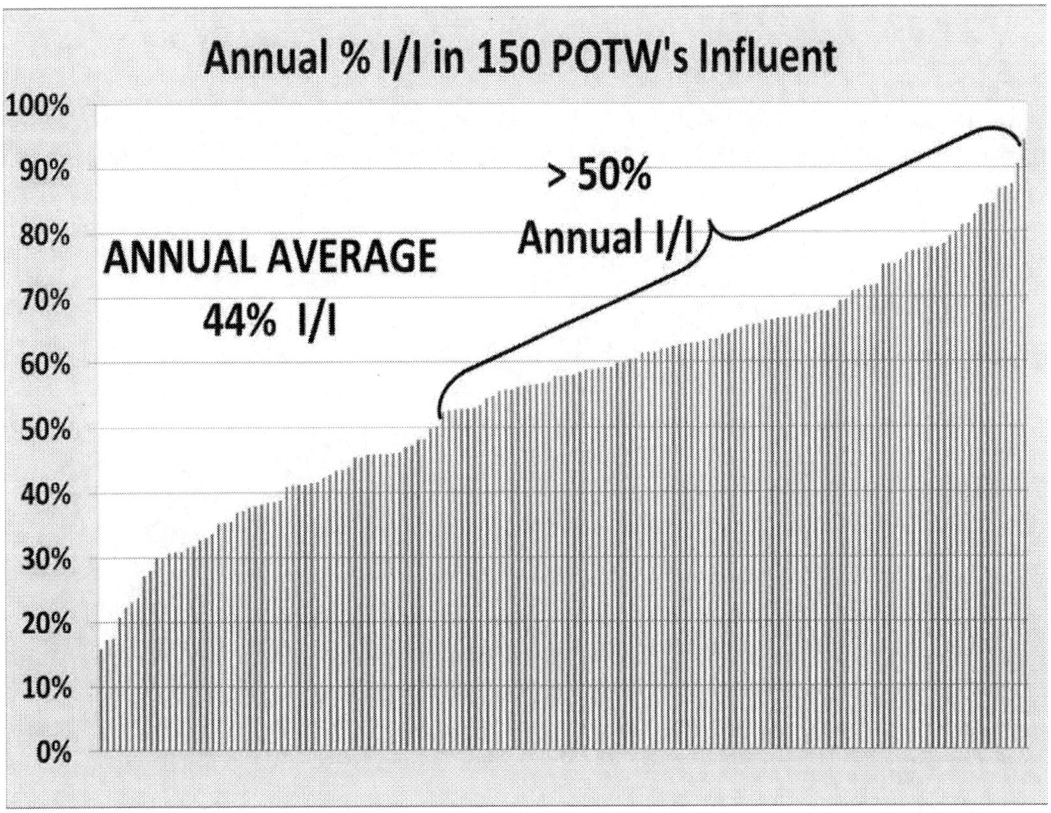
Annual % I/I in 150 POTW's Influent
100%
90%
80%
70%
60%
50%
40%
30%
20%
10%
0%
ANNUAL AVERAGE
44% I/I
> 50%
Annual I/I

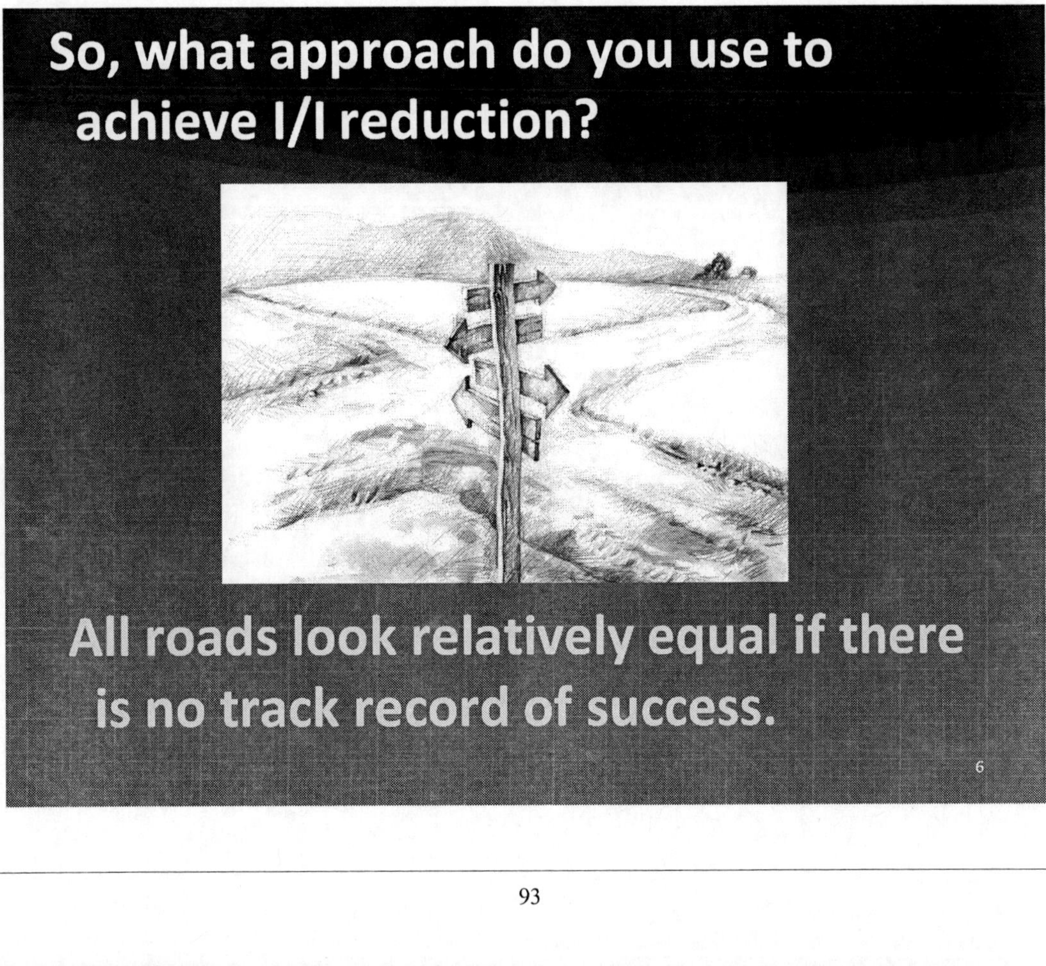
So, what approach do you use to achieve I/I reduction?
All roads look relatively equal if there is no track record of success.
6

Successful Sewer Rehabilitation

- Based on actual field results in Nashville & Brentwood
- Largest published database for measured I/I reduction in the US
- Analyzed 126 miles of rehabilitation (282 miles total - ~ 11% system)
- I/I cut in half
- 123 overflows eliminated
- EPA commends stream improvements

7

REHABILITATION EFFECTIVENESS

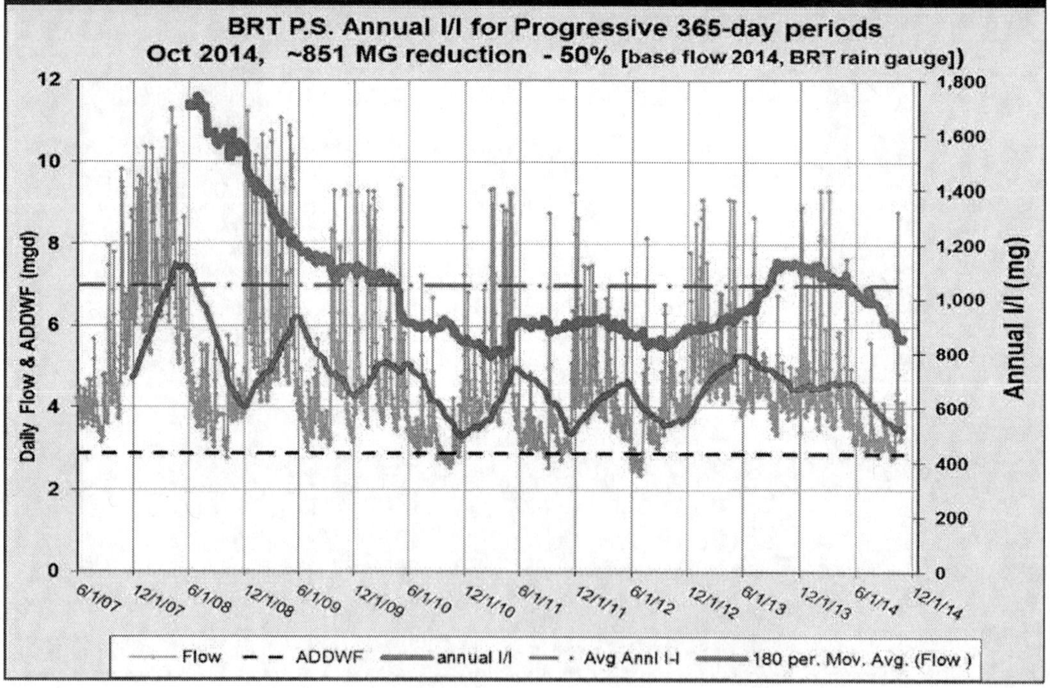

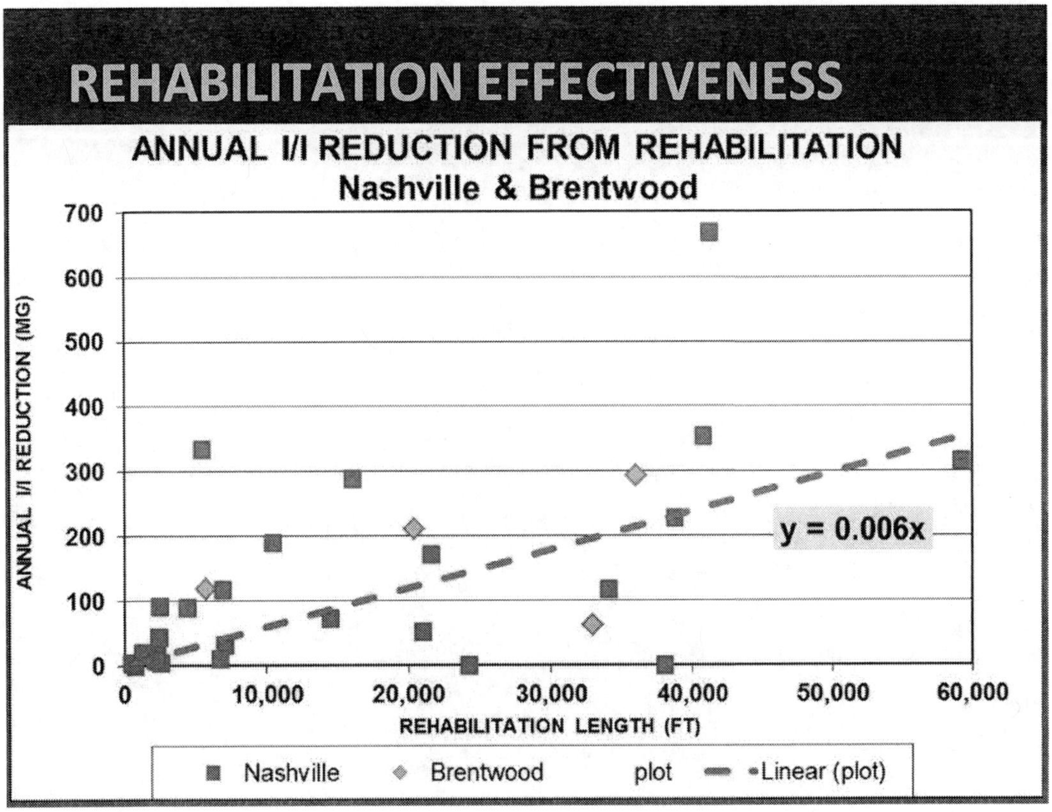
REHABILITATION EFFECTIVENESS
ANNUAL I/I REDUCTION FROM REHABILITATION
Nashville & Brentwood
ANNUAL I/I REDUCTION (MG)
700
600
500
400
300
200
100
0
0
10,000
20,000
30,000
40,000
50,000
60,000
REHABILITATION LENGTH (FT)
y = 0.006x
Nashville
Brentwood
plot
Linear (plot)

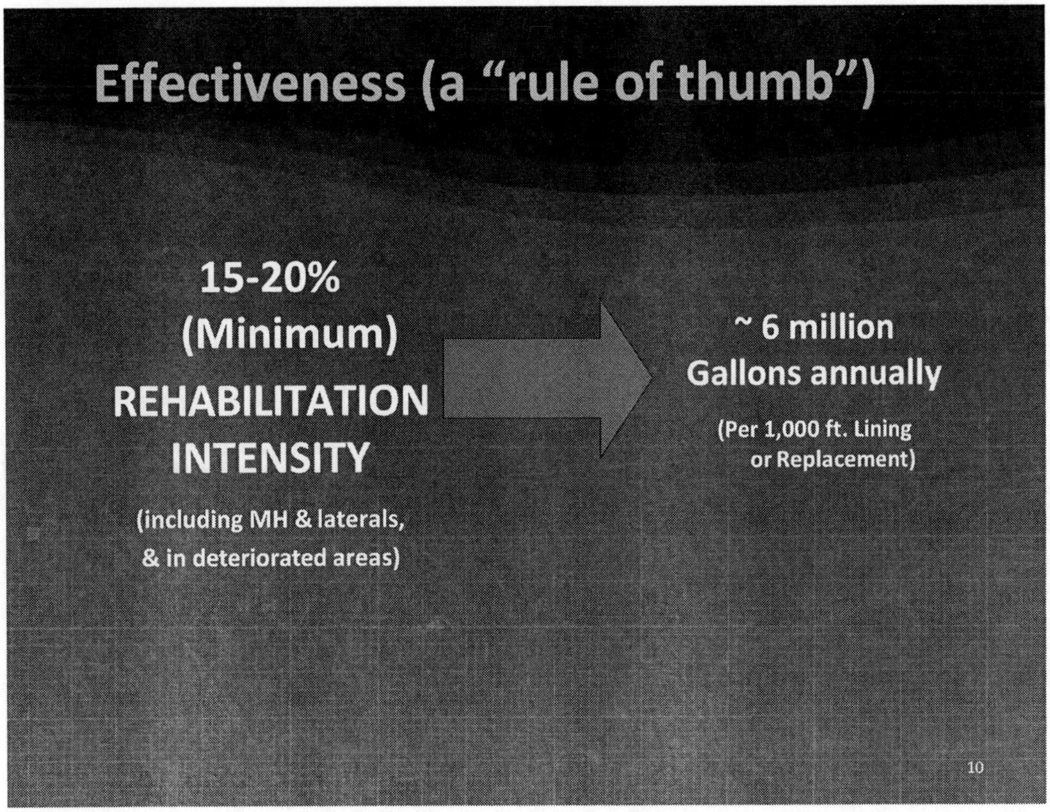
Effectiveness (a "rule of thumb")
15-20%
(Minimum)
REHABILITATION
INTENSITY
(including MH & laterals,
& in deteriorated areas)
~ 6 million
Gallons annually
(Per 1,000 ft. Lining
or Replacement)
10

Successful Rehab Factors

- Define goals
- Extensive flow monitoring & standard procedures for analysis
- System approach – lateral & manhole rehabilitation
- "Targeting" – stop water migration
- Accountability – verify desired results

11

Ten Step Strategy

- Identify Goals
- Select Target Area
- Quantify Problem
- Locate Defects
- Select Pipe Segments
- Estimate Cost-Benefit
- Design & Install
- Verify Performance
- Follow-up Flow Monitoring
- Calculate O & M Savings

12

1 – Identify Community Goals

"Eliminate overflows and basement backups"

- Period of time: 2 years?, 5 years?
- Relate time to rainfall event return interval
- No overflows legally sanctioned

13

2 – Select (and Characterize) Target Area

- Flow monitoring network (~100,000 L.F.) – subdivide the system
- Identify capacity problems
- Calculate observed & potential I/I
- Hydraulic model
- **Prioritize tributary areas**

14

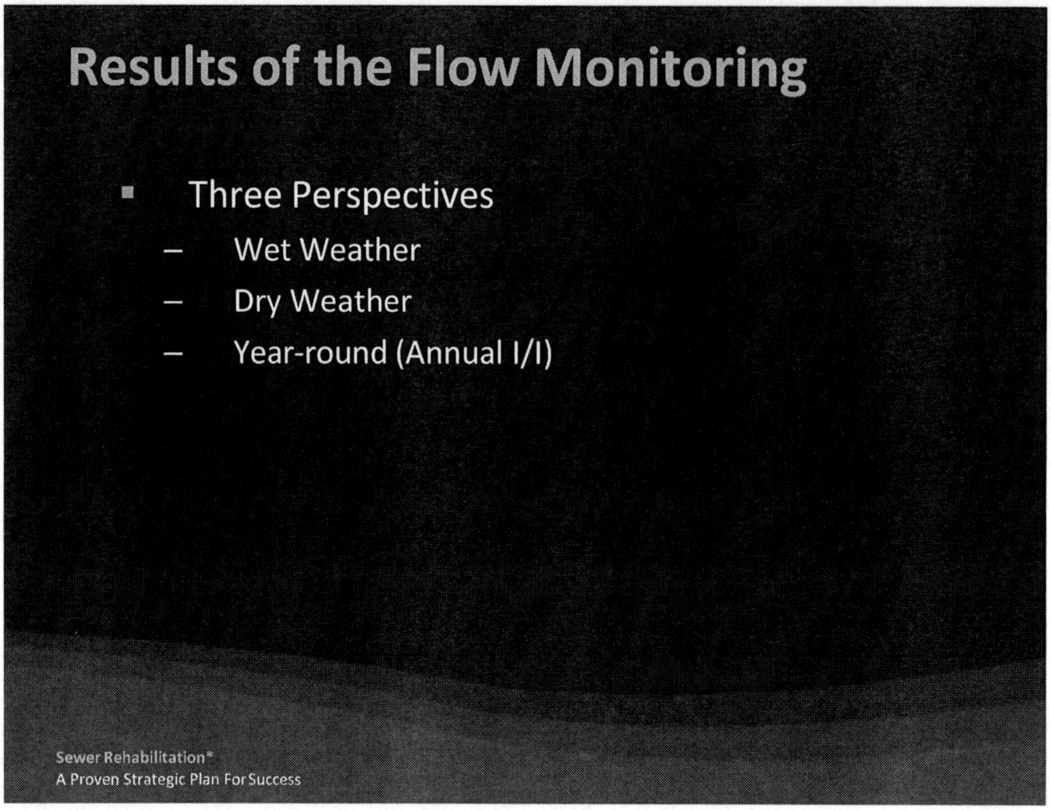
Results of the Flow Monitoring
Three Perspectives
– Wet Weather
– Dry Weather
– Year-round (Annual I/I)
Sewer Rehabilitation*
A Proven Strategic Plan For Success

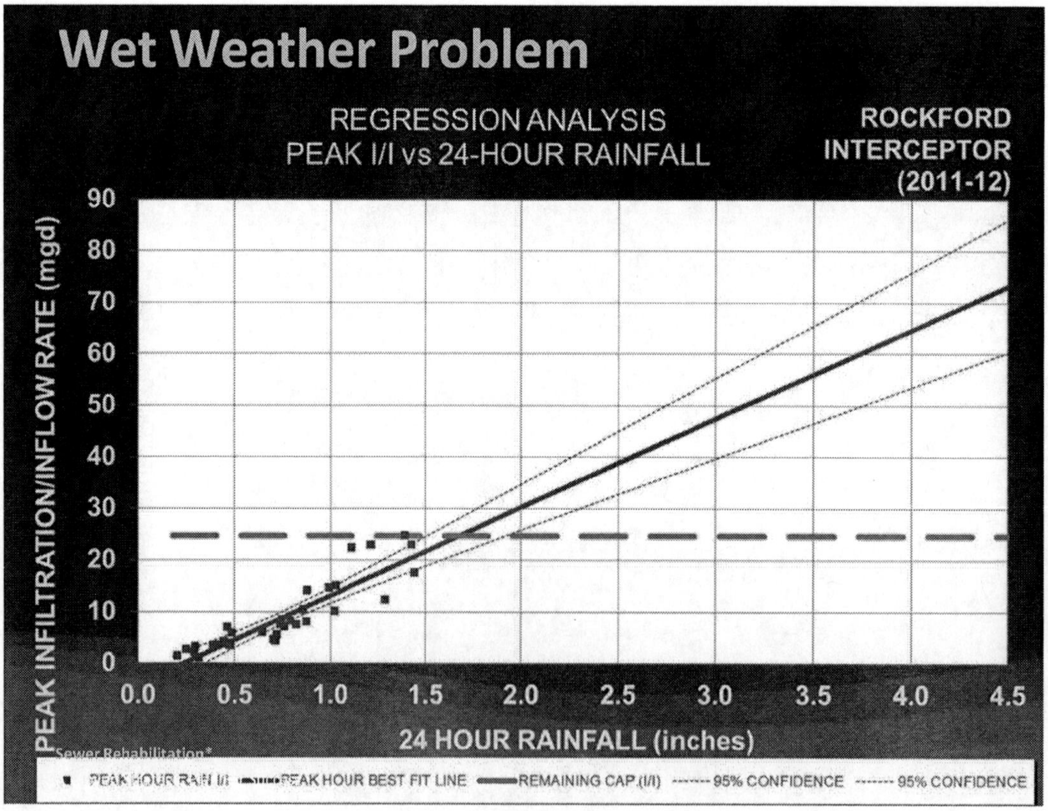
Wet Weather Problem
REGRESSION ANALYSIS
PEAK I/I vs 24-HOUR RAINFALL
ROCKFORD
INTERCEPTOR
(2011-12)
PEAK INFILTRATION/INFLOW RATE (mgd)
0
10
20
30
40
50
60
70
80
90
0.0
0.5
1.0
1.5
2.0
2.5
3.0
3.5
4.0
4.5
24 HOUR RAINFALL (inches)
Sewer Rehabilitation*
PEAK HOUR RAIN I/I
PEAK HOUR BEST FIT LINE
REMAINING CAP.(I/I)
95% CONFIDENCE
95% CONFIDENCE

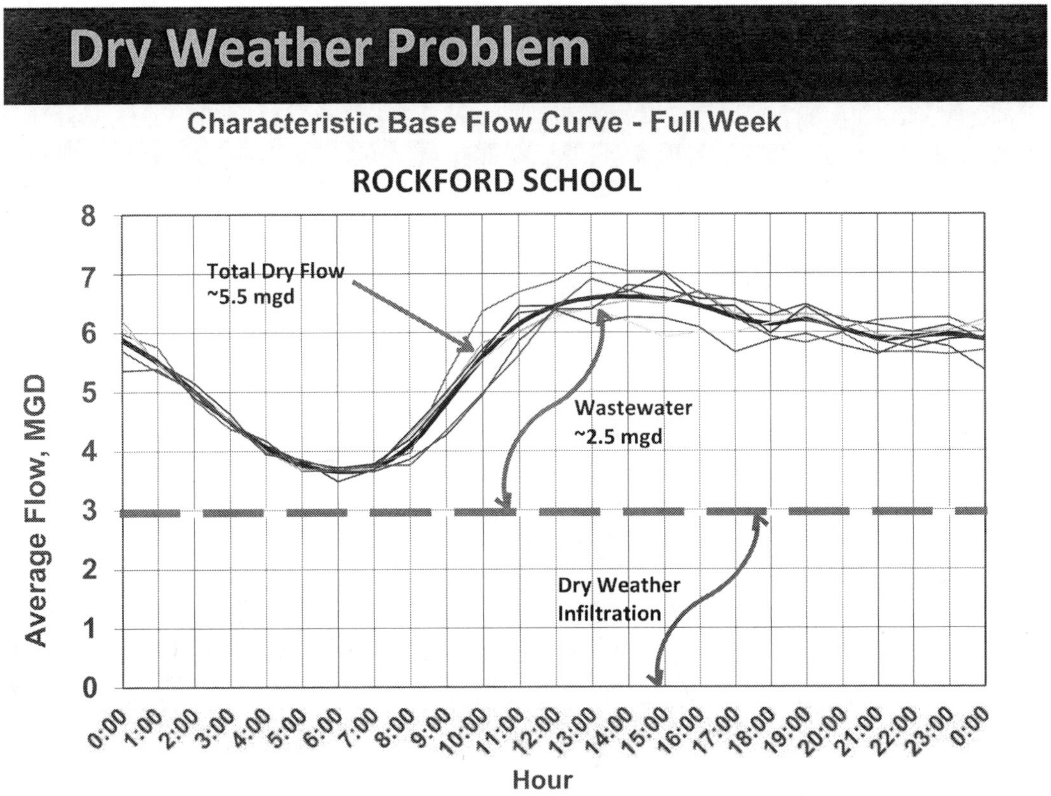
Dry Weather Problem
Characteristic Base Flow Curve - Full Week
ROCKFORD SCHOOL
Total Dry Flow
~5.5 mgd
Wastewater
~2.5 mgd
Dry Weather
Infiltration
Average Flow, MGD
0
1
2
3
4
5
6
7
8
0:00
1:00
2:00
3:00
4:00
5:00
6:00
7:00
8:00
9:00
10:00
11:00
12:00
13:00
14:00
15:00
16:00
17:00
18:00
19:00
20:00
21:00
22:00
23:00
0:00
Hour

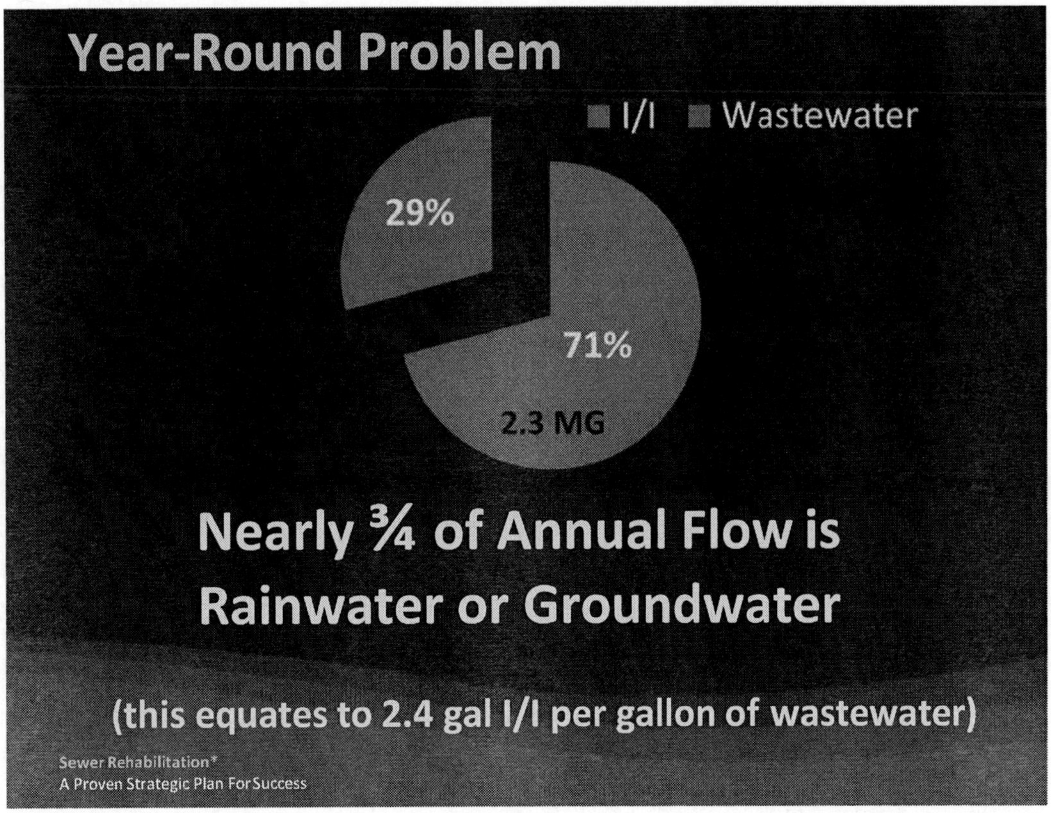
Year-Round Problem
I/I
Wastewater
29%
71%
2.3 MG
Nearly ¾ of Annual Flow is
Rainwater or Groundwater
(this equates to 2.4 gal I/I per gallon of wastewater)
Sewer Rehabilitation*
A Proven Strategic Plan For Success

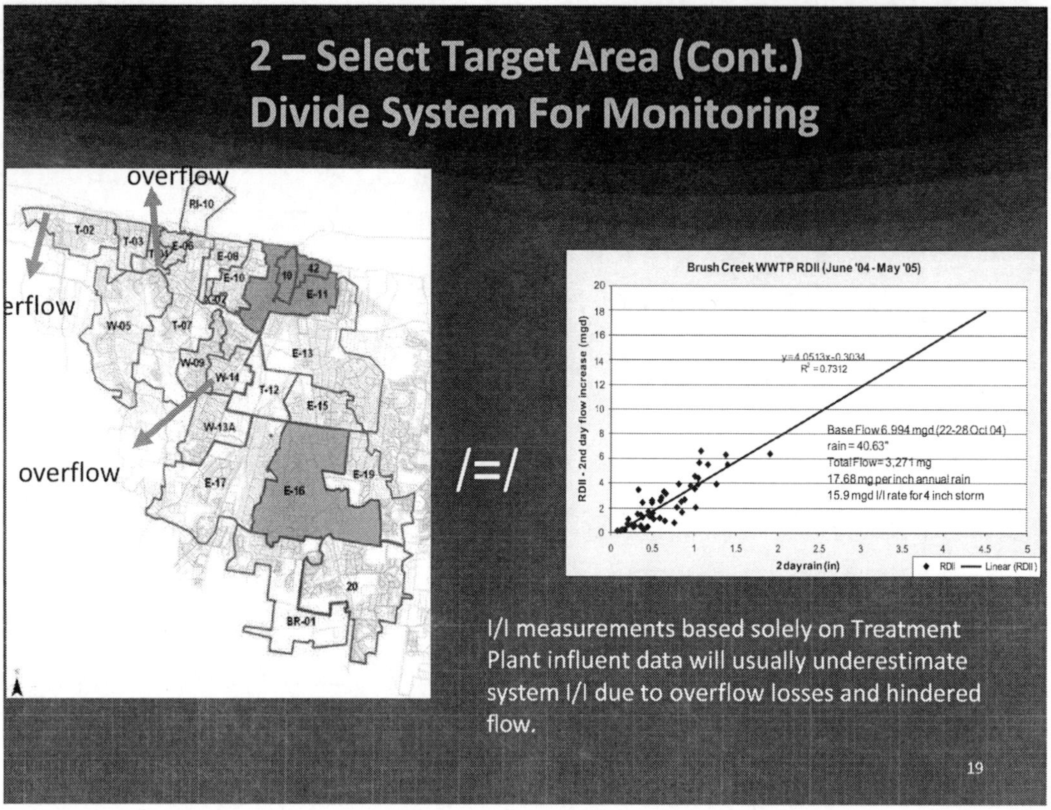
2 – Select Target Area (Cont.)
Divide System For Monitoring
overflow
erflow
overflow
RI-10
T-02
T-03
E-06
E-08
E-10
10
42
E-11
W-05
T-07
W-09
W-14
E-13
T-12
E-15
W-13A
E-17
E-16
E-19
20
BR-01
/=/
Brush Creek WWTP RDII (June '04 - May '05)
RDII - 2nd day flow increase (mgd)
y = 4.0513x - 0.3034
R^2 = 0.7312
Base Flow 6.994 mgd (22-28 Oct 04)
rain = 40.63"
Total Flow = 3,271 mg
17.68 mg per inch annual rain
15.9 mgd I/I rate for 4 inch storm
2 day rain (in)
RDII
Linear (RDII)
I/I measurements based solely on Treatment Plant influent data will usually underestimate system I/I due to overflow losses and hindered flow.
19

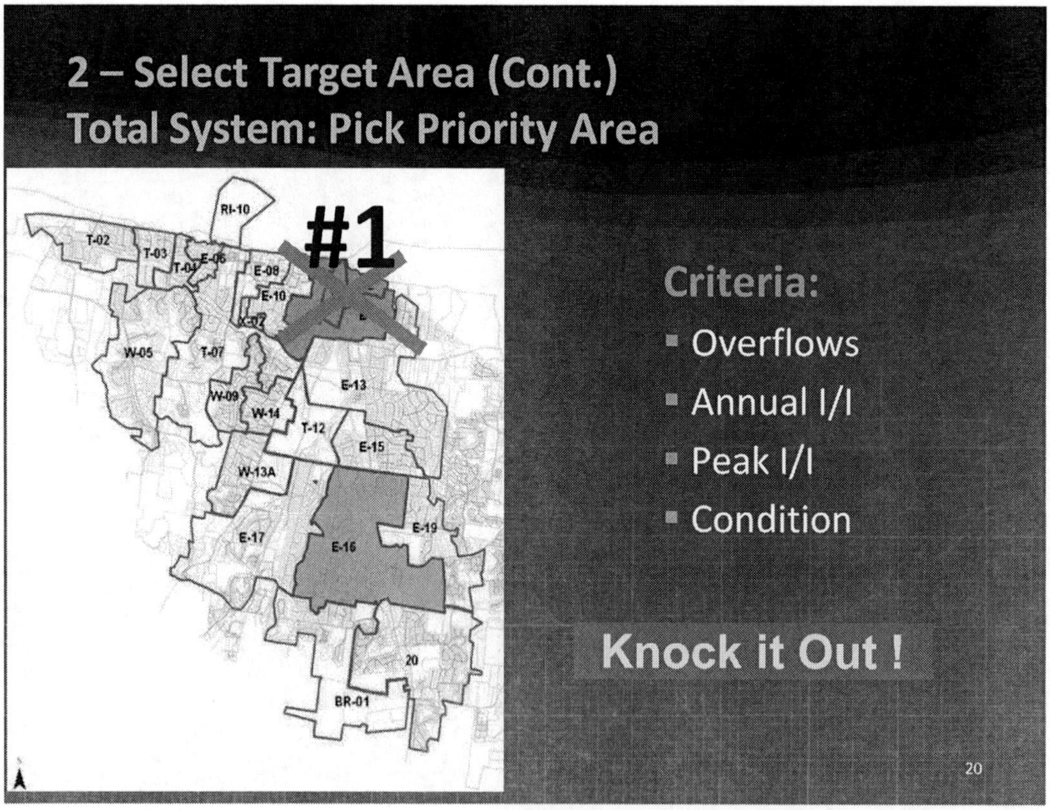
2 – Select Target Area (Cont.)
Total System: Pick Priority Area
RI-10
#1
T-02
T-03
E-06
E-08
E-10
W-05
T-07
W-09
W-14
E-13
T-12
E-15
W-13A
E-17
E-16
E-19
20
BR-01
Criteria:
Overflows
Annual I/I
Peak I/I
Condition
Knock it Out !
20

3 - Quantify Problem Conditions
(refine the process for the target areas)

- Intensive monitoring in top priority tributary areas (8,000 - 15,000 LF)
- Observed and potential I/I
- Additional capacity problems

21

3 - Quantify Problem Conditions (cont.)
Hydraulic Capacity Analysis

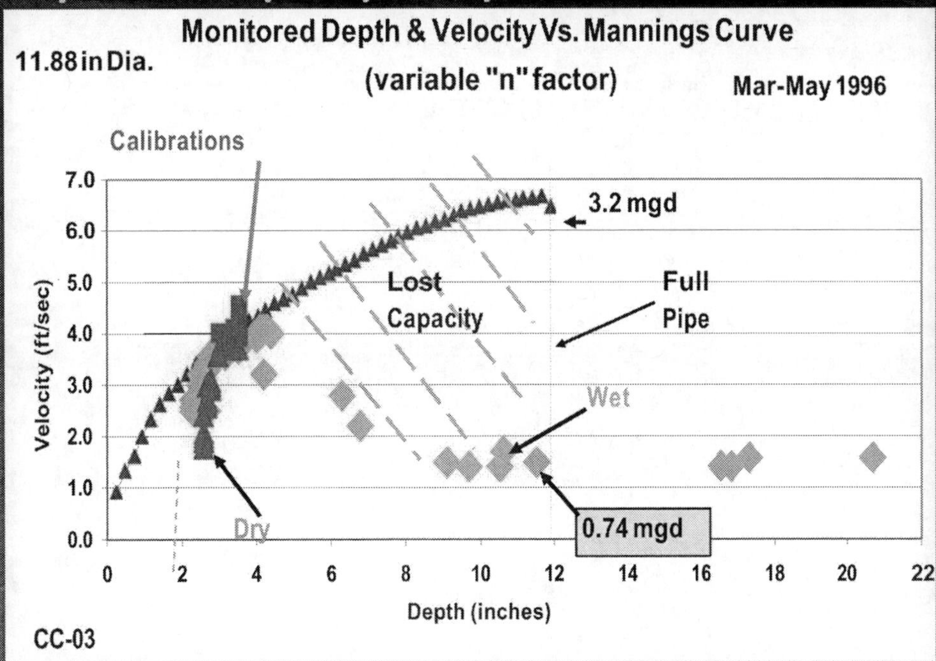

Quantify the I/I (Observed and Potential)

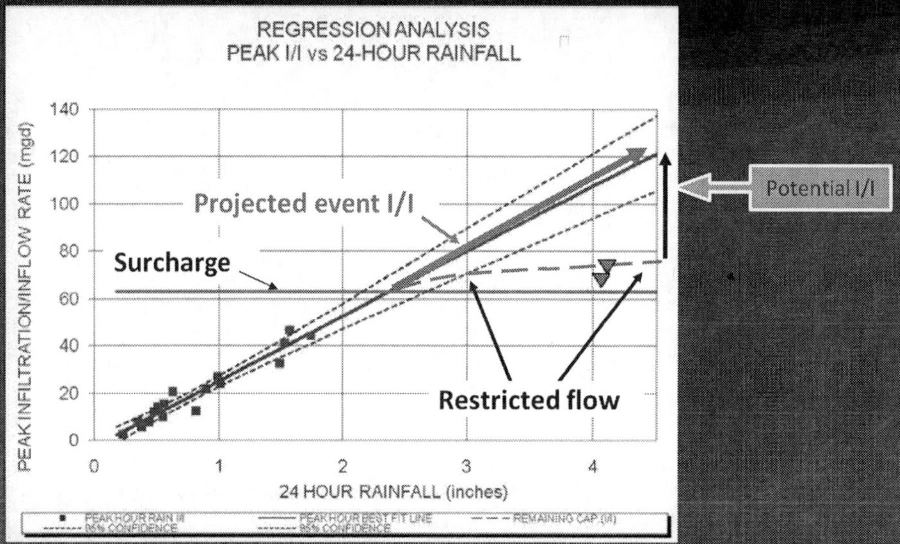

r =0.97, 95% CONF. = 26%

23

3 - Quantify Problem Conditions (cont.) "Potential" I/I

- I/I which cannot enter the sewer because the pipe is already overloaded!
- Obscures overall I/I removal goals
- Monitor depth & velocity
- Extrapolated

Monitoring in the upper reaches of a basin (upstream of significant hindered flow conditions) allows a more realistic estimate of I/I !

24

Data Interpretation

- **Need to standardize criteria**
  - 24-hour rainfall more reliable than peak hour rain for predicting peak design I/I
  - **AMC –** Antecedent Moisture Condition is critical for selecting valid rainfall events
  - **Hindered flow -** Potential I/I There are ways to correct for this, however the analyst must be aware of this condition
  - Underestimating the peak flow can result in the inadequate design of new facilities

25

4 - LOCATE & IDENTIFY DEFECTS

- Televise target area system (may be concurrent with monitoring)
- Categorize defects with respect to I/I potential

26

4 - Locate & Identify Defects (Cont.)

- "Invisible" defects – electric field leak detection, segmental isolation
- Gross inflow (roof drains, etc.)

27

Electroscan

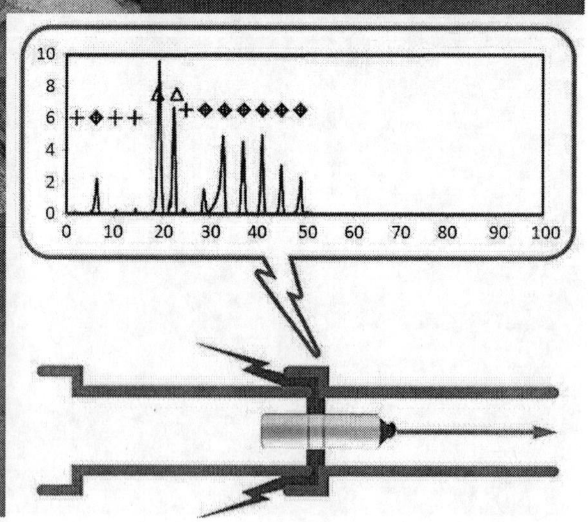

5 - Select Segments For Rehabilitation

- Categorize & color code lines
  - 3 or more major defects
  - 1–2 major defects
  - No major defects
- "3 or more" – renew!
- Check adjacent segments
- Renewal "intensity" – range of 15–20% (**<u>or greater</u>**) in first round

29

Connect The Dots

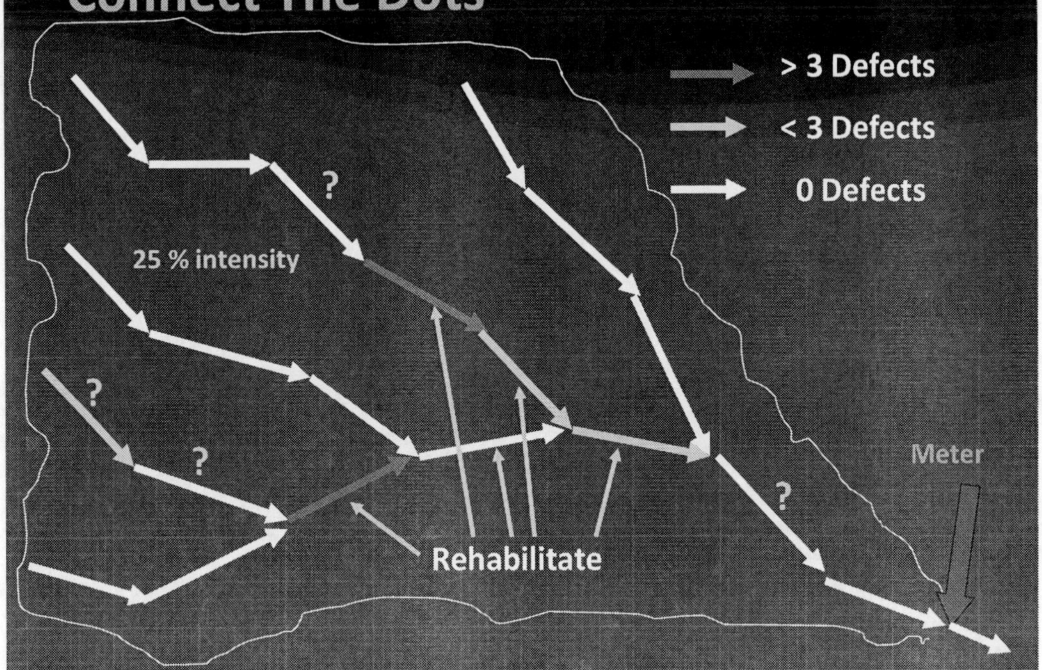

30

Sewer Rehab Strategy: Halt Migration!

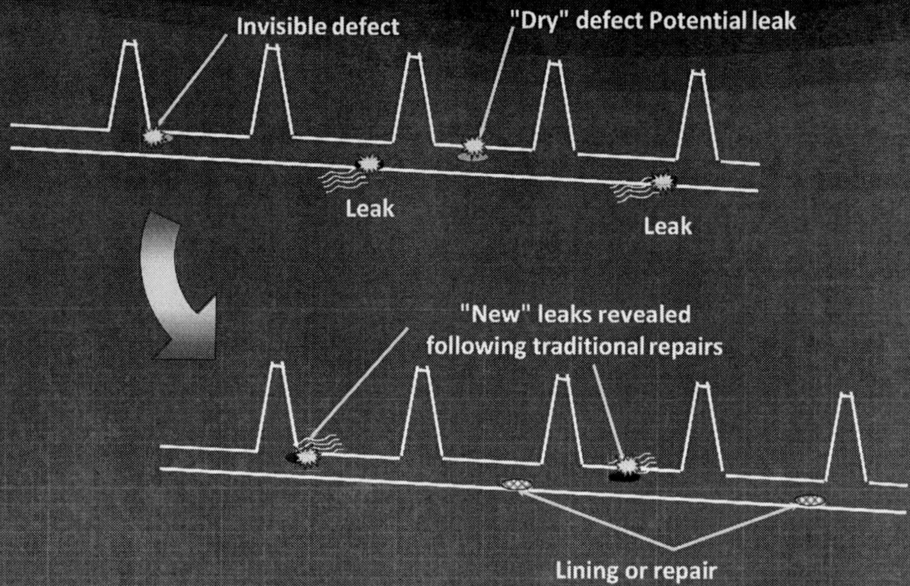

31

6 - ESTIMATE COST-BENEFIT

- Compare renewal costs to: O & M costs ($1.73 – $1.87/ 1,000 gal)
- At least 50% I/I removal
- Costs:
  - Lining (8–10" cipp) ~ $43 / lf
  - Laterals ~ $2,500 ea. (1/ 200 lf)
  - Manholes ~ $1,000 –1,300 ea. (1/200 lf)
  - Engineering ~ 12% – 15% of total
  - Owner's expenses (admin, etc.)

(Gross= ~$100 to $132/ft rehab)

32

7 - Design and Install Rehab

- Halt migration from outside pipe
- Halt migration ("tracking") inside pipe
- Provide seal at manhole junction
- Renew service laterals

Over 10,000 service laterals rehabilitated in Nashville's Program

33

Peak Hour I/I Reduction with Lateral Rehab

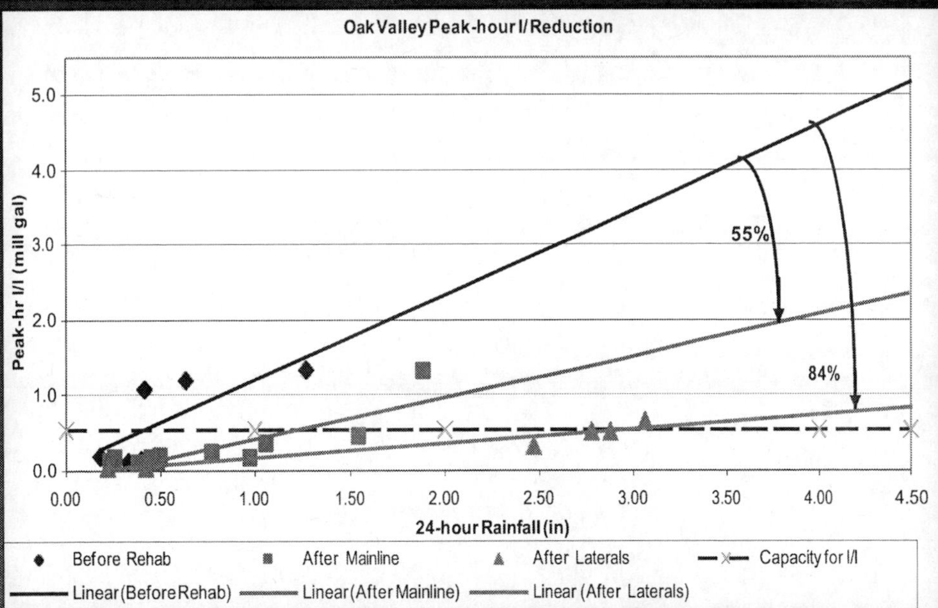

34

8 - Performance Testing

- Air-test sewer service connection!
  - Most vulnerable part
  - Not accepted until performance verified

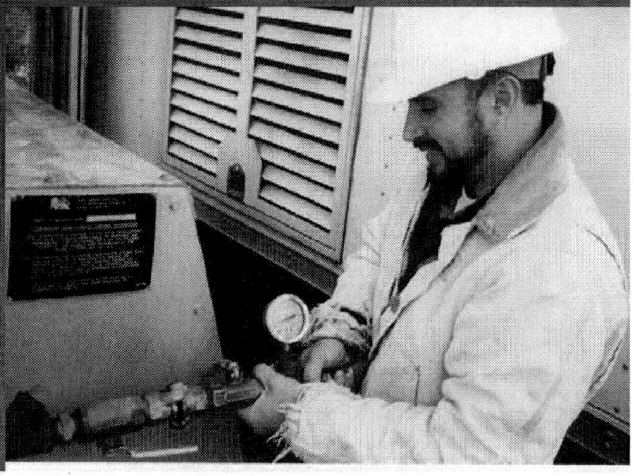

9 - Follow-up Flow Monitoring

- Quantify I/I reduction
- Standardized I/I analysis
- TV during wet weather
- Rerun hydraulic model
- Determine if design goals met!

36

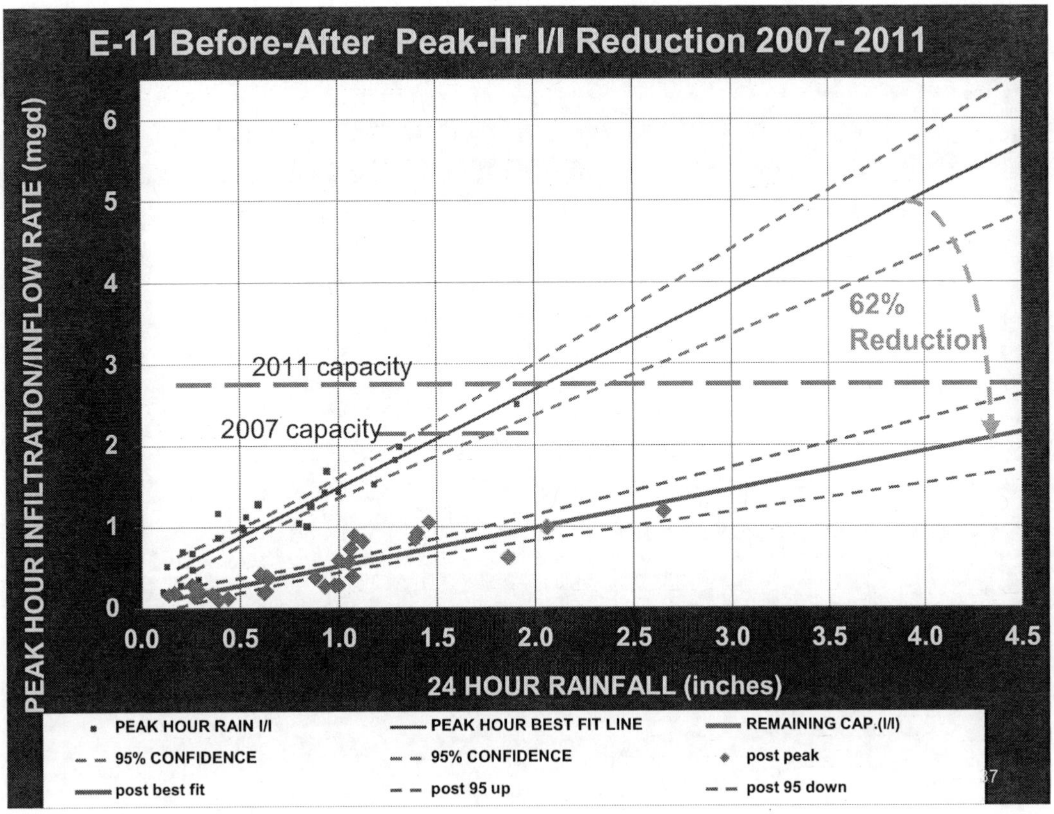

10 - CALCULATE O&M SAVINGS

- Possible 10-13 Year payback (on installation, design, investigation costs – TOTAL PROGRAM)
- Provides data for future program planning
- Accountability to community

- Brentwood is saving ~ $1.6 million/year by eliminating 851 million gallons of I/I annually – pays for the program in 13 years

38

Recap: Successful Rehab Factors

- Extensive flow monitoring
- Lateral renewal to easement
- "Targeting"– lining selected by observed defects, age, proximity, migration potential, surface water
- Performance (air) test line and lateral

39

Questions ?

40

THE TOTAL SOURCE

educational sessions

NASHVILLE'S REHAB PROGRAM*

A strategic plan removes 3.6 billion gallons of I/I

PRESENTED BY

George Kurz, P.E., DEE

George.kurz@comcast.net

** A significant portion of this work was conducted with CTE-AECOM as part of the Nashville Overflow Abatement Program 1991-2005*

? ? ?

?

IS SEWER REHABILITATION EFFECTIVE ?

? ? ?

? ?

2

MEASURES OF EFFECTIVENESS

- Reduction of I/I
- Reduction of Overflows
  - (and concurrent water quality improvement)

3

NATIONAL TRENDS

WERF* - study of trends of I/I reduction in 44 utilities:

"Unfortunately, none of the information between projects was comparable..."

** Water Environment Research Foundation*

4

LACK OF INFORMATION

- Holds our industry back
- Discourages investment in infrastructure renewal
- Hinders environmental improvements

5

Successful Trenchless Rehabilitation

- Based on **actual field results** in Nashville, TN
- **Largest published database** for measured I/I reduction in the US
- Analyzed **94 miles of rehabilitation** (250 miles total - ~ 10% system)
- **I/I cut in half**
- **123 overflows eliminated**
- EPA commends **stream improvements**

6

Strategy for Application of Products is as Important as Product Quality

- Product Effectiveness "***Better Be***" Equal (no leaks ***where applied***)
- Therefore: Concentrate Effort to Defeat Migration
- Dry Defects on Video May be Wet Weather Leaks
- Some Defects May Not be Visible on Video
- Pipe Segments: "Connect-the-Dots"

7

System Approach

Ten steps:

1. Identify Goals
2. Select Target Area
3. Quantify Problem
4. Locate Defects
5. Select Pipe Segments
6. Estimate Cost-Benefit
7. Design & Install
8. Verify Performance
9. Follow-up Flow Monitoring
10. Calculate O & M Savings

8

Nashville Design Criteria -- Total System Approach

Goal:

"Containment" for flows from 5-year, 24-hour rainfall

Policy:

- All service laterals connected to the rehabilitated pipes will be renewed to the easement line or the property line.
- All manholes connected to rehabilitated pipes will be renewed.

9

THREE TYPES OF I/I MEASUREMENTS USED FOR NASHVILLE PROGRAM

- Annual I/I
  - (normalized for a year with average rainfall)
- I/I Rates Projected For a Standard Storm Event
  - 24 Hour I/I from 24 Hour Rainfall event
  - Peak Hour I/I rate from 24 Hour Rainfall Event

10

Evaluation of Work In Nashville

- 27 areas analyzed
- 94 miles rehab'd
- (1/3+ of total work in Nashville)
- + 50% I/I reduction (in the project areas)

Project	Length	Annual I/I Reduction		5-Year, 24-Hour I/I Reduction		5-Year, Peak-Hour I/I Reduction	
	(ft)	(mg)	(%)	(mg)	(%)	(mgd)	(%)
Berwick Trail	2290	15	52				
Brookwood	2,550	4	8				
Cleeces Ferry (15)	2,143	9.3	50	0.151	66	0.426	62
Clifton Park	38,744	228	62	5.25	43	10.788	40
Foster Ave.	10,445	189	52	3.998	56	7.561	57
Gibson Cr. (GC4 & DR6)	38,006	18	33	1.848	54	2.385	42
Hermitage Hills	34,100	116	47	2.536	46	2.93	32
Hopedale	16,084	289	72	1.376	49	0.271	8
Kenner	2,463	91	88	0.508	31		
LE Nashville (5&6)	7,077	32	11	0.687	56	0.809	22
Madison Heights	716	0.2	7	0.147	98	0.513	96
Nolensville Rd	2,391	43	37	0.884	25	1.016	
OakValley	4,400	89	76	3.868	90	4.331	84
Osage	5,440	334	52			1.98	18
Paragon Mills(P1&2,S3&4)	21,584	171	20	4.119	39	30.853	66
Post Rd	6,931	116	70	3.79	67	5.438	69
Rainbow Terr.	515	4	98	0.087	89	0.235	91
River Dr.	1,259	20	50	1	71	3	83
Riverside Dr.	2,285	7	47	0.463	40	1.841	63
Shelby Park A-1`	40,830	353	73	3.71	77	6.05	71
Smith Springs	59,178	316	38	8.924	51	16.772	48
Sugartree	41,255	668	37	15.38	49	17.27	35
Vandiver	14,500	72	44	2.74	40	4.45	57
Wallace Rd	24,267	0		8.65	69	9.41	55
West Linden	21,005	51	45	0.575	17	0.456	7
Whites Creek	88,000	400	33	12.018	41.4	19.469	42
Williamson Ferry	6,719	11	8	0.4	13	1.8	30

12

Successful Rehab Factors

- Define goals
- Extensive flow monitoring & standard procedures
- Lateral rehabilitation
- "Targeting" – stop water migration
- Accountability – verify desired results
- Planning for a <u>comprehensive</u> program
  - **Piecemeal or "find & fix" approaches are not adequate to stop migration !!**

13

Total System: Pick Priority Area

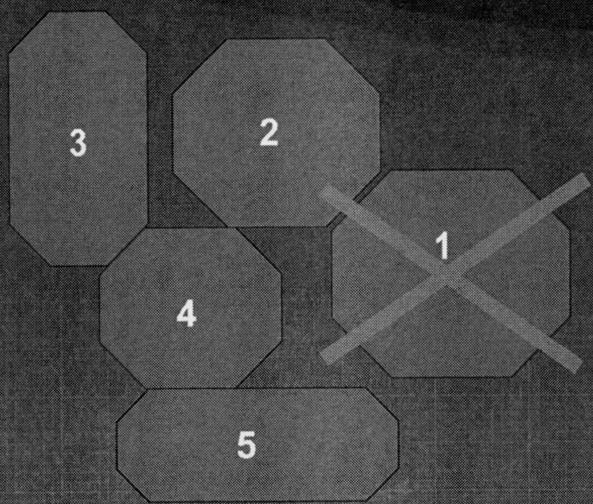

Work on the top priority area -

- AND KNOCK IT OUT !

14

Lateral Rehabilitation Effectiveness

- Downington, PA -area 5
  - 38 laterals in 1,513' CIP - 100% treated
  - 97% reduction - 22 mg annually
- Stege San. Dist, CA - subarea N
  - 111 laterals in 13,400' Slip-PE & MH - 100%
  - 86% reduction
- Nashville, TN - Oak Valley
  - 63 laterals in 4,400' CIP - 41% treated
  - 77% reduction (20%-laterals) - 67 mg annually

Over 10,000 service laterals rehabilitated in Nashville's Program

15

Flow & Rain Monitoring Is A Key Tool

- Target and prioritize basins
- Correctly interpret hydraulic conditions
- Conduct model calibration
- Monitor post-rehabilitation to verify project effectiveness

16

Documentation

- Reliable Flow Monitoring Data (USEPA – ETV verified equipment)
- Intensive Flow & Rainfall Monitoring (90+ long-term, 700+ temp, 20 rain gauges)
- Standardized (non-proprietary) Analytical Methods for I/I Analysis
- Standardized Criteria for Reporting I/I

17

Before-after Project Evaluation

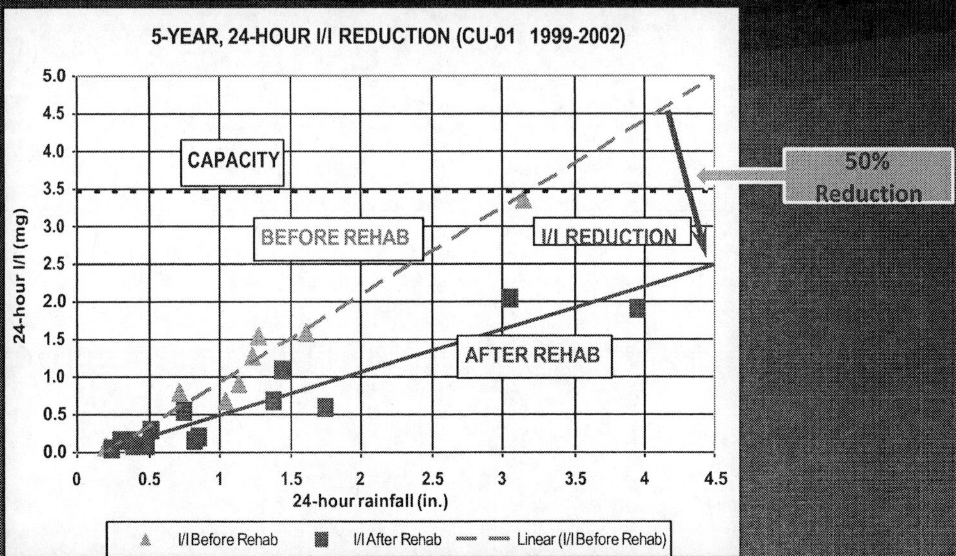

18

Proper Interpretation Considering Site Hydraulics

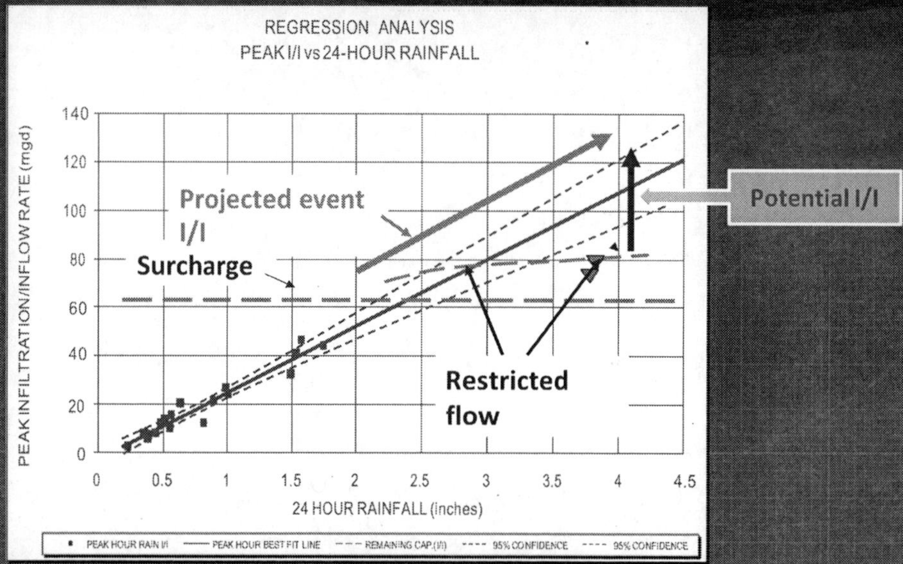

R=0.97, 95% CONF. = 26%

19

STANDARDIZED STATISTICAL METHODS

- Use statistical tests to evaluate the analytical methodology & results
- r value: coefficient of linear correlation (r *not related to RTK method)*
- r^2 value: per cent of variation of I/I attributable to 24-hour rainfall
- 95% confidence interval

20

Standardized I/I Analysis

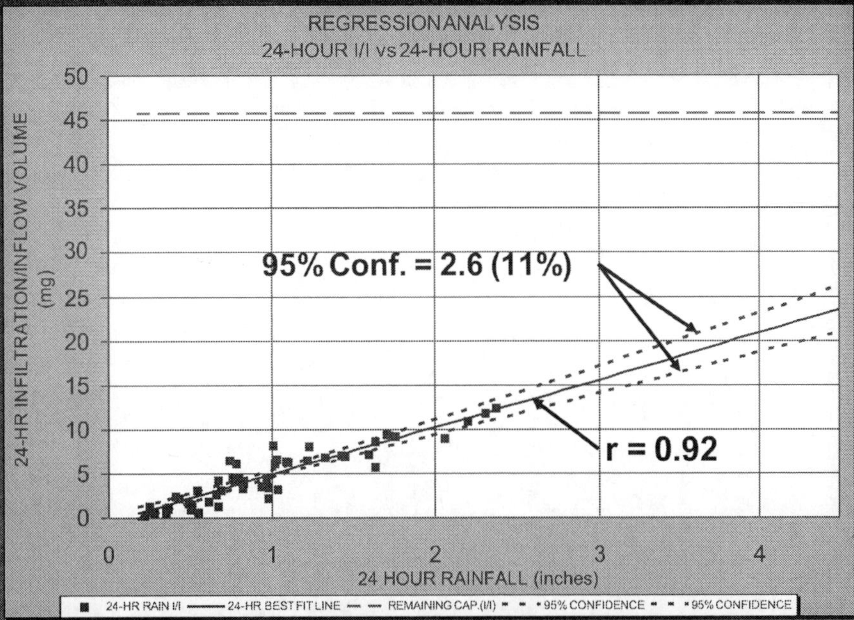

21

Standardized Approach (summary highlights)

- Design storm (e.g., 5-year, 24-hour event)
- Statistical criteria for QA/QC
- Define rainfall event (e.g., 10 hrs dry prev.)
- Use maximum number of events in period
- Minimize analyst bias (selectivity)

22

Nashville measured results

For the 27 areas (94 miles) analyzed so far:

- I/I eliminated annually: 3.647 billion gal.
- Annual I/I eliminated: 49.6%
- 24-hour, 5-year I/I reduction: 53%
- Peak-hour, 5-year I/I reduction : 52.2%

23

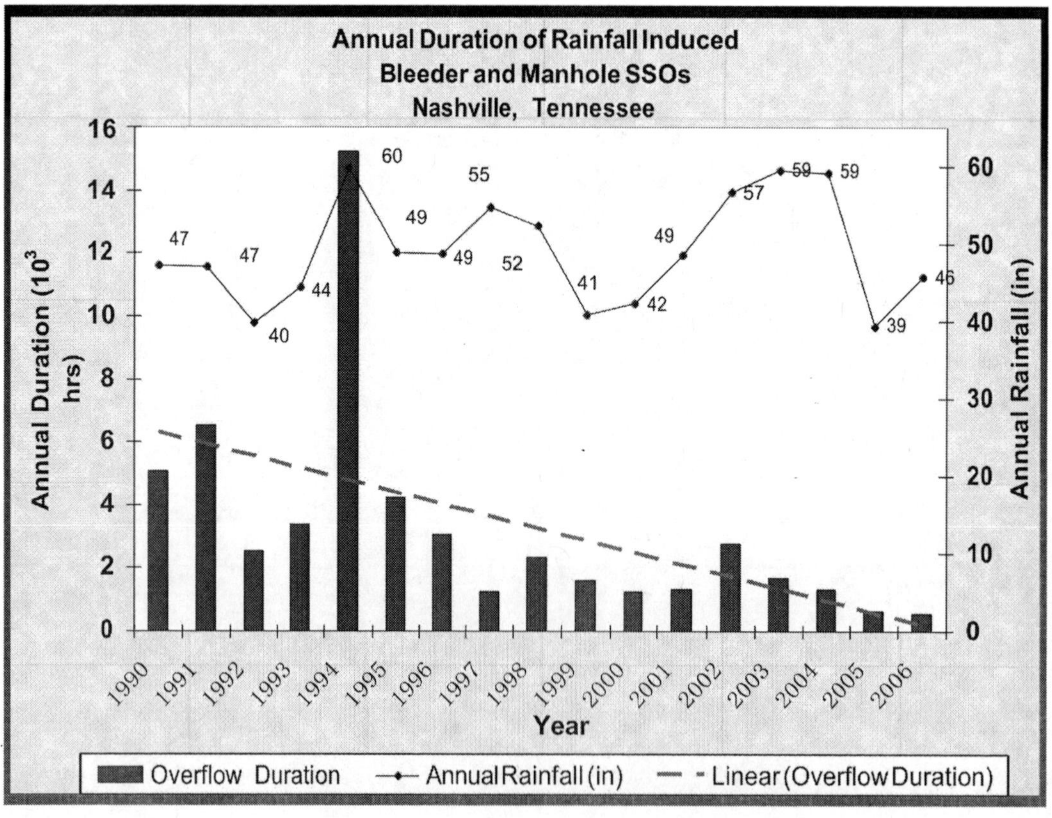

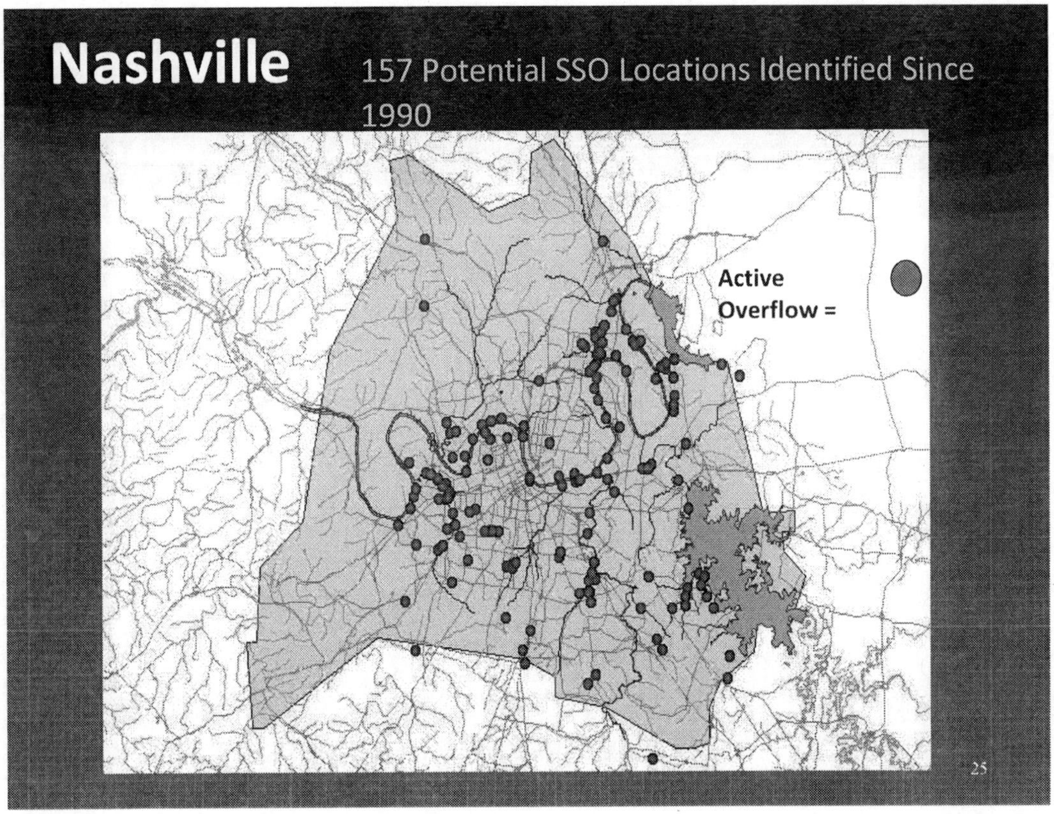
Nashville
157 Potential SSO Locations Identified Since 1990
Active Overflow =
25

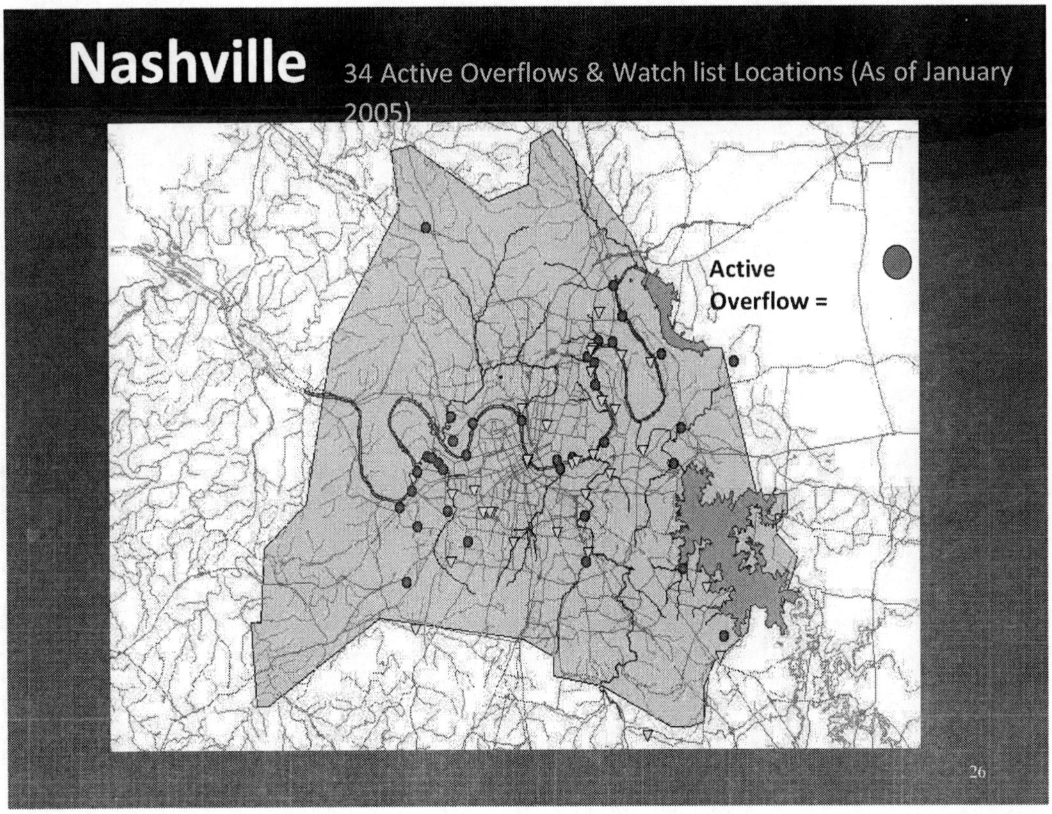
Nashville
34 Active Overflows & Watch list Locations (As of January 2005)
Active Overflow =
26

Governor's Ceremony 11-25-2002

Cleaning up the Cumberland

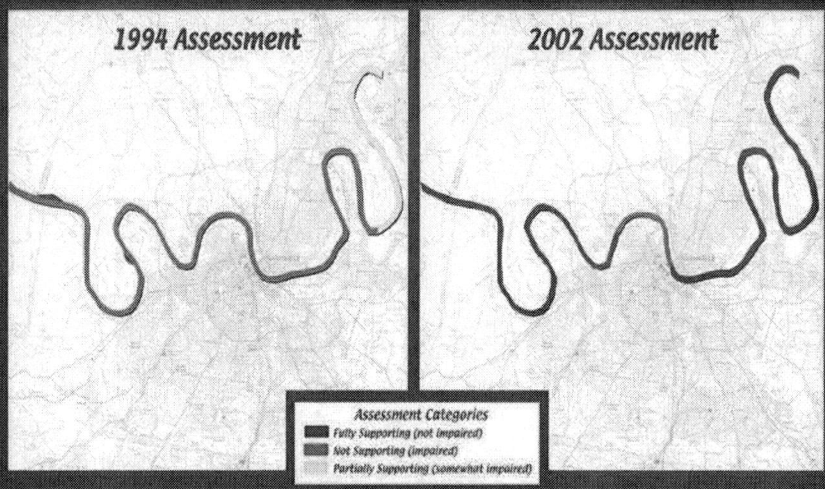

Graphic Courtesy of Tennessee Department of Environment and Conservation

27

EPA & Governor's Deposting Ceremony

November 2002

- 33 miles "de-posted"
- Many segments on 303d list removed or no longer attributed to collection system failure

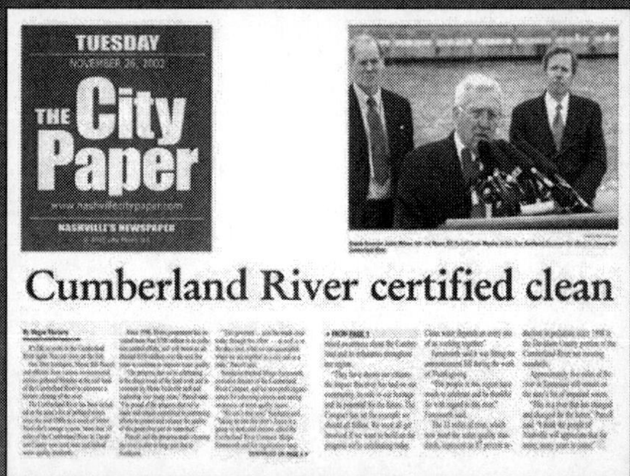

TUESDAY
NOVEMBER 26, 2002
THE City Paper
www.nashvillecitypaper.com
NASHVILLE'S NEWSPAPER

Cumberland River certified clean

28

Strategic goals met

- I/I reduction
- SSO reduction
- Stream improvement

29

Questions ?

NASHVILLE'S REHAB PROGRAM
A strategic plan removes 3.6 billion gallons

30

THE TOTAL SOURCE

educational sessions

Rehabilitation of Sewer Service Laterals

Crucial to program success!

PRESENTED BY

George Kurz, P.E., DEE

(615) 714-6120 | george.kurz@comcast.net

** A significant portion of this work was conducted with CTE-AECOM as part of the Nashville Overflow Abatement Program 1991-2005*

Summary

- Why renew laterals ?
- Sparse project information
- Methods
- Example pilot project
- Verify effectiveness

2

Why Is Sewer Lateral Renewal Needed?

- Serviceability of lateral for customer
- Structural integrity
- I/I reduction

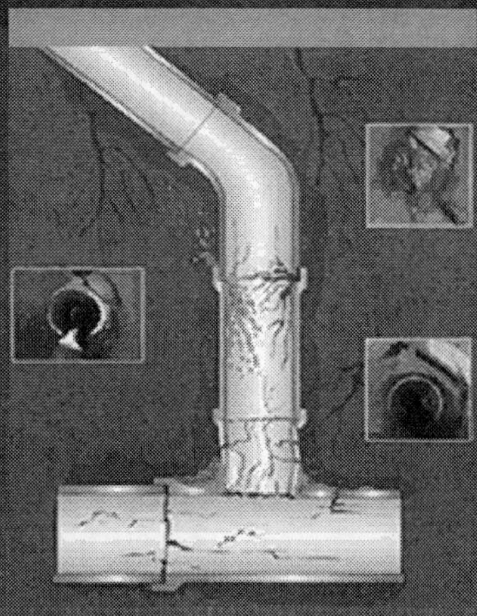

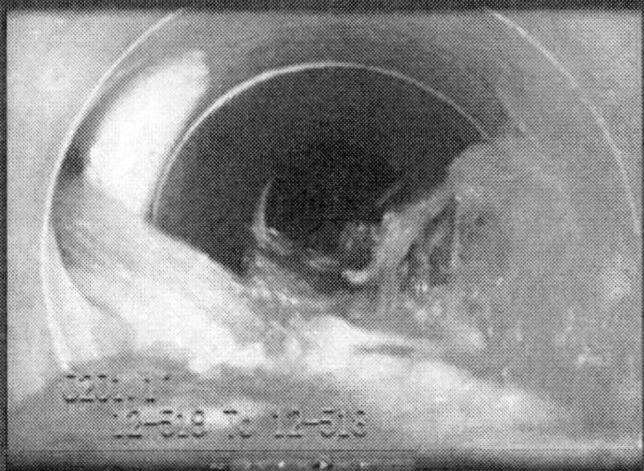

3

Private Sector I/I

- 1983 RJN report to EPA
- Study of 19 municipalities – Illinois
- 63% "inflow" from private sector (weighted average)
- May be based on source estimates
- Probably includes downspouts & foundation drains etc.

4

I/I Removal with Various Lateral Treatment

LOCATION	METHOD	mgd	%Red.	Footage	MG/1000-ft
Pembroke Pines, FL	cip & ff 3 gr lt	0.33	21	8,770	13.7
Dania, FL	cip & ff 122 gr lt	0.97	60	48,100	7.4
Miramar, FL	cip & ff 31 gr lt	0.81	39	22,780	13
Hollywood, FL	cip & ff 158 gr lt	3.53	33	98,710	13.1
Colonial Heights, VA	Cip- lateral cut	0.2	-	5,500	13.3
El Paso, TX	Cip- lateral cut	0.071	51	10,000	2.6
Lynn, MA-35%	Cip- lateral cut-214, MH-20	2.8	-	~26,000	39.3

5

Lateral Rehabilitation Methods

- Replacement and point repairs
- Fusion weld
- Lining – (cured-in-place)
- Grouting
- Flood Grouting
- Bursting
- Robotics
- Vacuum Excavation – Clean-out installation

6

Dig and Replace
7

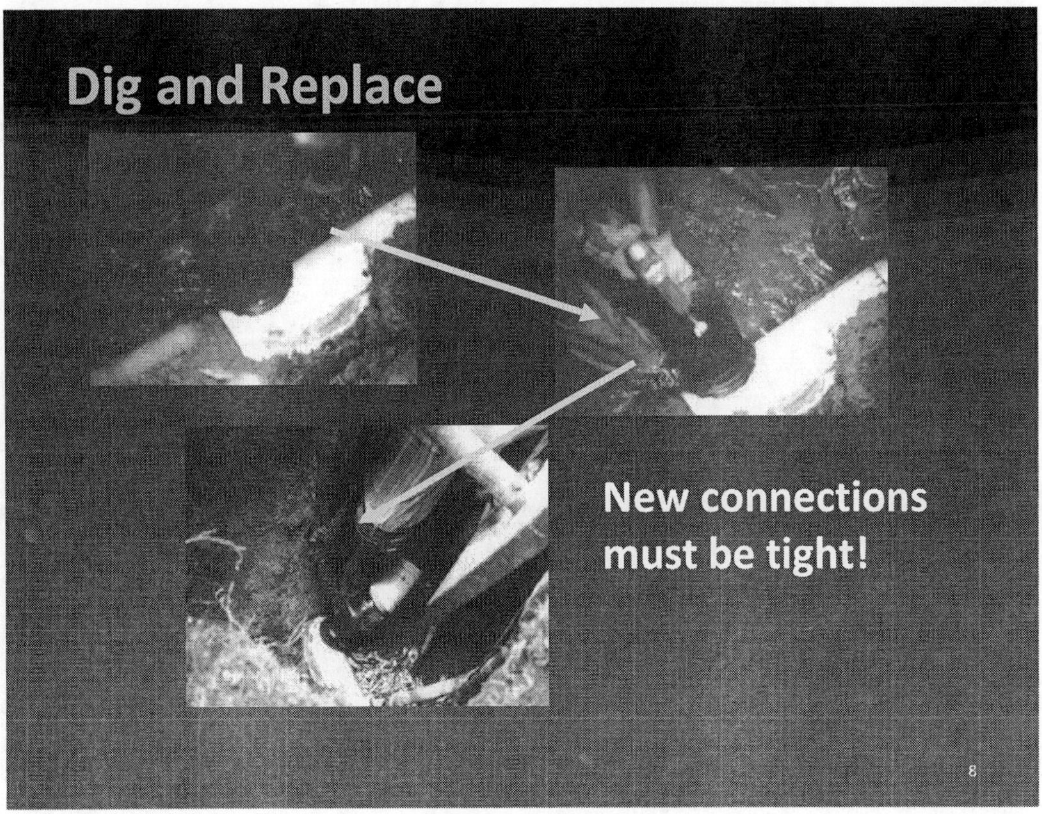
Dig and Replace
New connections
must be tight!
8

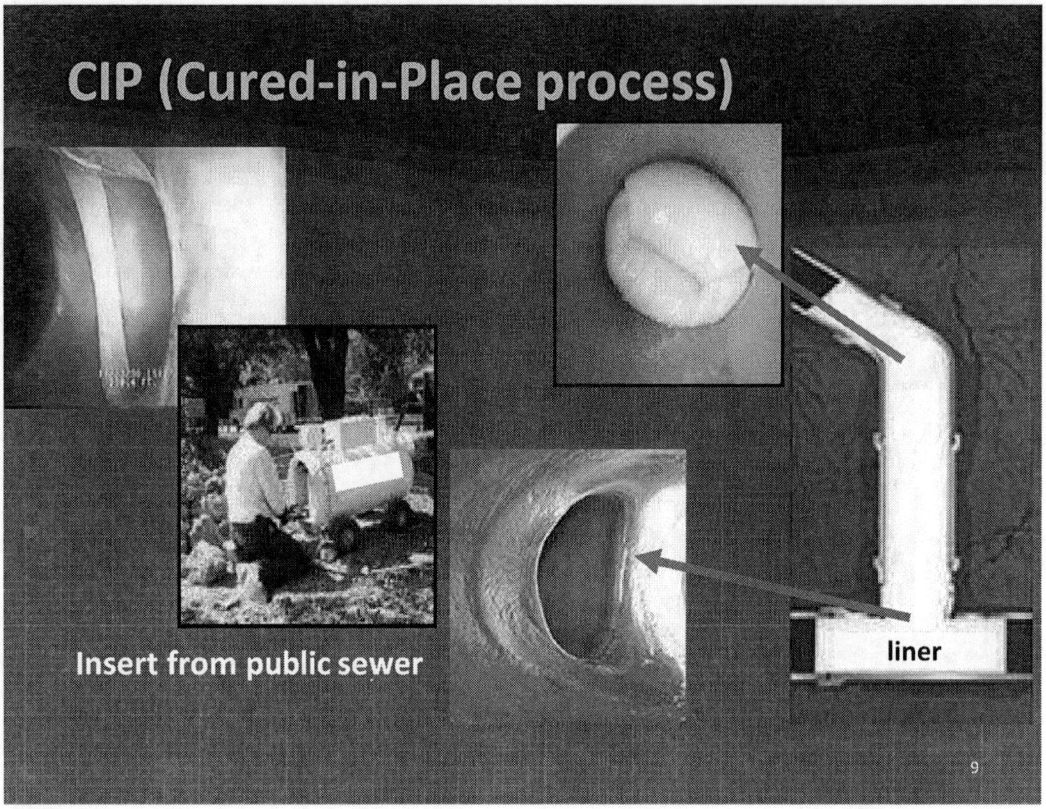
CIP (Cured-in-Place process)
Insert from public sewer
liner
9

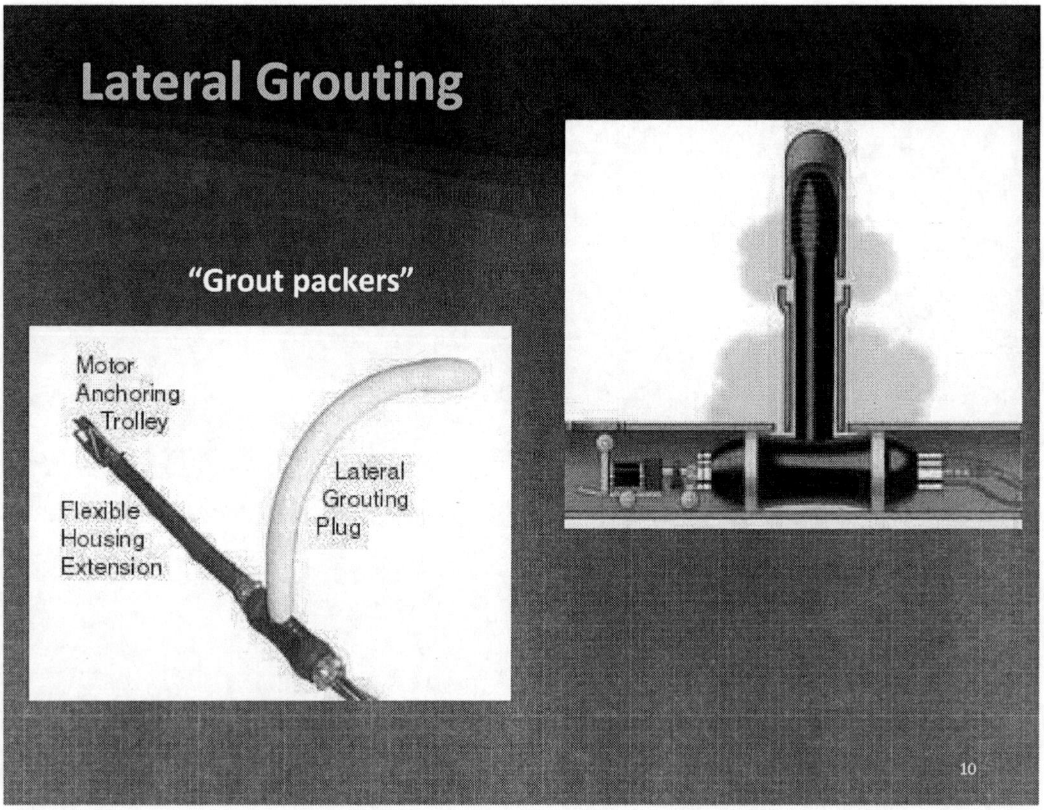
Lateral Grouting
"Grout packers"
Motor
Anchoring
Trolley
Flexible
Housing
Extension
Lateral
Grouting
Plug
10

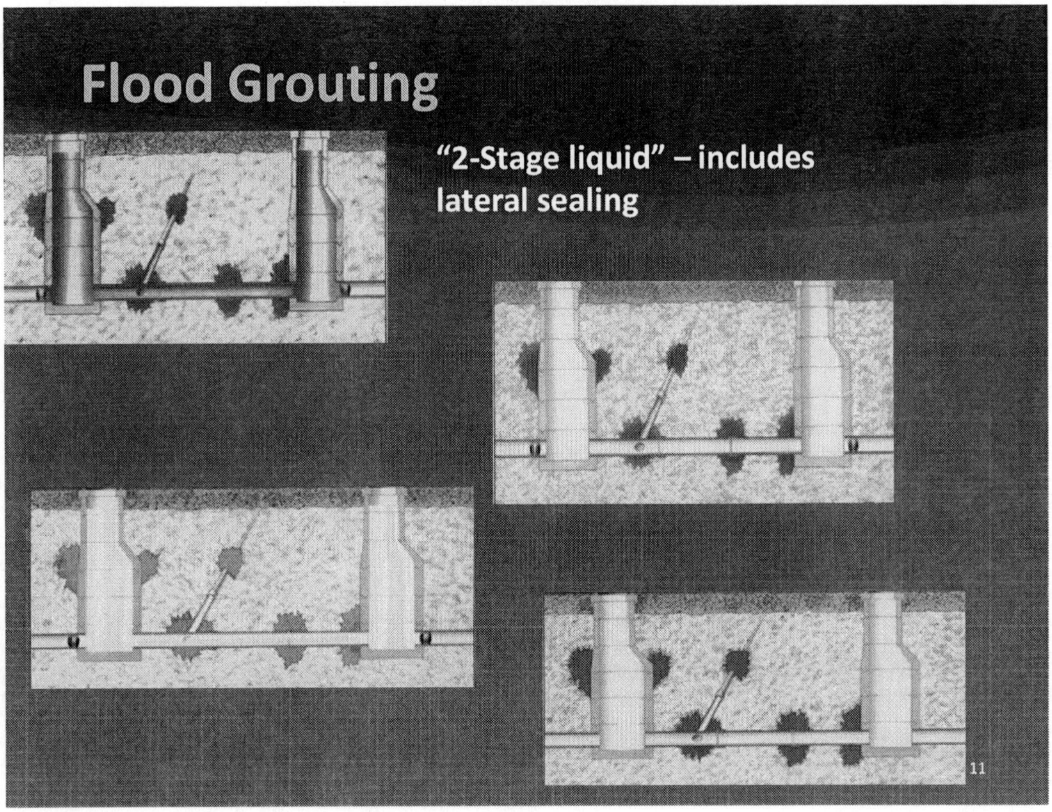
Flood Grouting
“2-Stage liquid” – includes lateral sealing
11

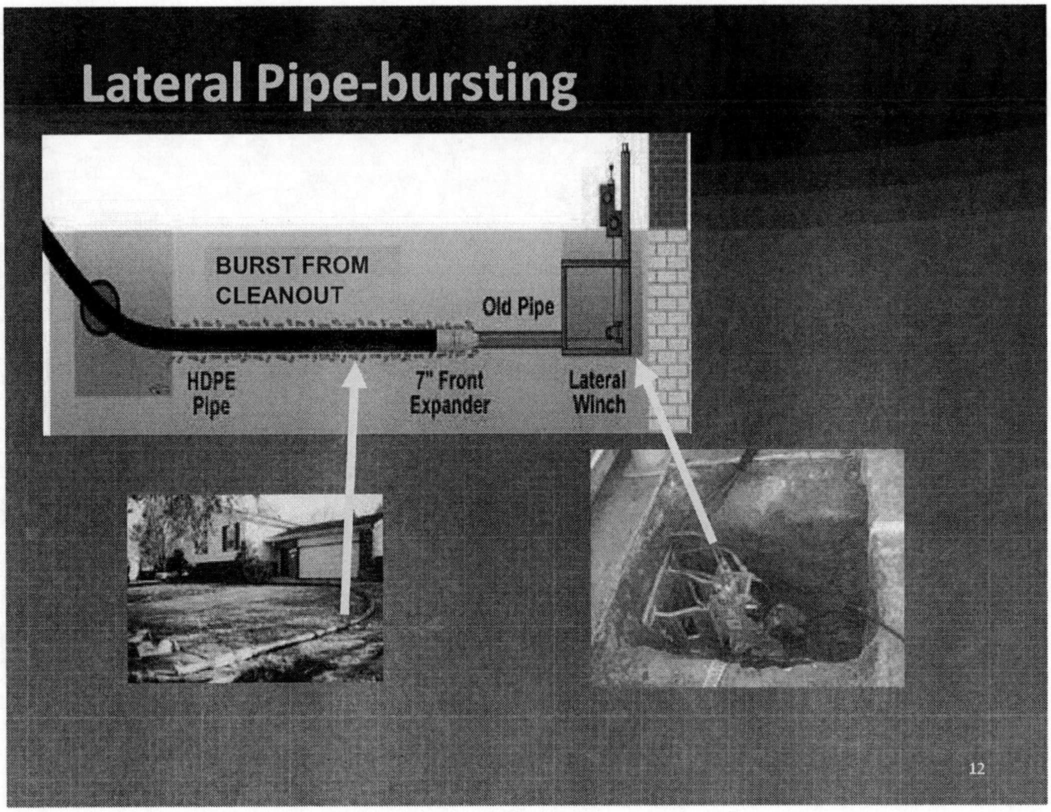
Lateral Pipe-bursting
BURST FROM CLEANOUT
Old Pipe
HDPE Pipe
7" Front Expander
Lateral Winch
12

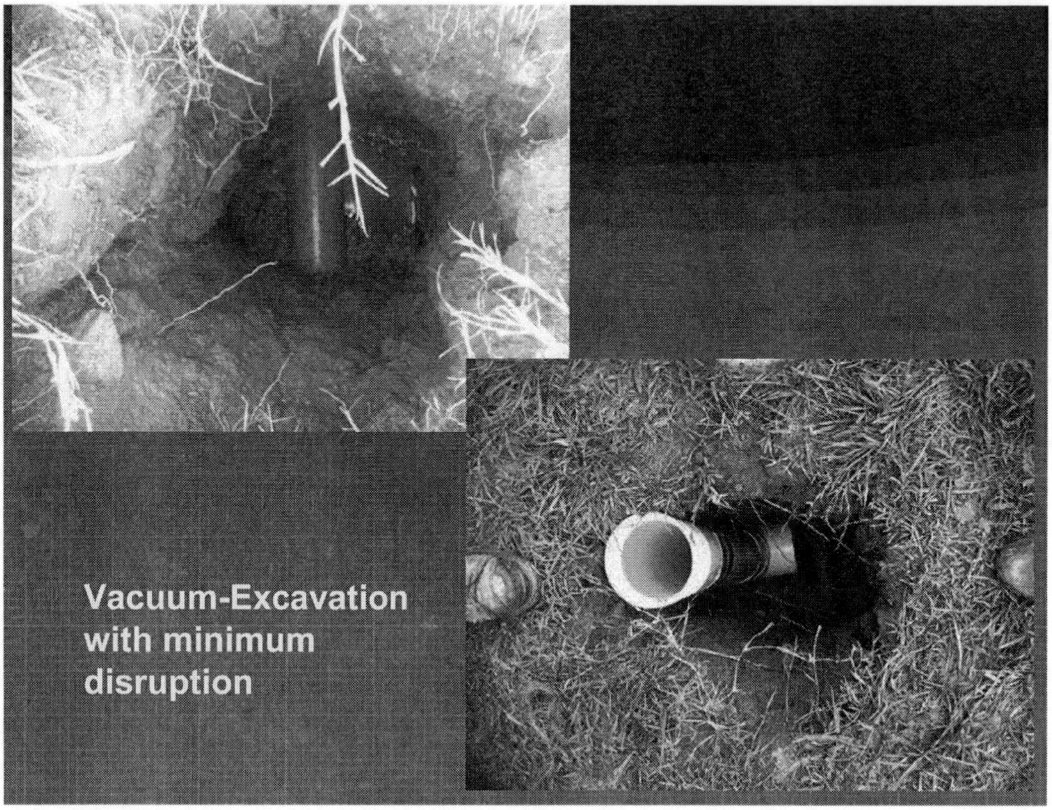
Vacuum-Excavation
with minimum
disruption

Vacuum-Excavation Clean-Outs
SPECTRASHIELD

Lateral Pilot Project (Oak valley – Nashville)

- Multiple Phases
- Initial Flow Monitoring – quantify I/I, TV for design – segment selection
- Main Line Rehabilitation only
- Flow Monitoring to quantify results
- Service Lateral Rehabilitation – CIP liner
- Flow Monitoring to measure additional removal

15

1.67" Rainfall – Before Renewal

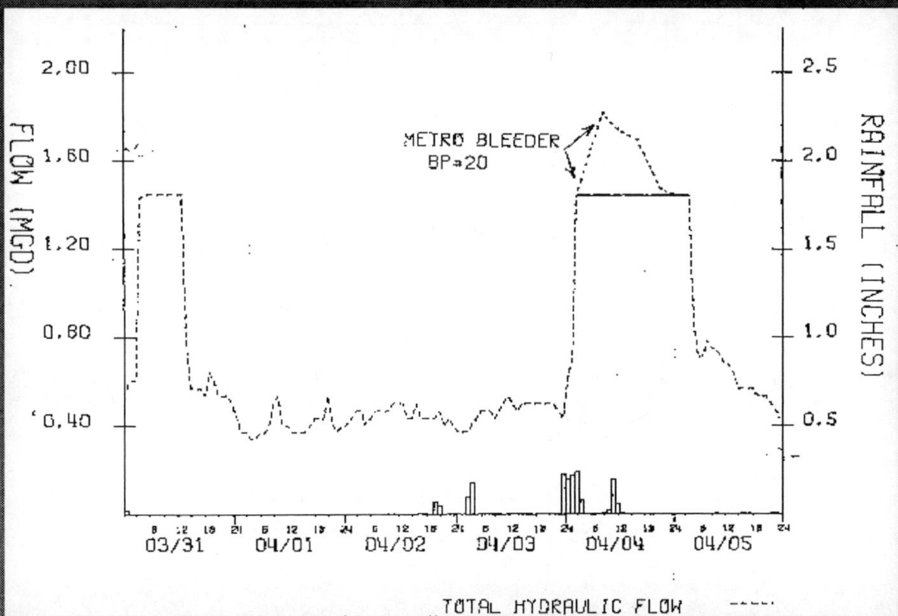

16

Projected I/I Before Renewal (30 days)

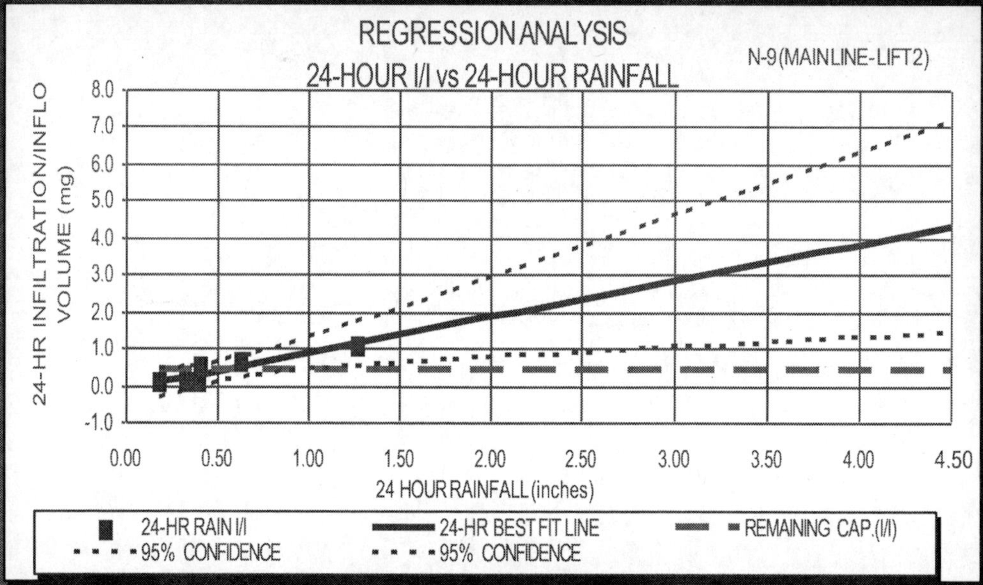

17

1.58" Rainfall After Pipe Lining

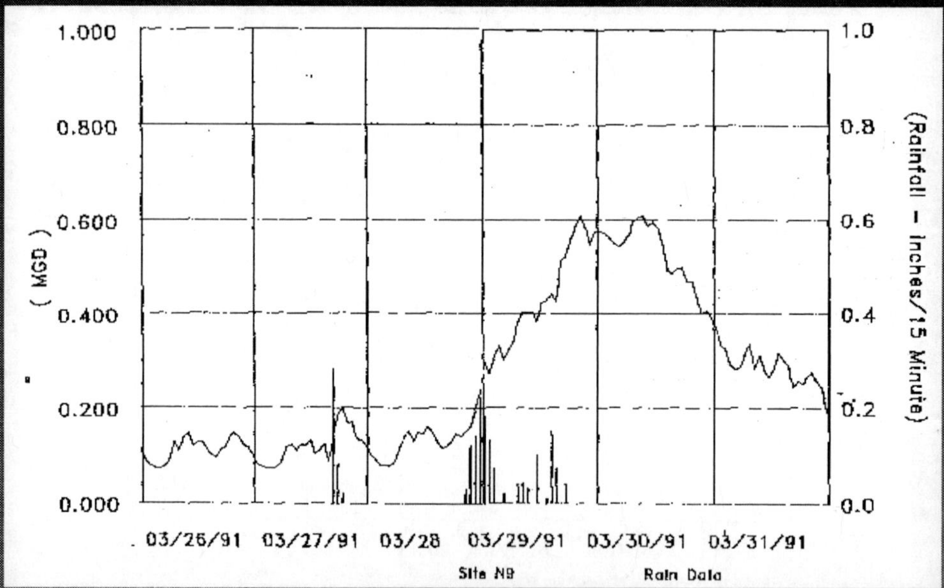

18

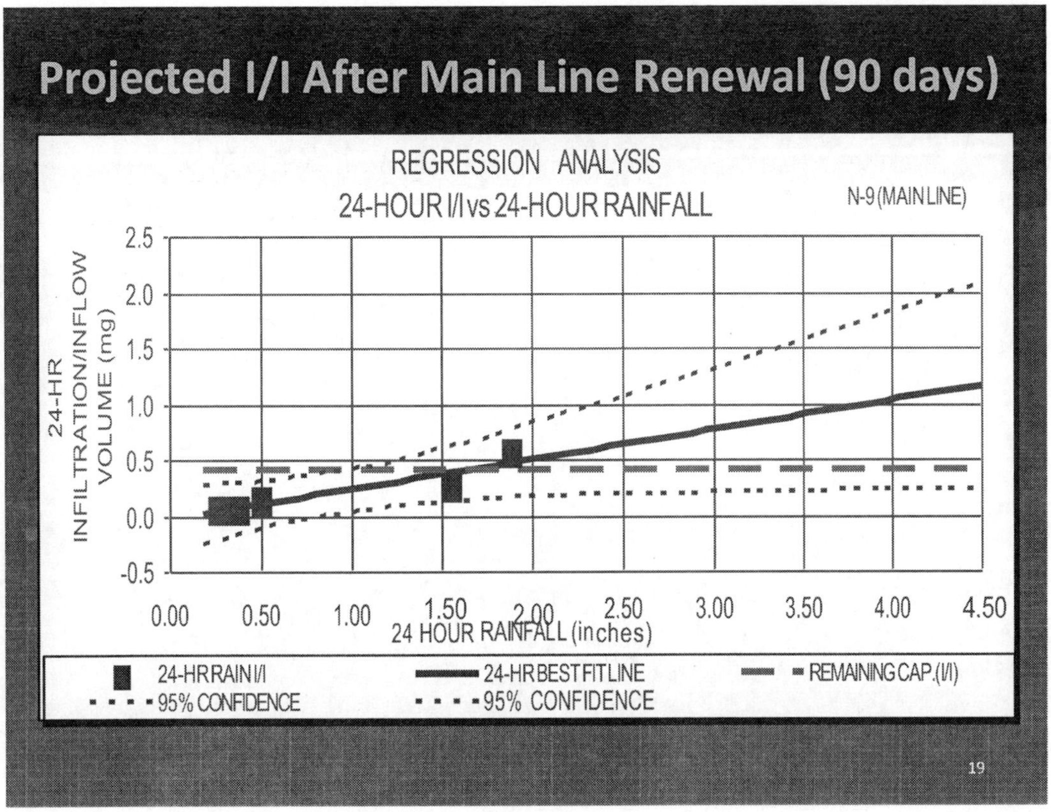
Projected I/I After Main Line Renewal (90 days)
REGRESSION ANALYSIS
24-HOUR I/I vs 24-HOUR RAINFALL
N-9 (MAIN LINE)
24-HR INFILTRATION/INFLOW VOLUME (mg)
2.5
2.0
1.5
1.0
0.5
0.0
-0.5
0.00
0.50
1.00
1.50
2.00
2.50
3.00
3.50
4.00
4.50
24 HOUR RAINFALL (inches)
24-HR RAIN I/I
24-HR BEST FIT LINE
REMAINING CAP.(I/I)
95% CONFIDENCE
95% CONFIDENCE
19

Insert CIP Liner from Cleanout
20

Observe Insertion to Mainline

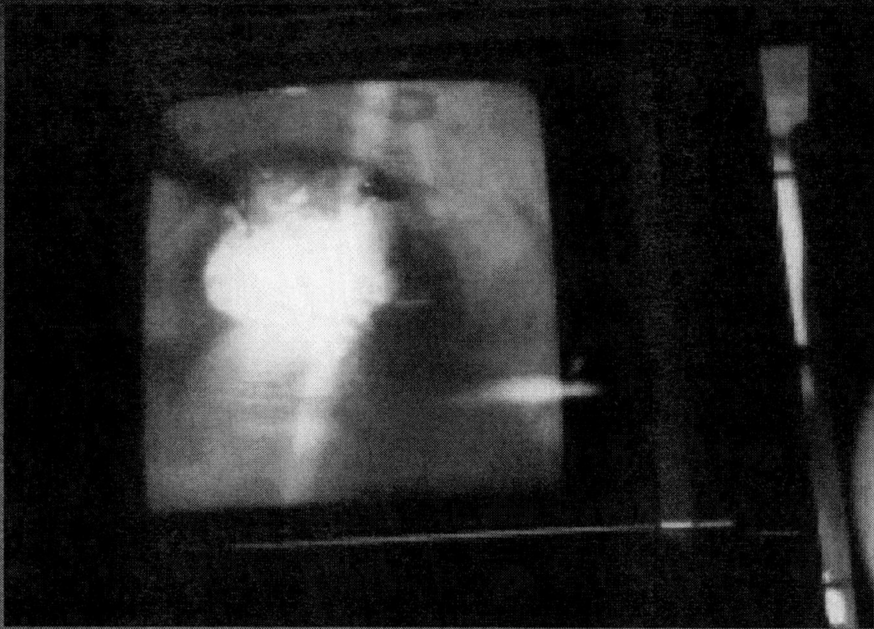

21

Cure the Liner with Hot Water

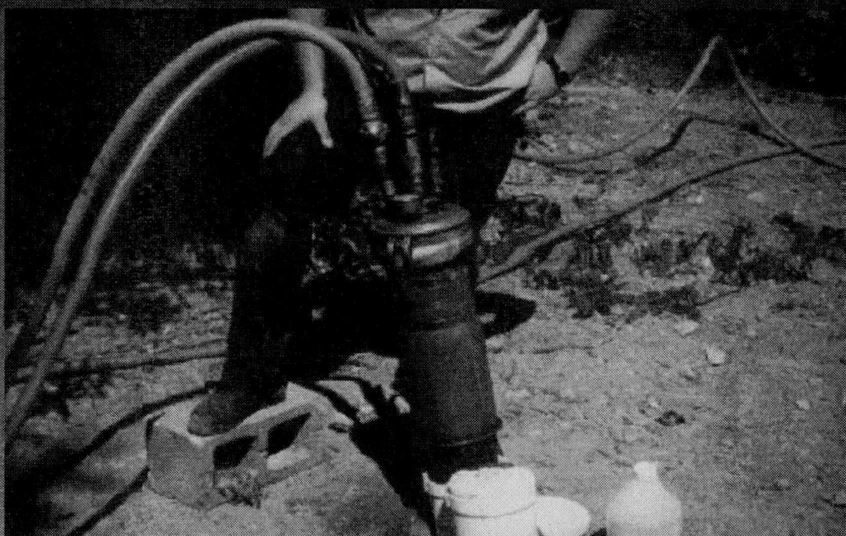

22

Verify Liner "Locked" to Public Sewer Liner

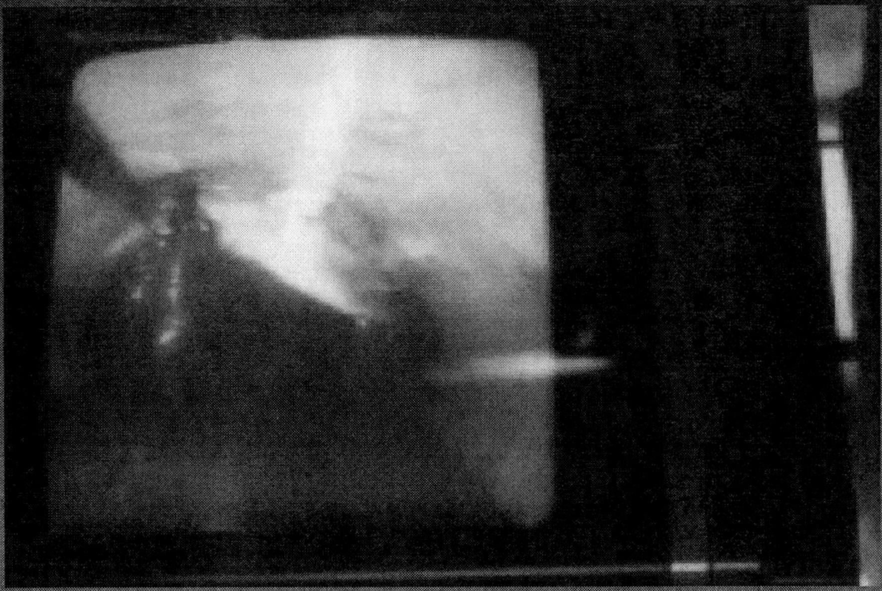

23

2.46" Rainfall After Lateral Lining

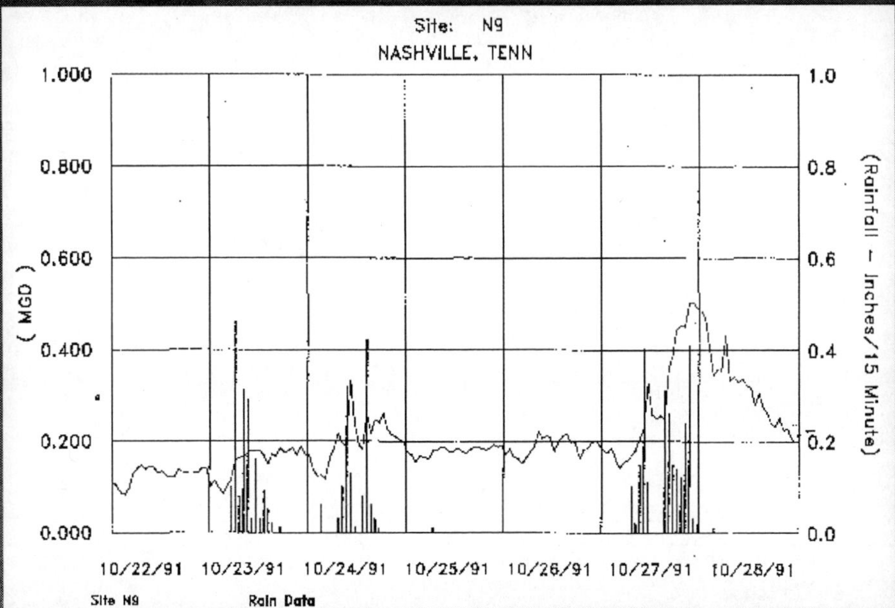

24

Oak Valley Lateral Renewal

	Before	After M.L.	After Lat.
Peak Q (mgd)	1.8	0.6	0.5
ADF (mgd)	0.2	0.13	0.13
SSO (mgd)	0.16	0	0
24 hr rain (in)	1.67	1.58	2.46
Prev. 48 hr.(in)	0.4	0.38	0
Prev. 21 days (in)	2.11	2.32	3.53

25

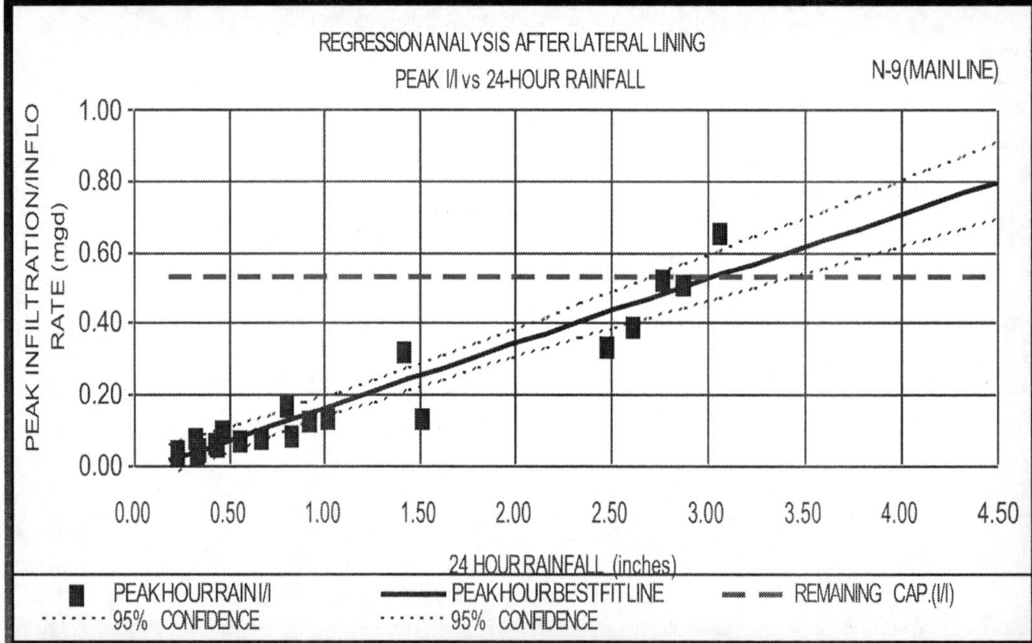

26

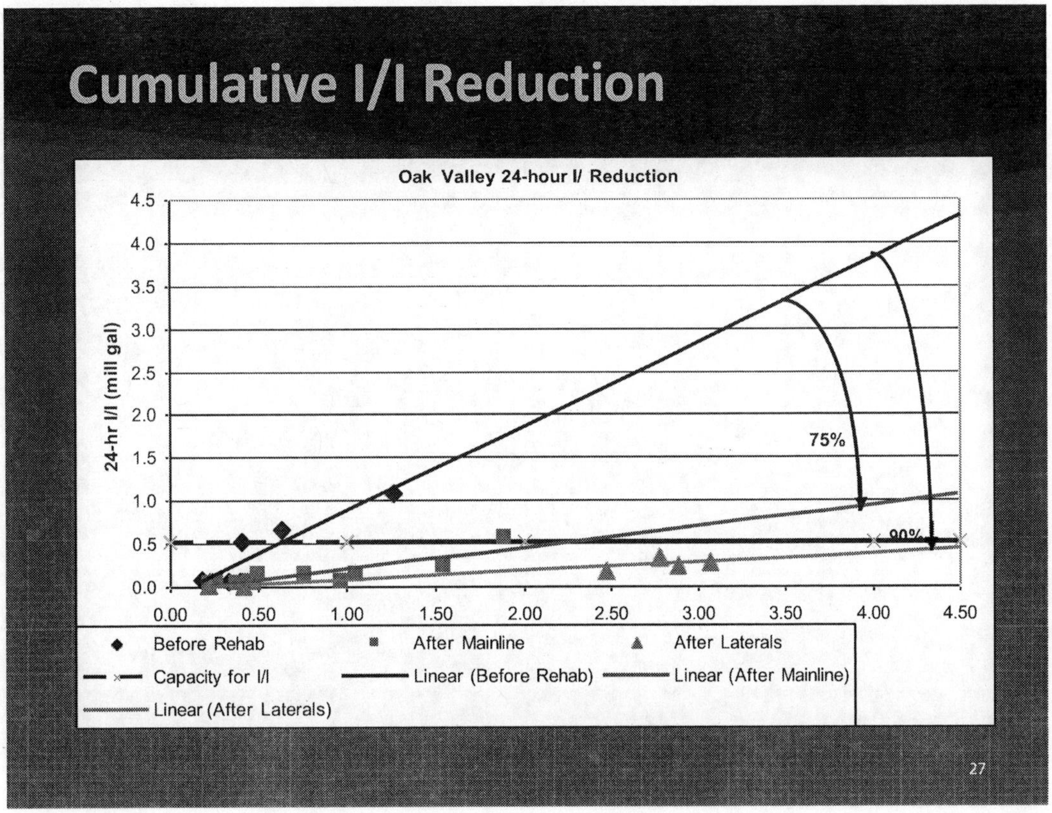
Cumulative I/I Reduction
Oak Valley 24-hour I/ Reduction
24-hr I/I (mill gal)
4.5
4.0
3.5
3.0
2.5
2.0
1.5
1.0
0.5
0.0
0.00
0.50
1.00
1.50
2.00
2.50
3.00
3.50
4.00
4.50
75%
90%
Before Rehab
After Mainline
After Laterals
Capacity for I/I
Linear (Before Rehab)
Linear (After Mainline)
Linear (After Laterals)
27

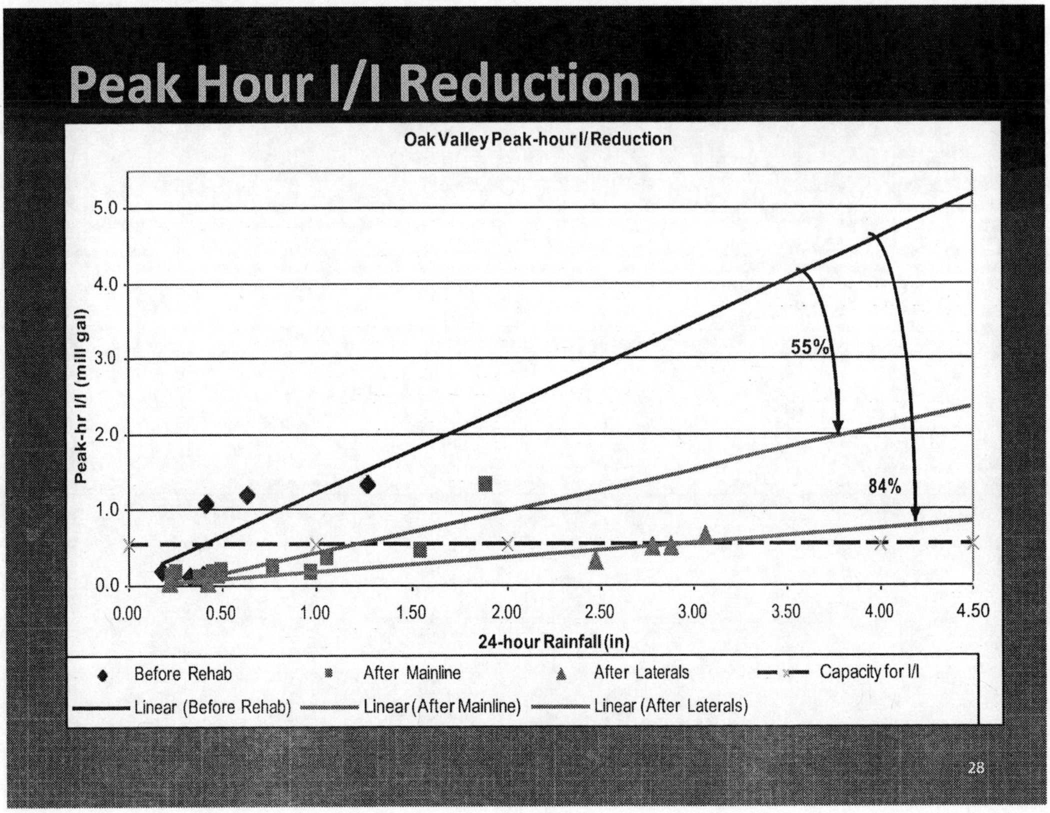
Peak Hour I/I Reduction
Oak Valley Peak-hour I/I Reduction
Peak-hr I/I (mill gal)
5.0
4.0
3.0
2.0
1.0
0.0
0.00
0.50
1.00
1.50
2.00
2.50
3.00
3.50
4.00
4.50
55%
84%
24-hour Rainfall (in)
Before Rehab
After Mainline
After Laterals
Capacity for I/I
Linear (Before Rehab)
Linear (After Mainline)
Linear (After Laterals)
28

Bang for the Buck

Result:

- ~ 20% additional I/I removal for ~ 10% contract cost

29

Pilot Project Results

Nashville rehabilitation policy *

- All laterals connected to pipes being rehabilitated or replaced shall be rehabilitated or replaced to the property line or easement line.

OVER 10,000 SERVICE LATERALS HAVE BEEN REHABILITATED OR REPLACED !

*** Nashville OAP program 1990-2005***

30

Application – Design Considerations

- Integration with overall sewer system rehabilitation (think: system strategy!)
- Cleaning – root removal
- Problems (Resin slugs, etc.)
- Lateral configuration (bends, transitions, defects, size, etc.)
- Inspection – verification (air test)

31

Follow-up Monitoring

- Flow metering – verify effectiveness
- Direct observation during wet weather – double stack clean-out

32

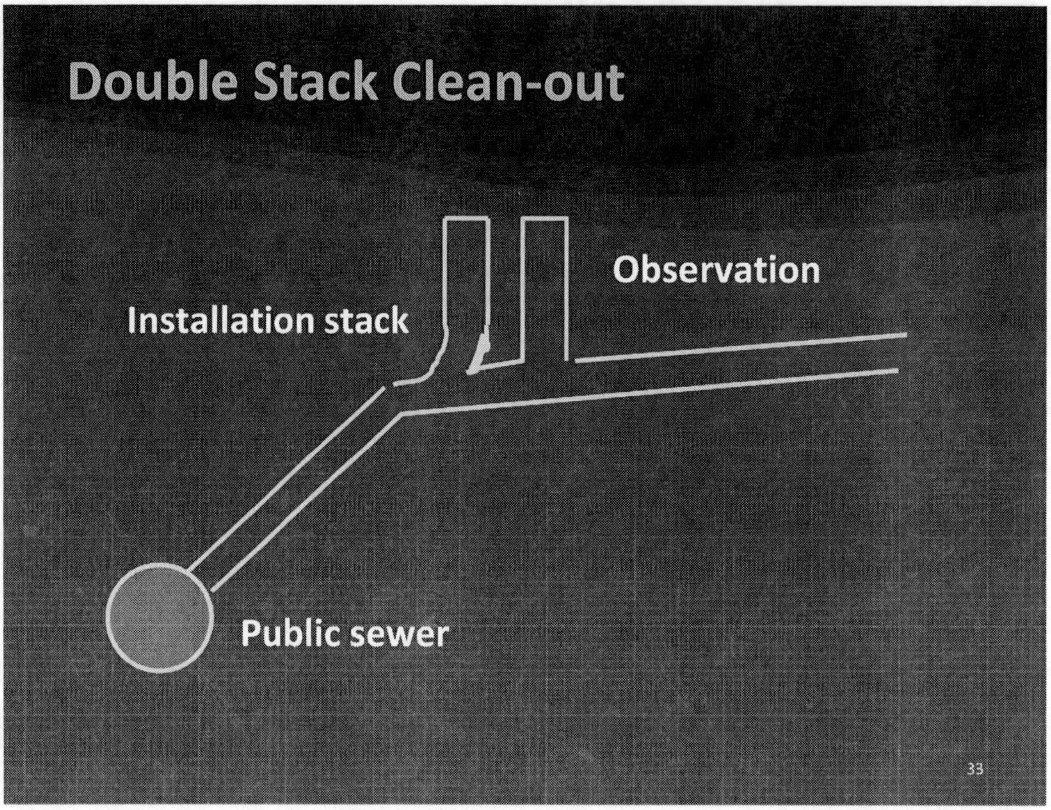
Double Stack Clean-out
Observation
Installation stack
Public sewer
33

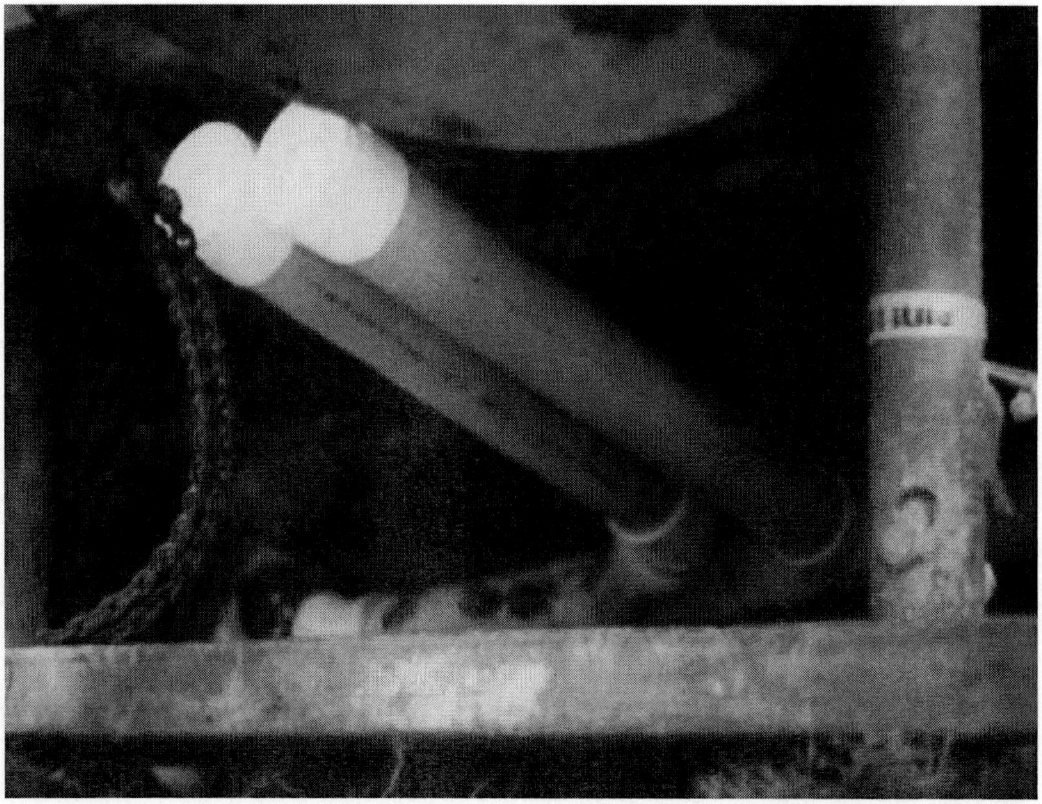

Lateral Air Testing

Test after lateral opening cut !

- Test full length line (manhole to manhole and plugs in laterals)
- Test each individual lateral - plug lateral and 2 plugs in pipe

35

Air Testing the Repaired Lateral Connections

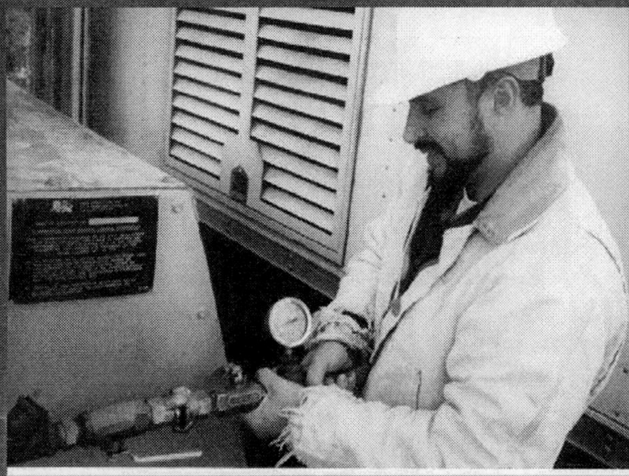

39

Questions ?

40

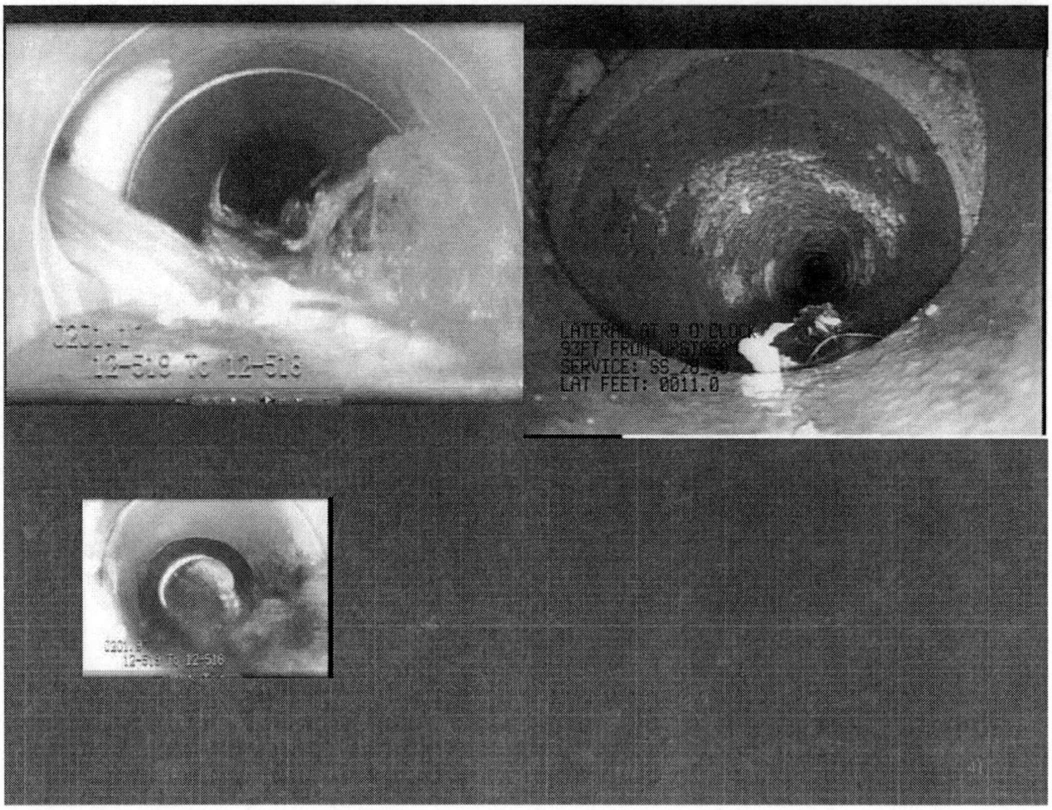
12-519 To 12-518
LAT FEET: 0011.0

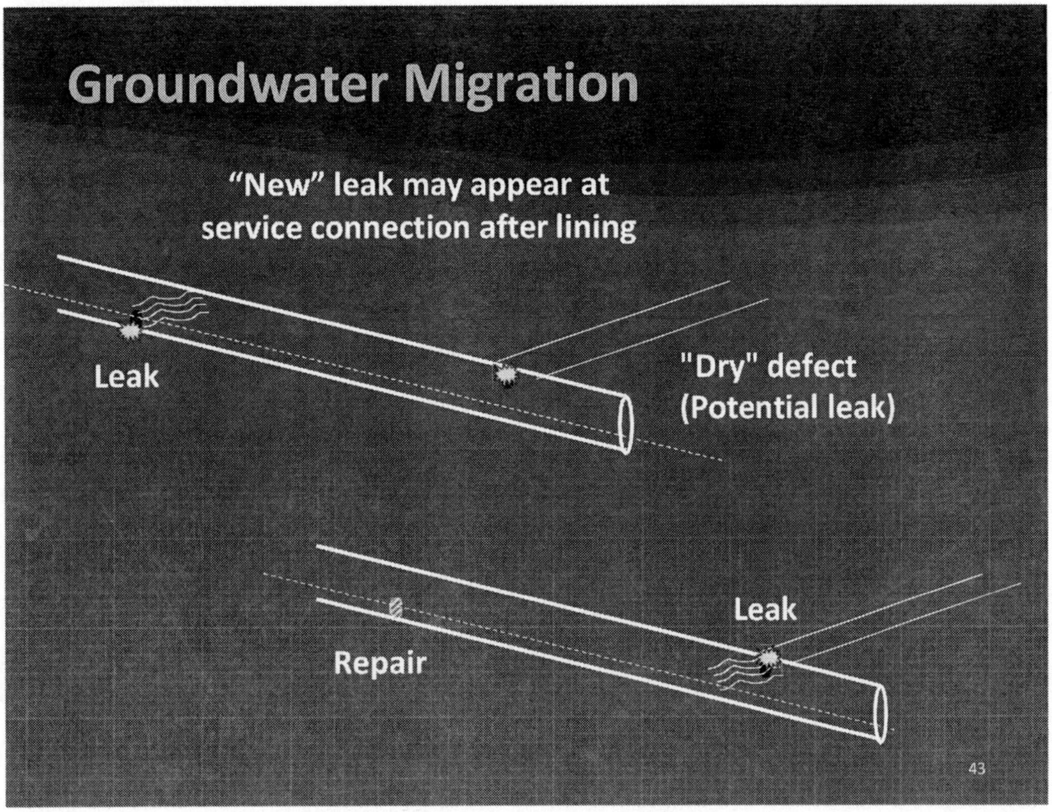
Groundwater Migration
"New" leak may appear at service connection after lining
Leak
"Dry" defect (Potential leak)
Repair
Leak
43

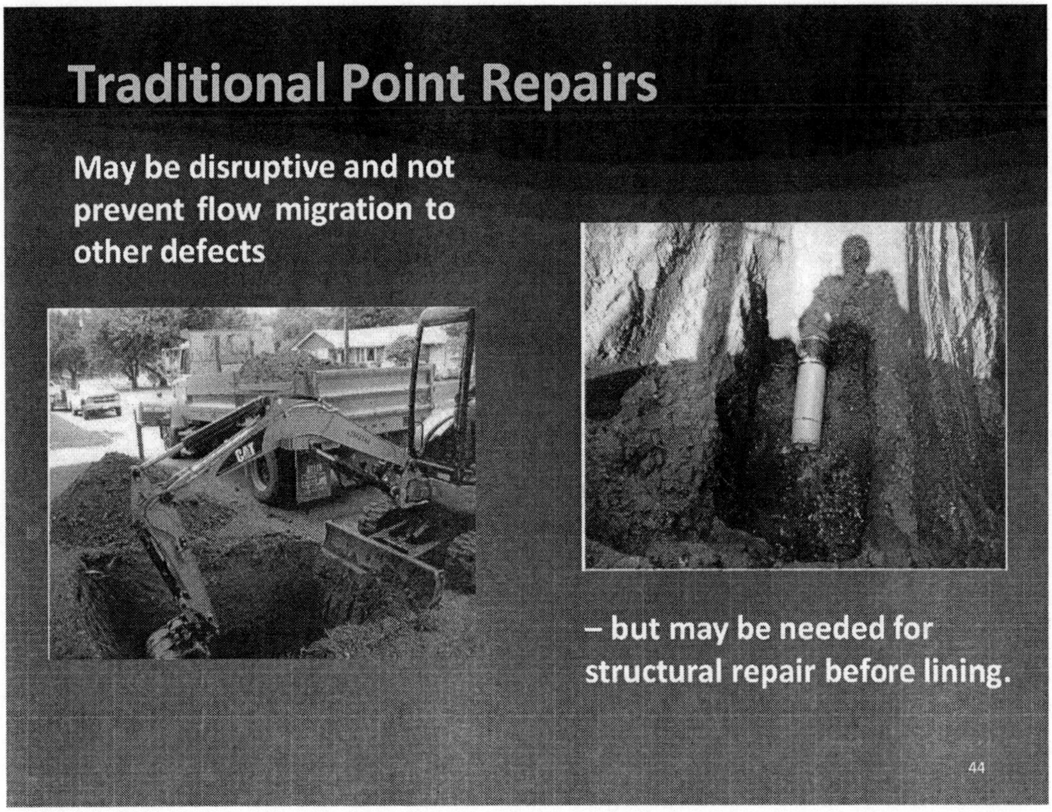
Traditional Point Repairs
May be disruptive and not prevent flow migration to other defects
CAT
– but may be needed for structural repair before lining.
44

Acrylate Grout Systems: Their characteristics, the role of reactants and field conditions on their performance

Shiva Sunder,
Material Specialist Engineer,
Western Technologies Inc.
3737 E. Broadway Road, Phoenix, AZ 85040
Michael Vargo,
Sr. Technical Consultant,
Prime Resins Inc.
2291 Plunkett Road NW, Conyers, GA 30012

Grouts and Grouting

- Grout:

= Liquid → Solid

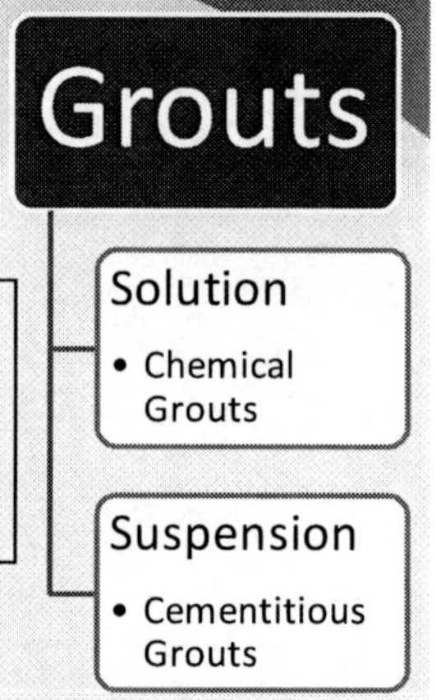

- Grouting:

Grouting is effective for sands With permeability as low as 10^{-4} cm/s (Christopher et al, 1989).

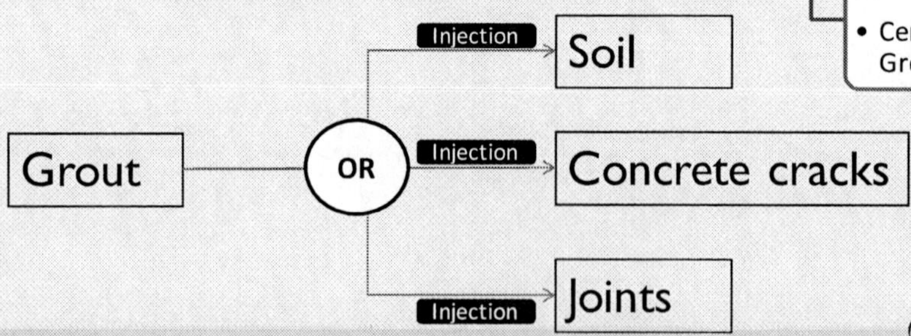

(Sunder, 2012)

SOLUTION GROUT TYPES

SOLUTION GROUTS

- ACRYLIC
  ACRYLATE
  ACRYLAMIDE
- EPOXY
- SODIUM SILICATE
- POLYURETHANE

(Vipulanandan, 2012)

ACRYLATE GROUTS

- Typically a **2-component** grout with 3 reactants.
- The aqueous-based acrylate grout systems undergo **radical polymerization**
- The acrylate monomer resin, an initiator and an activator – primary ingredients
- **The initiator** – responsible for initiating the reaction by creating reactive radicals. Example: persulfates of sodium and ammonium etc.
- **Activator** – responsible for the consistency and quality of the polymerization reaction. Example: triethanol amine, diethanol amine etc.

Acrylate - Polymer Chemistry

Monomer:
- A small atom or a molecule
- Active to react to form large molecules.

Polymer:
- A large molecule that is composed of small repeating units – MONOMERS.

Polymerization:
It is the chemical reaction where,
- The monomers react under different conditions
- Occurrence of a 3 dimensional structure formation, POLYMER.

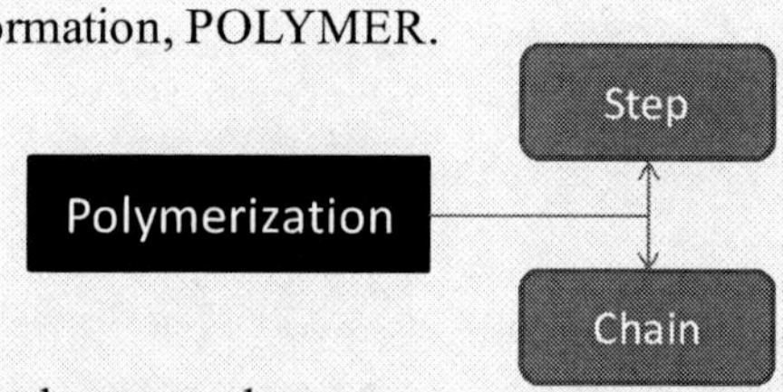

GELLING TIME – time at which the grout undergoes a phase transformation from liquid to solid

(Sunder, 2012)

FREE RADICAL POLYMERIZATION

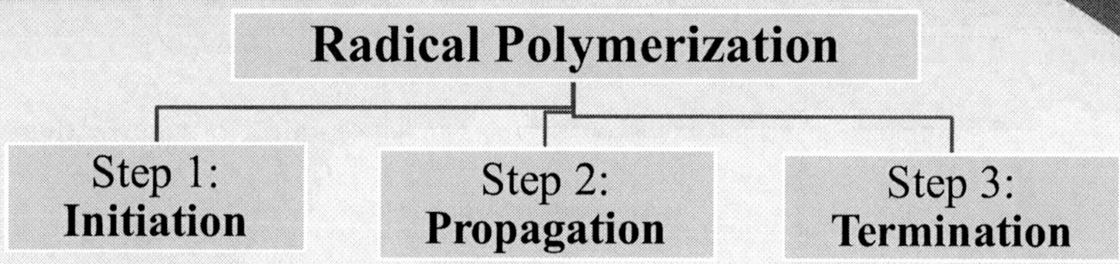

- A type of addition polymerization
- the reacting species is a long chain free radical
- initiated by the attack of free radicals

Typical Gel Time Curve of Acrylate Grout by Radical Polymerization

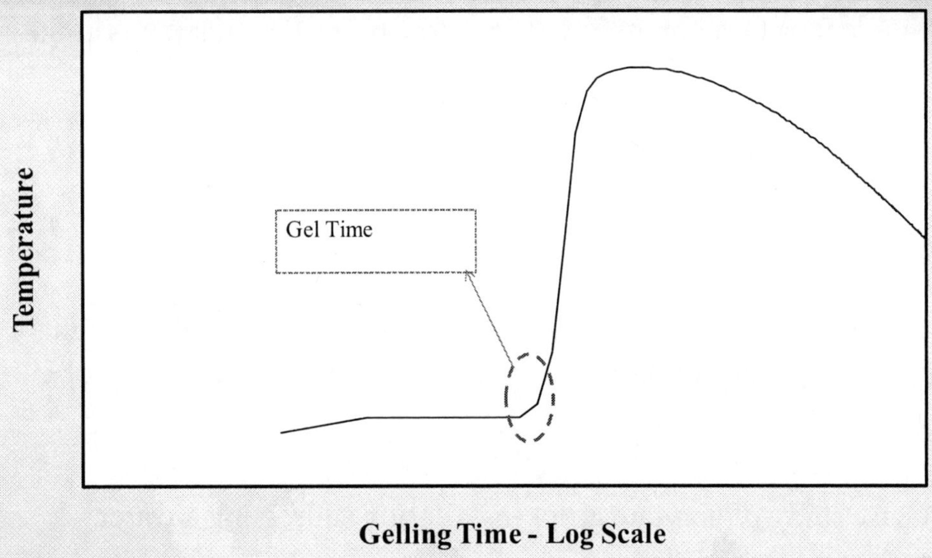

Problem statement

Acrylate Grout Systems

- Leak sealing applications
- Soil stabilization applications

Selection of the Grout:

- Resin concentration
- **Composition of the reactants** attributing to the reaction

Environmental Conditions:

- **Temperature** and temporal variations
- Peripheral hurdles
- Severity of the problem

This Presentation Addresses…

A laboratory program to investigate the role of the reactants in the grout system

- Resin
- Activator
- Initiator

A laboratory program to analyze the effect of temperature on the grout performance evaluated from the gelling characteristics of the grout

A case study of acrylate grout installation addressing a water leaking issue

The Effect of Material Temperature on Gelling Time of Acrylate Grout

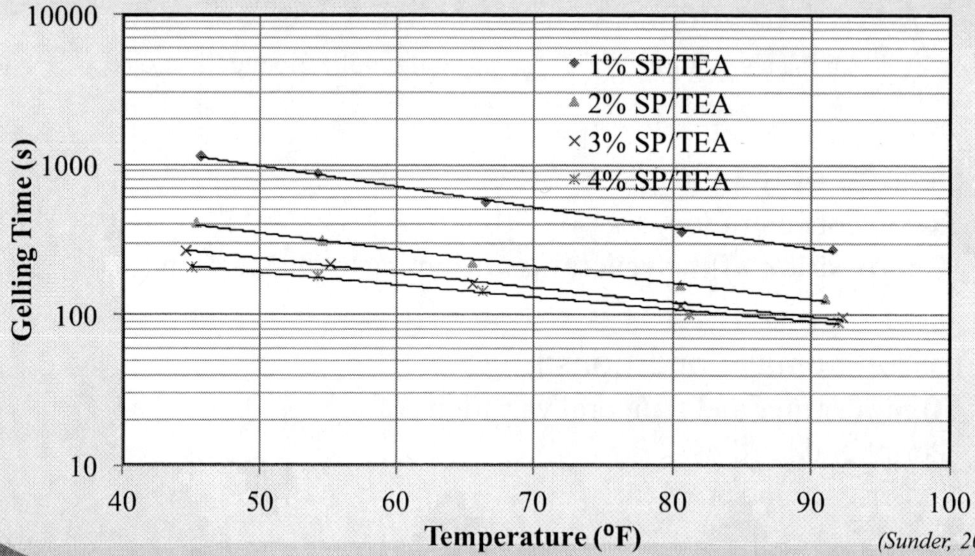

(Sunder, 2015)

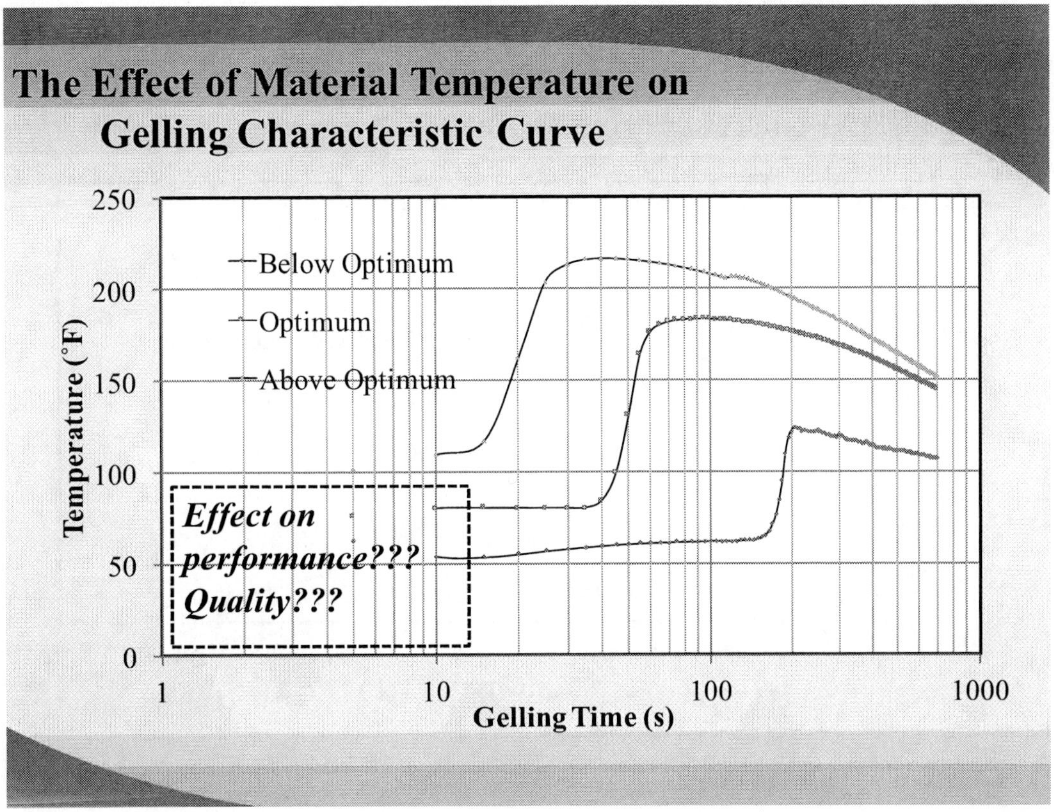
The Effect of Material Temperature on
Gelling Characteristic Curve
Temperature (°F)
250
200
150
100
50
0
Below Optimum
Optimum
Above Optimum
Effect on
performance???
Quality???
1
10
100
1000
Gelling Time (s)

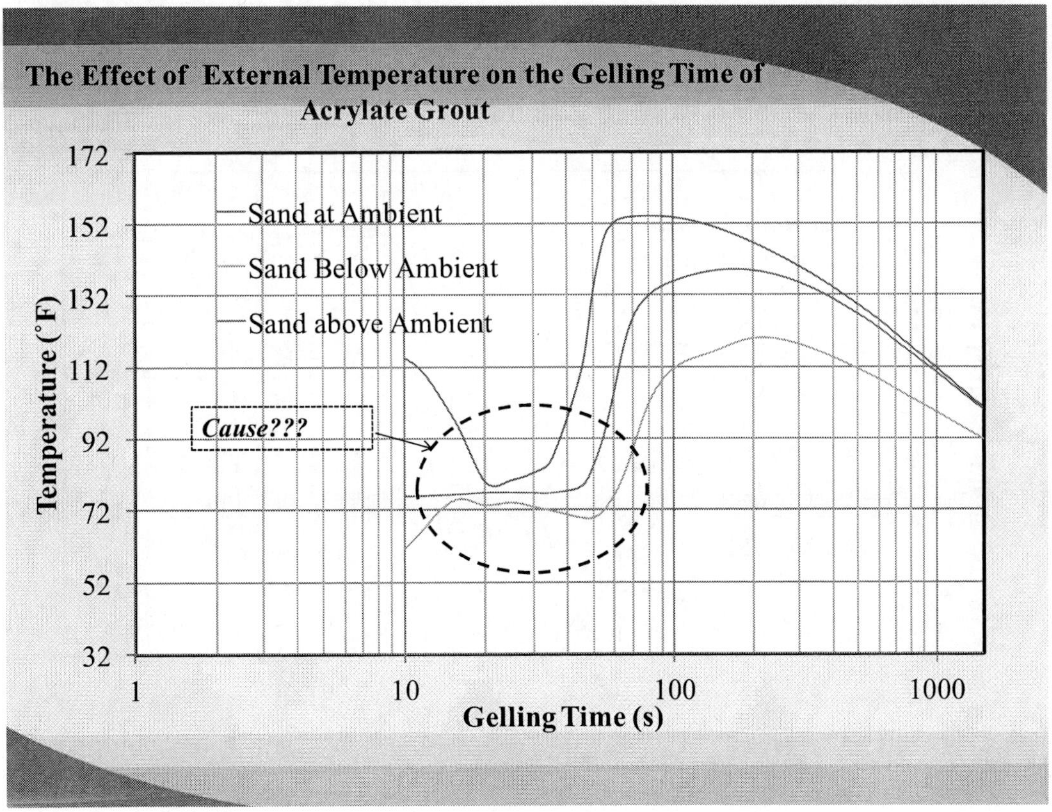
The Effect of External Temperature on the Gelling Time of
Acrylate Grout
Temperature (°F)
172
152
132
112
92
72
52
32
Sand at Ambient
Sand Below Ambient
Sand above Ambient
Cause???
1
10
100
1000
Gelling Time (s)

The Effect of Resin Content (Reactant) on the Gelling Time of Acrylate Grout

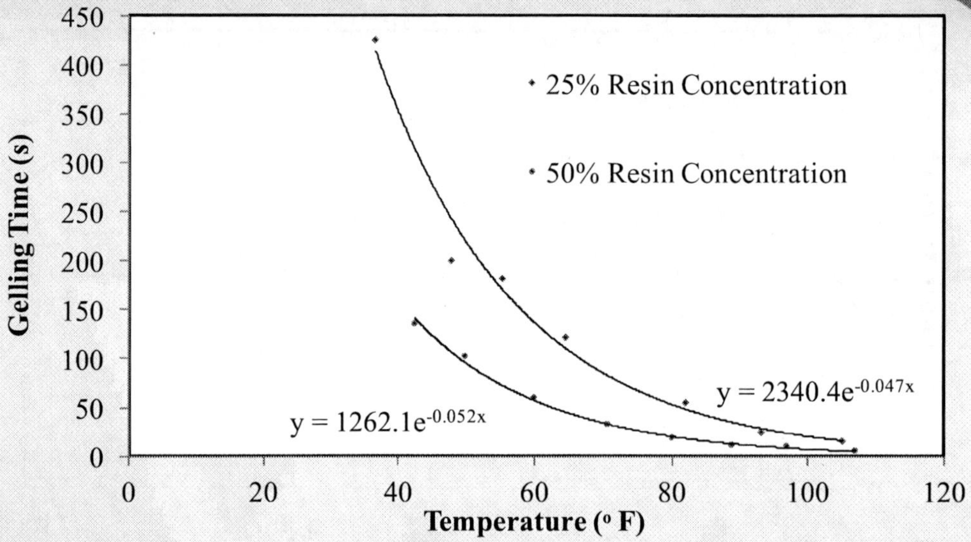

The Effect of Initiator and Activator (Reactants) on the Gelling Time of Acrylate Grout

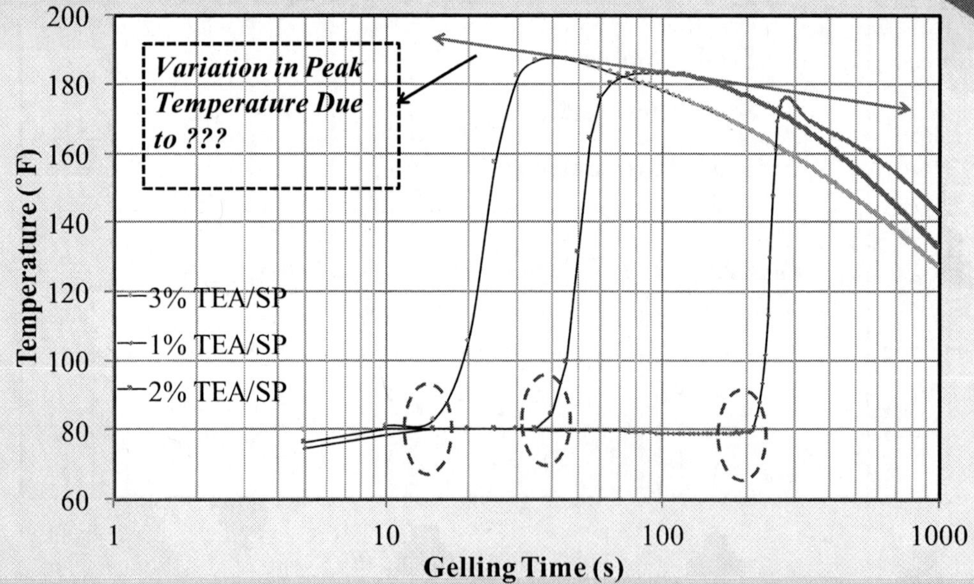

Case Study

Sealing water leaks

17-story apartment complex

Bronx, NY

In-Situ Condition

Basement condition prior to grouting

Ground water and soil contaminated with heating oil

Severe water leak through cracks

Large temporal variation

Existing oil tanks not to be crushed/punctured

Non-expansive sealing grout required

Grouting with AR800 acrylate system

Adjustable gel time depending on dilution

Ability to gel under saturated conditions

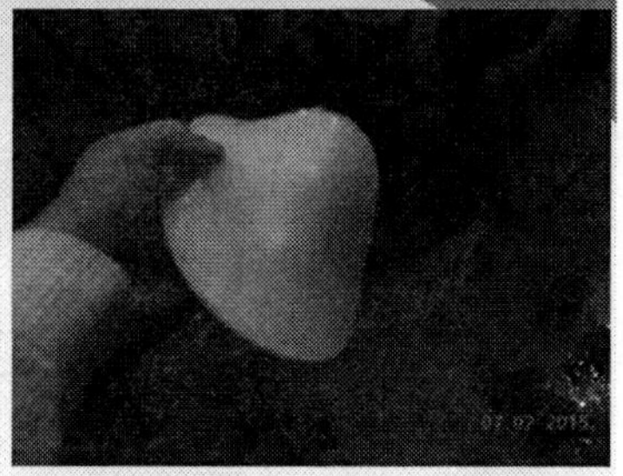

Non-expansive compared to polyurethane systems

Low-Viscosity Grout System

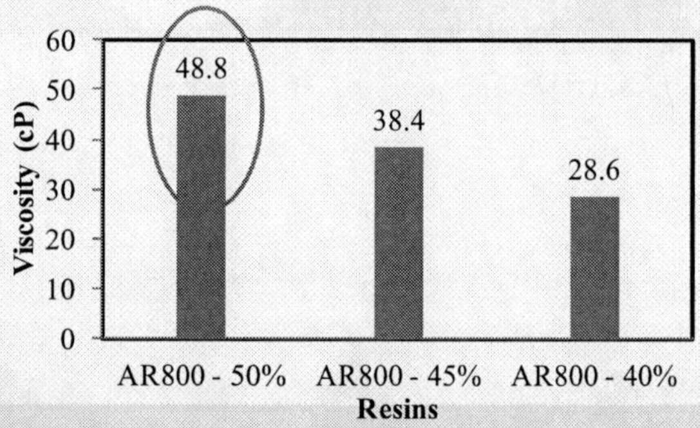

Grouting process

Grouting Parameters

- **Grouting Process:** 2.5 week duration
- **Selected Grout:** AR 800 acrylate grout system
- Recommended **premix ratio:** A:B = 1:1
- **Catalyst/activator percentage:** 2
- **Gel time:** 30 seconds (considering the in-situ temperature and dilution due to contaminated water leak as a contributing factor in the selection)
- **Total material injected:** 950 gallons (A = 475 gallons + B = 475 gallons)
- **Result:** *Successful grouting process resulted in a leak-free environment.*

After Grouting

Conclusion

- Environmental conditions are a major consideration for the selection of a grout.
- The reactant composition and the pre-mix temperatures play vital roles beyond just being used to modify the gelling temperatures.
- Safety considerations – NSF certification.
- **Acrylate grout systems** with adjustable parameters without compromise in its physical, mechanical and working properties is the best solution.

Acknowledgements

Acknowledgements for their help and/or being a part of this project.

- David L. Barton, President, Prime Resins Inc., Conyers, GA.
- Michael Vargo (co-author) of Prime Resins Inc.
- Tom Alexander and the research department of Prime Resins Inc.
- Western Technologies Inc., Phoenix, AZ

PRIME RESINS

THE TOTAL SOURCE

educational sessions

Alabama & Gulf Coast Railway (AGR) – Reline Under an Active Railroad

Monroeville, Alabama

AGR – Reline Under a Live Railroad

Alabama & Gulf Coast Railway (AGR)

- AGR began operations in Monroeville, AL in 1997 from Kimbrough, AL to Pensacola, FL from BNSF
- In 2003, AGR expanded from Kimbrough, AL to Columbus, MS, with operations to Mobile, AL via NS and Amory, MS via BNSF
- Operates total of 442 miles of track through MS, AL and FL
- Major commodities include lumber, paper products, chemicals, wood chips, metal and crude oil
- Interchanges with BNSF, CAGY, CSXT, CN, GTRA, KCS, MNBR, NS and TASD

AGR – Reline Under a Live Railroad

Reline Location Details

- Monroeville, AL, aka Fountain, is located at MP 814.2, headquarter location for the AGR
- Located centrally on the Pensacola Subdivision, servicing Georgia Pacific at MP 813.5
- Site location in question located at MP 823.5, 132"x86' multi plated corrugated pipe bottom section having deteriorated

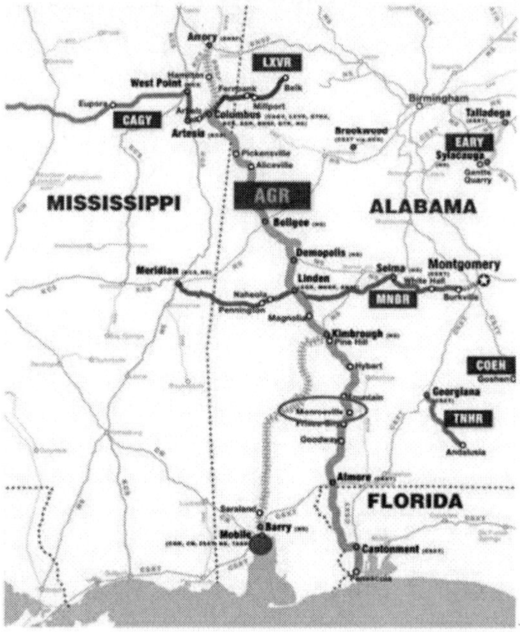

AGR – Reline Under a Live Railroad

Site Background

- Deteriorating culvert experienced major encroachment on the toe of the slope after a major rain event and catastrophic washouts
- Immediate Repair/Replacement was required
- Consulted three local contractors for proposed solutions / cost

AGR – Reline Under a Live Railroad

Condition / Assessment

- Options were to repair or replace
- Relining provided substantial cost benefits
- Structurally sound solution was required
- Avoid open cut and closure to rail line above
- 20-25 foot of cover

AGR – Reline Under a Live Railroad

Solution &
Technical Description:

- Smooth Cor™ (double wall corrugated steel pipe – ALT shell with Polymer-Coated liner) 108-in. diameter, 106 LF
- Tunnel Liner Plate – 10.5 LF, 7 GA

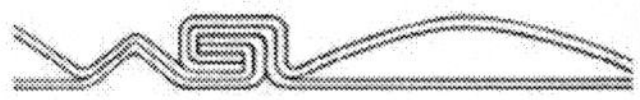

AGR – Reline Under a Live Railroad

Relining Process

- Insertion of Smooth Cor™ on struts from the south end.
- Partial collapse required hand lining with Tunnel Liner Plate for a short section
- Installation occurred from January to April 2015

AGR – Reline Under a Live Railroad

Relining Process, cont.

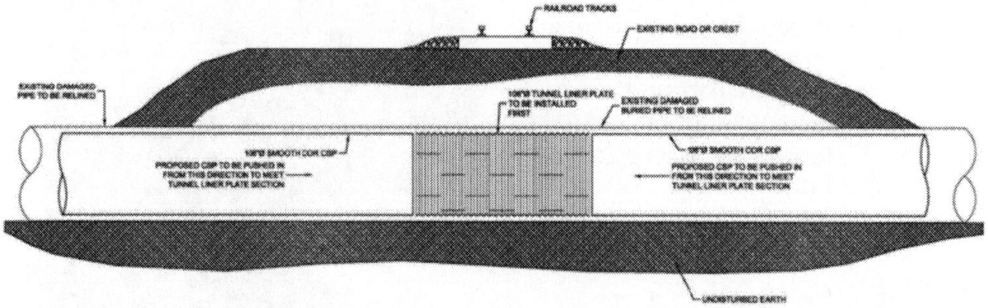

AGR – Reline Under a Live Railroad

Grouting Process

- Brace & block new pipe
- Multiple grouting stages
    - Through headwall/end ports and grout ports
- Non-shrink grout
- Low maximum grouting pressure maintained < 5 PSI

AGR – Reline Under a Live Railroad

Cost Comparison

- Original pricing for complete replacement was upwards of $500,000
- Relining the host pipe cut estimated costs by roughly 2/3

AGR – Reline Under a Live Railroad

RESULT

- Fully structural solution at minimal cost and disruption to the rail line.

AGR – Reline Under a Live Railroad

QUESTIONS?

Don Herbert, Contech – dherbert@conteches.com
Chris Frank, Genesee & Wyoming – cfrank@gwrr.com

A Case Study for Advanced Geopolymer Pipe Lining Rehabilitation System

UCT 2016

Joseph Royer, PhD - Milliken Infrastructure Solutions, LLC
Collis Parish - Inland Pipe Rehab, LLC

MILLIKEN INFRASTRUCTURE
A Milliken COMPANY

What is a Geopolymer

Not a Plastic

- Not HDPE/PVC/Epoxy

Look and feels like cement

- Workability
- Material Properties
- Service Life

Chemical structure like natural stone

- Monolithic
- Durable
- Corrosion Resistant

Cement Chemistry Primer

Portland cement (OPC) was developed from natural cements made in England in the early part of the nineteenth century, and its name is derived from its similarity to Portland Stone, a type of lime stone that was quarried on the Isle of Portland in Dorset, England. OPC is mainly composed of calcium silicates.

Source - Portland Cement Association.

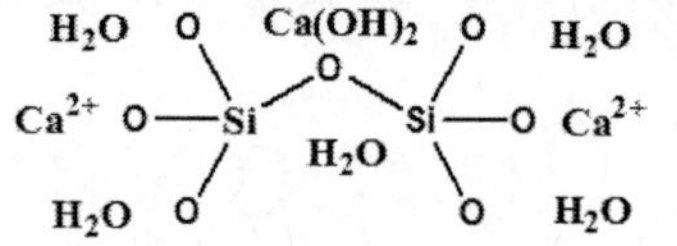

Representative Chemistry of OPC

Phase	OPC (%)
C_3S	50-70
C_2S	15-30
C_3A	5-10
C_4AF	5-15
CA	0

Chemical Composition of OPC

Cement chemistry notation is used C=CaO, S=SiO_2 A=Al_2O_3, and F=Fe_2O_3

Geopolymer Chemistry Primer

Geopolymers or alkali cements consist predominantly alumino-silicates or polymer Si-O-Al bonds and are similar in chemical make-up to natural Zeolites. Traditional starting materials for geopolymers cements are fumed silica, fly ash & metal slag.

- Geopolymers can be produced from commercially available industrial waste streams, creating a highly environmentally friendly product.
- The chemical reactions of geopolymer create only a fraction of the CO_2 emissions of standard cementitious materials.
- Geopolymers have excellent chemical and thermal resistance and can essentially be considered engineered stone.

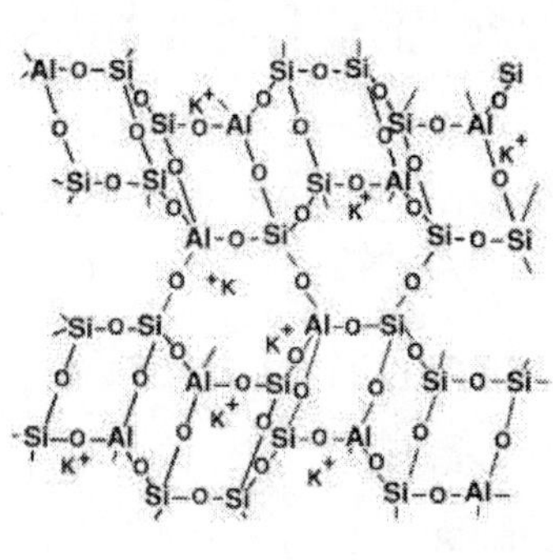

Typical Geopolymer Structure

Comparing Cementitious Materials

Geopolymer material solutions offer significant chemical, physical, environmental, and economic advantages over traditional materials

	Portland Cement	Geopolymers
Specific Strength	Variable	Excellent
Early Strength	Variable	Excellent
Acid Resistance	Poor	Very Good
Self-Adhesion	Low	High
% Recycled Content	<10%	>50%
CO_2 Emissions	High	Low
Single Pass Thickness	X	2X-3X
Total Installed Cost	$$	$

Geopolymer Advantages - Self Adhesion

Experimental Parameters:

- On the first day a series of 2" by 4" cylinders were cast half full and cured with on an approximately 45 angle.
- On subsequent days (1, 7, 14 and 28) the top half of the cylinders were cast and filled creating a 45 angled joint in the center of the cylinders
- Compressive strength tests were conducted 28 days after the top half of the cylinders were cast.

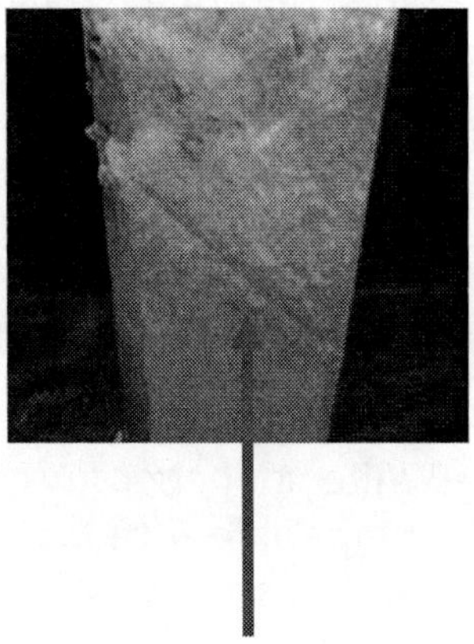

Joint in Tested Sample

Geopolymer Advantages - Self Adhesion

Results:

- Samples cast of geopolymer (both halves) when tested under compression did not break at the joint.
- The chemical nature of geopolymers allows for chemical bonding across the interface even with pours 28 days apart resulting in a monolithic structure.
- Samples cast of competitive cement mortars always broke along the angled joint essentially creating 2 separate layers.

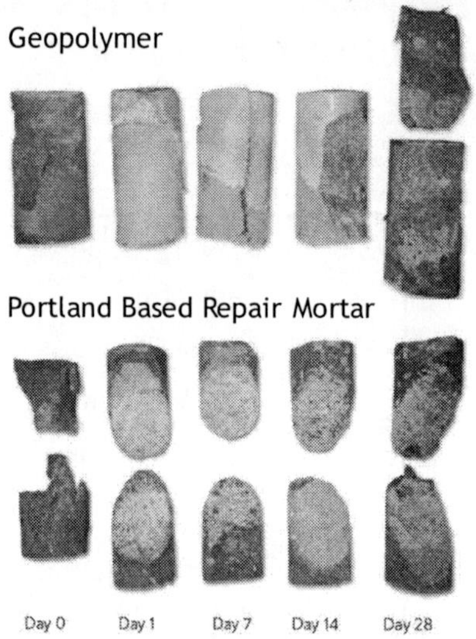

Centrifugal Casting of Geopolymers

- Rapidly distributes material within the pipe's internal circumference
- Forms a strong, low permeability, cement mortar lining
- Depending upon design & thickness, creates a "pipe within a pipe" and does not rely on the integrity of the host structure
- While not generally required it is compatible with most common reinforcement technologies

Centrifugal Casting of Geopolymers

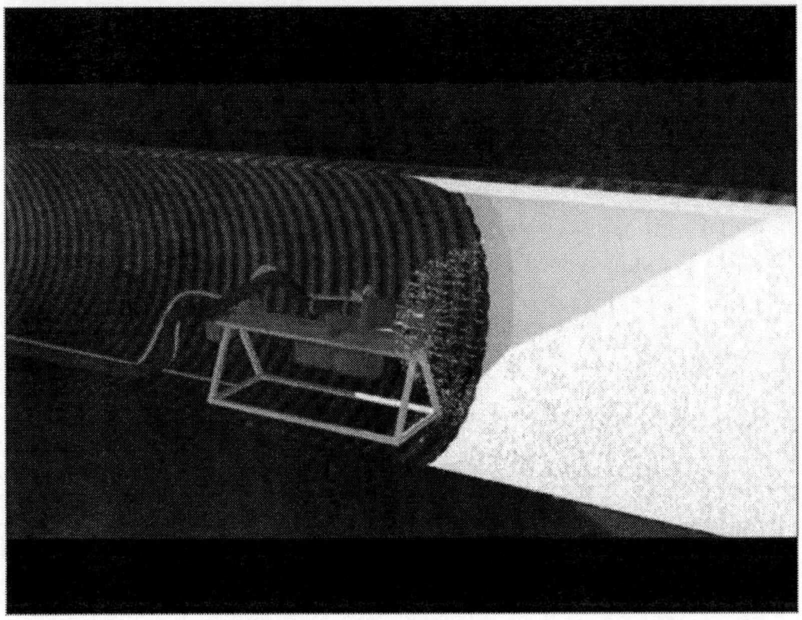

Centrifugal Casting of Geopolymers

- Shape
  - Non-round, elliptical, egg shaped
  - Manholes, wet-wells, junction boxes
- Length
  - Short lines, culverts, point repair
  - Longer sewer and storm segments
- Diameter
  - From 36 inches to 180 inch
  - Varying diameter in a single run
- Ovality
  - Greater than 10%
  - Box culverts and structures

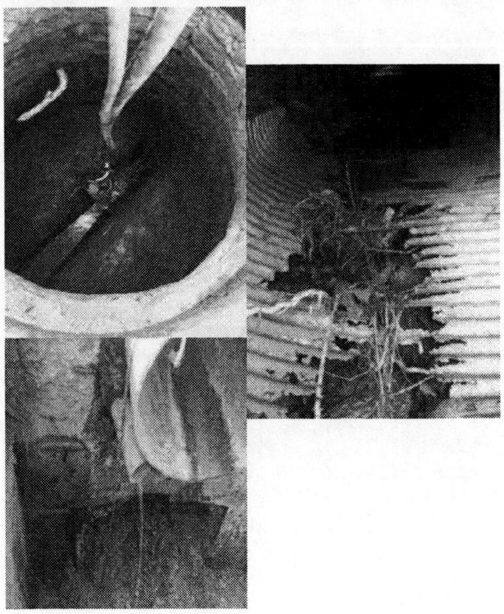

Centrifugal Casting of Geopolymers

Angles
- Bends, curves, and change in elevation

Laterals
- Monolithically tied to structure

Host pipe material
- Brick, Rock, Concrete
- Corrugated Metal or Cast Iron

Host pipe condition
- Slightly to fully deteriorated

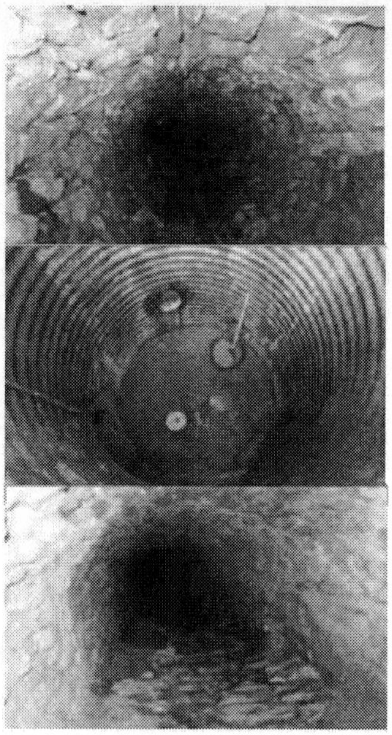

Case Study
McAllen Texas

Rado Storm Drain

Dual CMP Storm Sewer

Completed August 2014

MILLIKEN INFRASTRUCTURE
A Milliken COMPANY

Case Study - McAllen Texas

Project Details:

- Rado Drain McAllen, TX

- Storm Water CMP
  - 2 twin pipes
  - 4,400 linear feet total
  - 114 inches
  - 1,850,000 lbs of GeoSpray applied

- Contractor - IPR South Central
- Engineer - Fred Kurth with Meldon & Hunt

City of McAllen

Case Study - McAllen Texas

Project Background:

After severe drought prior to summer 2013, heavy rains in south Texas created a situation with the Rado drain. The soil became saturated resulting in heavy infiltration and ground water flows inflicted damage which included:

- Offset joints
- Partial collapse
- Heavy corrosion
- Damage to earlier repairs
- Sinkholes surrounding the pipes
- Small 80ft easement on local neighborhood
- Previous shot-crete repairs had failed

Case Study - McAllen Texas

Consulting Engineer Meldon & Hunt Considered & Concluded:

Do Nothing - Not an option with sinkholes & collapsed pipe sections.
Dig & Replace - Expensive due to easements and potentially dangerous given soil conditions.
CIPP - Expensive, Dramatic reduction in pipe size (106' was limit), Multiple access pits would have to be excavated, Easement issues.
Geopolymer (EcoCast™) - Cost Effective, Non-Invasive, Passive By-pass

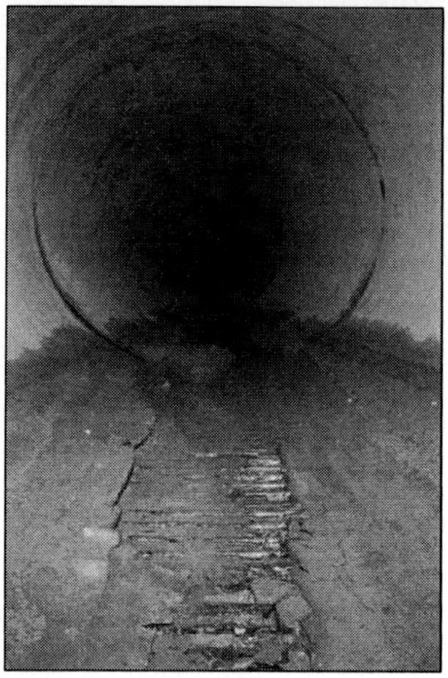

Case Study - McAllen Texas

Project Bidding

HGACBuy THE SMART PURCHASING SOLUTION

- The Houston-Galveston Area Council (H-GAC) is a regional council of governments operating under the laws of the State of Texas and governed by a board comprised of 35 elected officials from the 13 county region.
- The H-GAC Board awards contracts for the purchase of material and services, which can then be made available to local governments nationwide thru HGAC Buy
- The Texas Interlocal Cooperation Act permits joint participation by local governments, states, state agencies, and certain non-profit corporations.

Case Study - McAllen Texas

Project Bidding

- HGAC is active throughout the United States and provide more than 7,497 members with 40 major categories of products and services from more the 750 contractors.

- End users (state and local governments) can issue a purchase order based on competitively set prices for services and products without the need for a formal bid process from the local agency.

- HGAC Buy currently has over 2,700 registered end users in 48 states nationwide.

- This includes local municipalities and airports

Case Study - McAllen Texas

Project Outline - Steps to Complete the Repair

- **Pre-construction Inspection and Review**
- **Interior of the pipe pressure washed and cleaned - min 3500 psi.**
- **Fast setting geopolymer plug material use to stop active infiltration and stop ongoing leaks.**
- **Chemical grouts applied to junctures to stop heavy infiltration flows.**
- **Replace collapsed sections**
- **Joint prep**
  - Surrounding soil was "sugar-like" sand
  - Joints with failure or separation were Grouted, Sealed and reinforce with Carbon Mesh
- **Pipe repair - Centrifugal Casting of Geopolymer (Done in 150 LF sections). 1.5 inch liner was specified as repair mechanism.**
- **QC during repair.**
- **Post-construction inspection.**

Case Study – McAllen Texas

Milliken® Infrastructure Solutions
GeoSpray®
Geopolymer Mortar
STORM + SANITARY

Specification of the repair materials used in this study

Test Method	Duration	GeoSpray	Conventional Repair Mortar
Compressive Strength ASTM C-39/C-109	1 Day 28 Days	Min. 2,500 psi / 17 MPa Min. 8,000 psi / 55 MPa	 5000 psi / 34 MPa
Flexural Strength ASTM C-78	7 Day 28 Days	1100 psi / 7.6 MPa 1800 psi / 12.4 MPa	 500 psi / 34 MPa
Modulus of Elasticity ASTM C-469	1 Day 28 Days	3,000,000 psi / 20700 MPa 5,800,000 psi / 40000 MPa	 3,000,000 psi / 20700 MPa
Bond Strength to Concrete ASTM C-882	1 Day 28 Days	Min 900 psi / 6.2 MPa Min. 2,500 psi / 17 MPa	 N/A
Set Time ASTM C-807 Initial Cure Time	Initial Set Final Set	60 - 75 Minutes 90 - 110 Minutes	120 Minutes 300 minutes
Freeze Thaw Durability ASTM C-666	300 Cycles	100% Zero loss	80% to 90% 10% to 20% degradation
Shrinkage ASTM C-1090	28 Days	0.00% @ 65% R. H.	0.35% to 0.50% Shrinkage
Tensile Strength ASTM C-496	28 Days	Min. 800 psi / 5.5 MPa	400 psi / 2.7 MPa
Abrasion Resistance ASTM C-1138	5 Cycles @ 28 Day Maturity	2.7% Loss	4.7% Loss
Rapid Chloride Ion Permeability ASTM C-1202	28 Days	Very Low	N/A

Case Study – McAllen Texas

A proper, detailed construction specification is a must

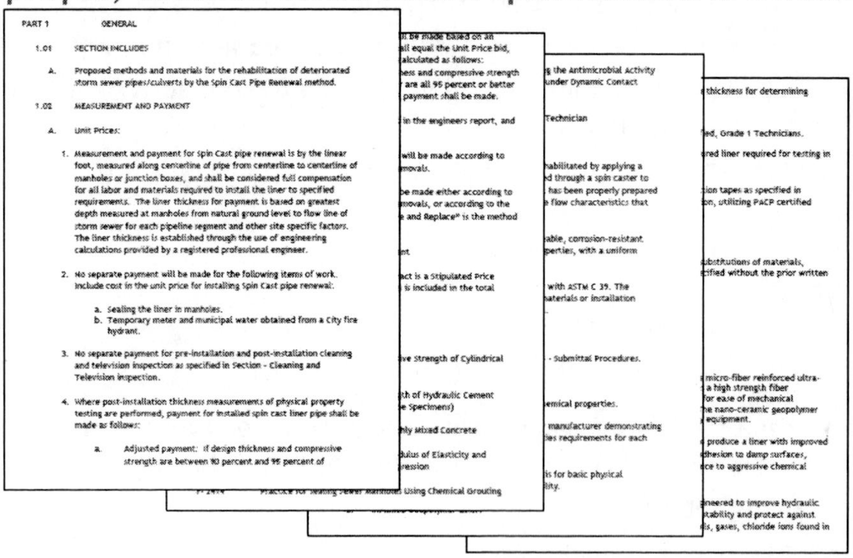
PART 1 GENERAL

1.01 SECTION INCLUDES

A. Proposed methods and materials for the rehabilitation of deteriorated storm sewer pipes/culverts by the Spin Cast Pipe Renewal method.

1.02 MEASUREMENT AND PAYMENT

A. Unit Prices:

1. Measurement and payment for Spin Cast pipe renewal is by the linear foot, measured along centerline of pipe from centerline to centerline of manholes or junction boxes, and shall be considered full compensation for all labor and materials required to install the liner to specified requirements. The liner thickness for payment is based on greatest depth measured at manholes from natural ground level to flow line of storm sewer for each pipeline segment and other site specific factors. The liner thickness is established through the use of engineering calculations provided by a registered professional engineer.

2. No separate payment will be made for the following items of work. Include cost in the unit price for installing Spin Cast pipe renewal:

   a. Sealing the liner in manholes.
   b. Temporary meter and municipal water obtained from a City fire hydrant.

3. No separate payment for pre-installation and post-installation cleaning and television inspection as specified in Section - Cleaning and Television Inspection.

4. Where post-installation thickness measurements of physical property testing are performed, payment for installed spin cast liner pipe shall be made as follows:

   a. Adjusted payment: If design thickness and compressive strength are between 90 percent and 95 percent of

Case Study - McAllen Texas

Project Timeline

- Project Awarded - December 2013
- Subcontract work on point repairs and collapse began - January 2014
- Geopolymer repair started, with by-pass using the twin pipes (one in operation - one being repair) - April 2014
- Completion and inspection - August 2014

Case Study - McAllen Texas

Located adjacent to upscale gated neighborhood

Case Study – McAllen Texas

A single utility truck and material loader is all the on-site equipment needed

Case Study – McAllen Texas

Typical joint failures (note previous shot-crete repair)

Case Study – McAllen Texas

Sections with heavy infiltration

Case Study – McAllen Texas

Invert erosion and corrosion due to poor bonding

Case Study – McAllen Texas

Material sections damaged from infiltration

Case Study – McAllen Texas

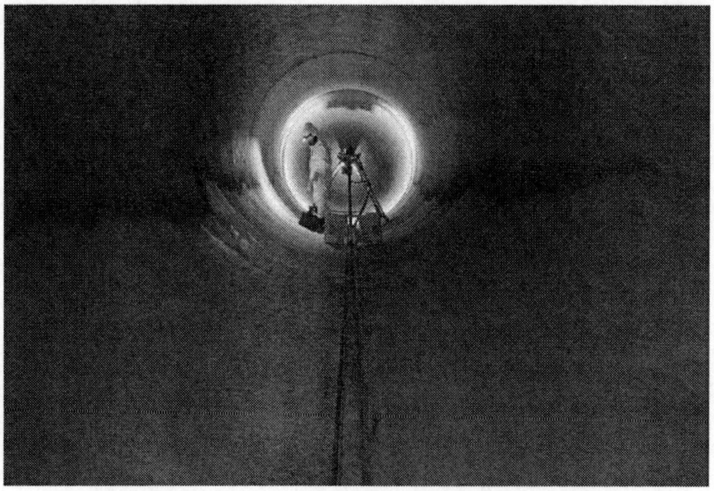

Geopolymer Spin Casting In process

Case Study – McAllen Texas

Up-close Repair

Case Study – McAllen Texas

Even closer view of the action

Case Study – McAllen Texas

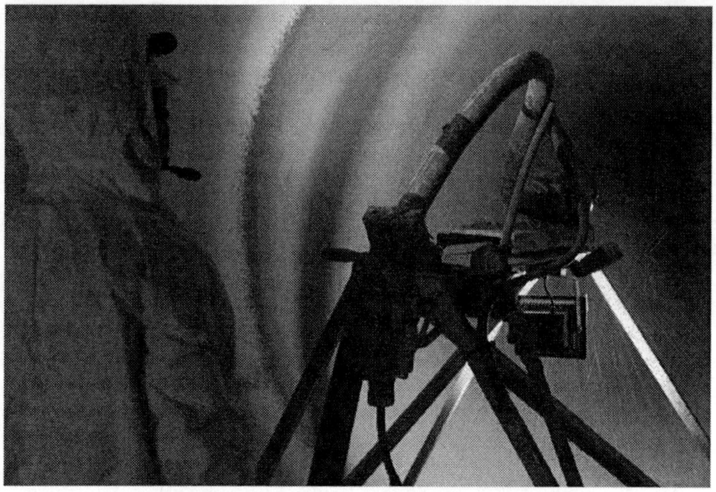

Geopolymer Spin Casting In Process

Case Study – McAllen Texas

View from the repaired side

Case Study – McAllen Texas

Completed Repair

Case Study – McAllen Texas

A view of the twin pipes near completion

Case Study – McAllen Texas

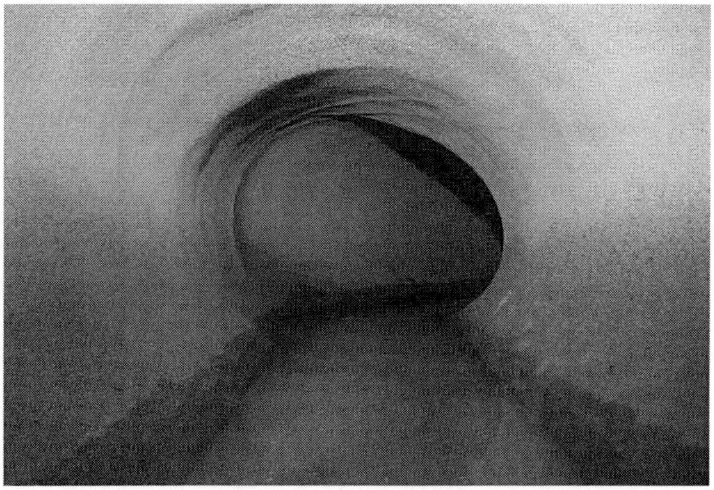

Monolithic liner applied around a 45° bend

Case Study – McAllen Texas

Completed Repair

Case Study – McAllen Texas

The Happy Repair Crew!

Case Study – McAllen Texas

QC/QA Testing

- **During the repair QC samples were taken at 32 different days.**
- **These materials were tested by and ACI accredited third party lab and reported to both the engineer and the asset owner in the final QC package.**

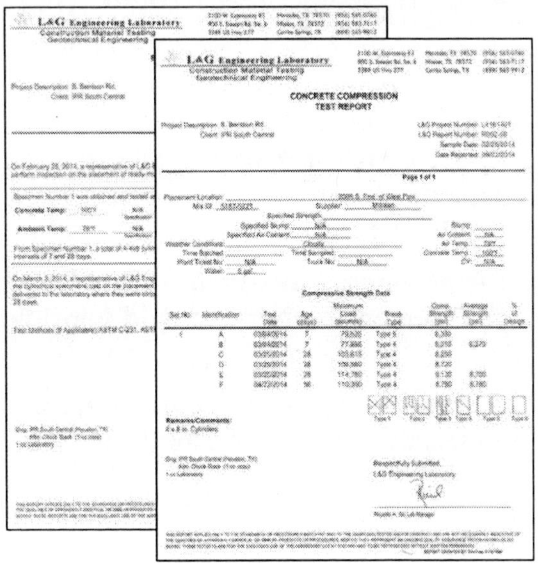

L&G Engineering Laboratory

CONCRETE COMPRESSION TEST REPORT

Compressive Strength Data

Case Study - McAllen Texas

Critical Aspects to Consider

- Environmental conditions - 100°F or greater ambient conditions can be an issue with other technologies but are compatible with geopolymers.

- Construction specs are critical to ensuring the contractor has a detailed plan of attack agreed to by all stake holders, especially with small easements in a residential area.

- QA/QC before, during and after the installation are critical to insuring you get what you paid for.

Case Study - McAllen Texas

Conclusions

Geopolymers are type of cementitious material made from Aluminosilicate.

Geopolymers have significant advantages as compared to standard OPC cements related to cold jointing and chemical resistance.

Geopolymers can be a cost-effective structural repair option, especially for large diameter structures.

Thank You
MILLIKEN INFRASTRUCTURE
A Milliken COMPANY

Formatting Issue - Best Available Paper/Graphic

THE TOTAL SOURCE

educational sessions

Evolution of PVC Restrained Joint Piping Systems

Continuous Improvement Pays Big Dividends

Presented by:
Joshua Clapper
North American Pipe Corporation

Mission/Abstract

Continuous Improvement Pays Big Dividends - Evolution of PVC Restrained-Joint Piping Systems

Segmented restrained-joint PVC pipe has been a well-accepted product as a result of the numerous advantages it offers to the Engineering and Contracting communities. These advantages include corrosion resistance, joint strength, flow performance, and ease of assembly with minimal right-of-way disruption. As more experience has been gained, spline-lock PVC pipe has been enhanced with a more robust sealing system for enhanced hydraulic integrity, and to better tolerate various installation conditions, including bending and angular deflection often seen on the job-site. The presentation will highlight the engineering design, development, and structural analysis of the enhanced system in C900 sizes. Segmented systems allow for quick installation, especially where assembly must take place within pits. Presentation will also focus on the rigorous product qualification process, which involved lab tests and customer trials.

Product applications include water transmission, sewer force mains, and storm sewers where Horizontal Directional Drilling and Static Pipe bursting methods are used. Products can also be used for conventional open-cut construction where there is a need to eliminate concrete thrust blocks.

Presentation will utilize case studies from several projects to illustrate the new technology solution.

Mission of the Session

This presentation will of interest to Consulting Engineers, Municipal Engineers, and Contractors, as the information to be presented will provide them with a way to confidently consider multiple materials for trenchless projects.

Agenda

- Introductions
- Segmented Restrained Joint Overview
  - Coupled vs Integral Bell Joints Comparison
- Review of Pipe Bending via Uni-Bell Guidance
- Review of Angular Deflection & ASTM D3139 Testing
- Job Site Challenges
- Design Optimization / Integral Bell Improvements

Segmented RJ Overview

- Segmented Restrained Joint PVC Pipe Advantages

Coupled vs Integral Bell

- Restrained joint systems can be assembled using couplings or via integral bells (IB or RJIB).
- Each design has advantages:

Coupled Design

- More deflection per joint
- Thicker coupling stock can provide improvements in strength/rigidity
- Greater field cutting/grooving flexibility

Integral Bell Design

- Ease/Speed of assembly
- Smaller OD
- Similar tensile strength
- More cost effective

PVC Pipe Bending

- PVC's response to bending is considered a significant advantage of PVC pipe in buried applications.
  - PVC bending can occur ***deliberately***, or from responses to various unplanned conditions:
    - Differential settlement of a manhole, valve or structure rigidly connected to the pipe
    - Uneven settlement of bedding
    - Ground movement
    - Erosion of bedding or foundation
    - Seasonal variations in soil
  - An allowed amount of bending is calculated to ensure safe wall stress based on pipe size and pressure class.

PVC Pipe Bending

- Uni-Bell provides recommendations for bending considerations in PVC

CHAPTER VIII - SPECIAL DESIGN APPLICATIONS

TABLE 8.2
ALLOWABLE LONGITUDINAL BENDING FOR PRESSURE CLASS PIPE
(AWWA C900) IN 20 FOOT LENGTHS
(S_b = 800 lbs/in², E = 400,000 lbs/in²)

Nominal Size, in	4	6	8	10	12
DR 14 (PC 200)					
D_o, in	4.800	6.900	9.050	11.100	13.200
t_{min}, in	0.363	0.523	0.683	0.840	0.999
D, in	4.07	5.86	7.68	9.42	11.20
I, in⁴	12.3	38.3	138	338	716
M, in-lbs	4,170	12,400	28,000	51,600	86,800
R_b, in (min)	1200	1730	2260	2780	3300
R_b, ft (min)	100	144	188	232	275
β, degrees	11.5	7.9	6.1	4.9	4.2
α degrees	5.7	4.0	3.0	2.5	2.1
A, in	24	17	13	10	9
P, lbs	26	77	170	320	340
Ratio R_b/D_o	250	250	250	250	250
DR 18 (PC 150)					
D_o, in	4.800	6.900	9.050	11.100	13.200
t_{min}, in	0.283	0.406	0.533	0.654	0.777
D, in	4.23	6.09	7.98	9.79	11.65
I, in⁴	10.3	43.8	130	293	586
M, in-lbs	3,420	10,200	22,900	42,300	71,100
R_b, in (min)	1200	1730	2260	2780	3300
R_b, ft (min)	100	144	188	232	275
β, degrees	11.5	7.9	6.1	4.9	4.2
α degrees	5.7	4.0	3.0	2.5	2.1
A, in	24	17	13	10	8.7
P, lbs	21	63	140	260	440
Ratio R_b/D_o	250	250	250	250	250

HANDBOOK OF PVC PIPE

FIGURE 8.1
PVC PIPE ALLOWABLE BEND
CALCULATIONS MADE AT 73° F (23° C)

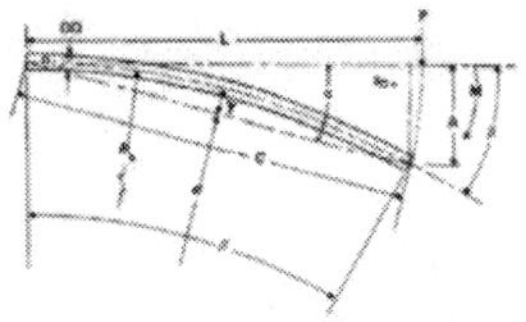

(EQUATION 8.3)

$$M = \frac{S_b I}{c}$$

(EQUATION 8.7)

$$\beta = \frac{360L}{2\pi R_b}$$

(EQUATION 8.8)

$$\alpha = \beta/2$$

(EQUATION 8.9)

$$A = 2R_b(\sin^2\beta/2) = C \sin\beta/2 - L \tan\alpha$$

EQUATION 8.11

$$L = \frac{R_b \pi}{90}\alpha$$

EQUATION 8.12

$$d = R_b \cos\beta/2$$

EQUATION 8.13

$$Y = R_b - d$$

EQUATION 8.14

$$C = 2R_b \sin\beta/2 + L$$

"Chapter 8." *The Uni-Bell PVC Pipe Association Handbook of PVC Pipe: Design and Construction.* 5th ed. Dallas, TX: Uni-Bell PVC Pipe Association, 2012. 282-84. Print.

PVC Pipe Bending

- Examples of max allowable bending:

6" C900
Bend Radius: 143.75 ft

12" C900
Bend Radius: 275 ft

24" C905*
Bend Radius: 537.5 ft
*Unibell suggests too difficult to bend

Segmented PVC Joint Deflection

- Segmented PVC joints provide a second element for direction change: Angular Deflection.
  - Uni-Bell describes best practices and ASTM defines a sealing test for performance validation.

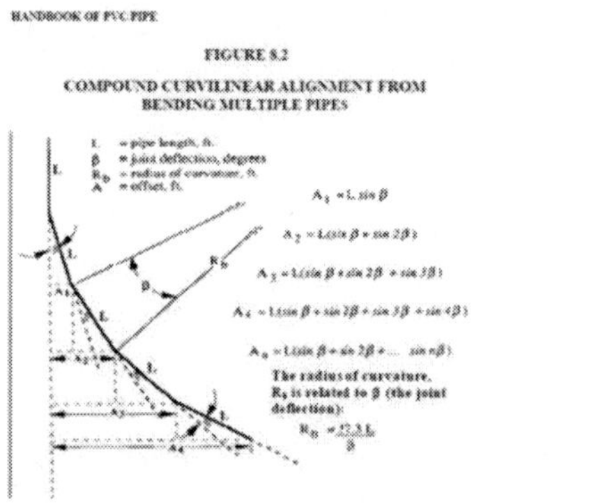

"Chapter 8." *The Uni-Bell PVC Pipe Association Handbook of PVC Pipe: Design and Construction*. 5th ed. Dallas, TX: Uni-Bell PVC Pipe Association, 2012. 282-84. Print.

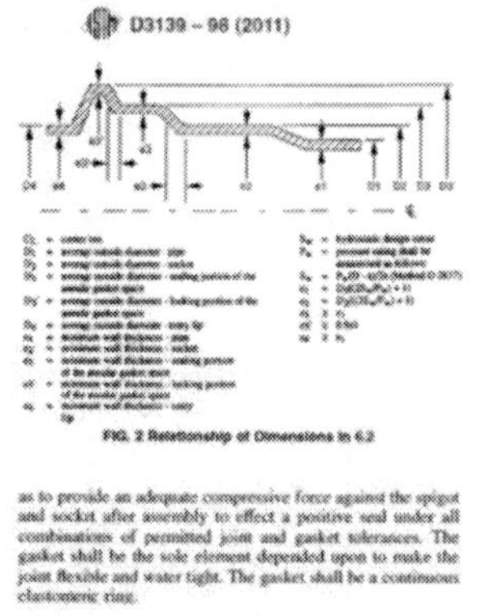

"Specification for Joints for Plastic Pressure Pipes Using Flexible Elastomeric Seals." ASTM D3139-98 (2011)

Segmented PVC Joint Deflection

- Examples of max allowable joint deflection:

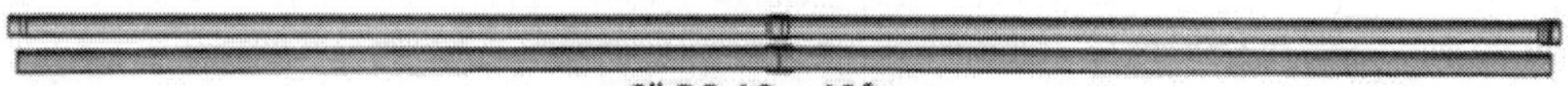

6" DR 18 – 40ft
RJIB: 0.3° / Coupled: 0.6°

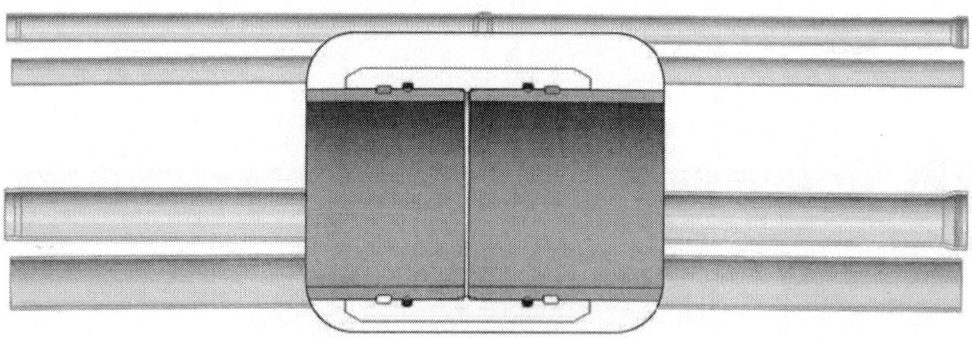

RJIB: 0.5° / Coupled: 1.0°

Segmented PVC Joint Deflection

- Uni-Bell does not recommend combining pipe deflection and pipe bending in segmented products.
  - Initial loads caused by direction change deflect the pipe within the compliance of the joint
  - Additional direction change applies loads/stress to pipe system causing pipe to bend
    - Secondary effect of pipe bending is deformation loads at each joint
    - Over deflection of Restrained Joints can have detrimental effects on the system

Job Site Challenges

- Deflection + bending occurs during some installations
  - Usually taken into consideration for bore paths
  - Often overlooked for pit design

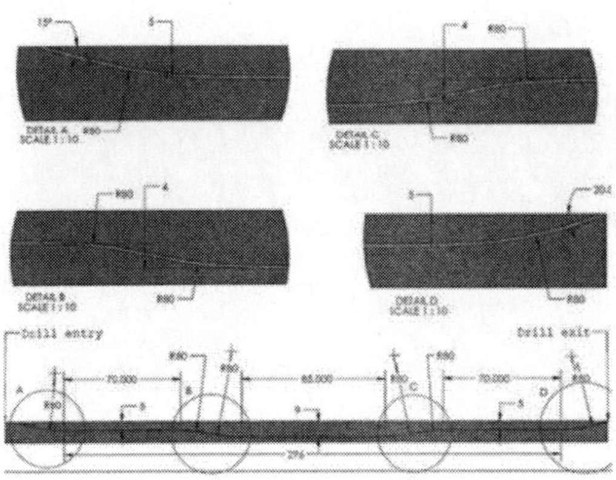

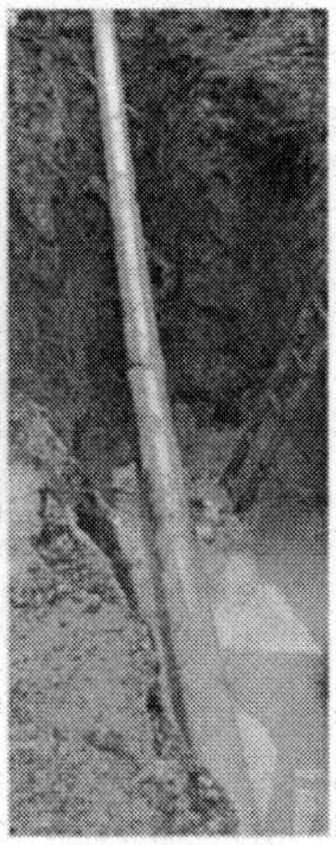

Engineering Investigation

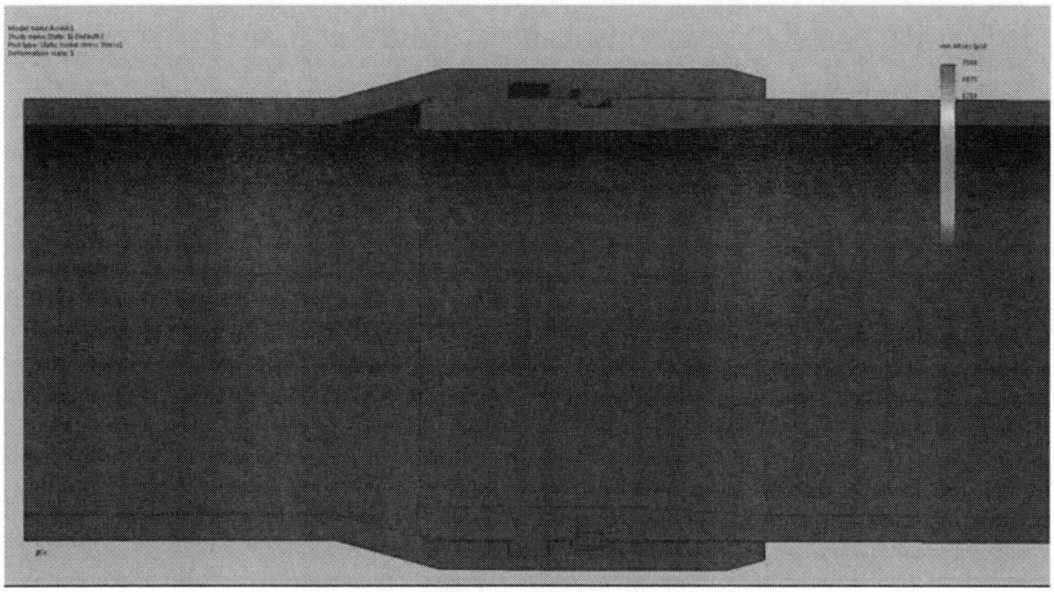

FEA of Coupled Pipe Under High Bending Loads

Engineering Investigation

- Coupled and Integral Belled products behave differently under deflection:
  - Joint deflection caused by unsupported cantilevered pipe

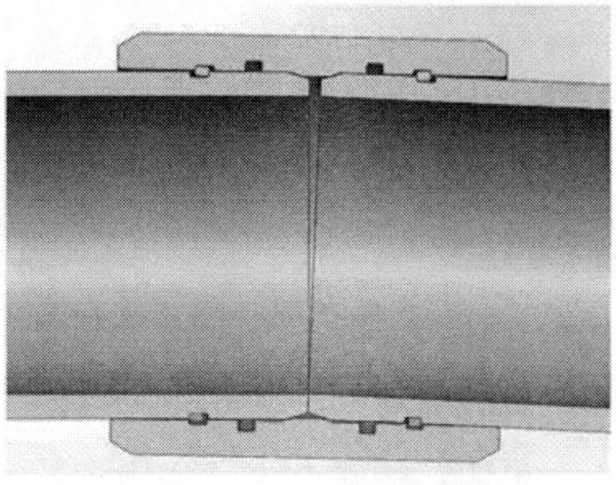

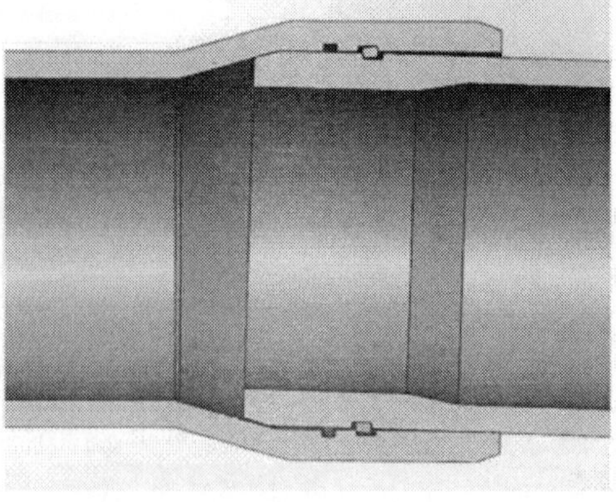

Engineering Investigation

- Excessive over-deflection will cause leaks
  - Seal compression can be reduced/eliminated by coupling/bell deformation

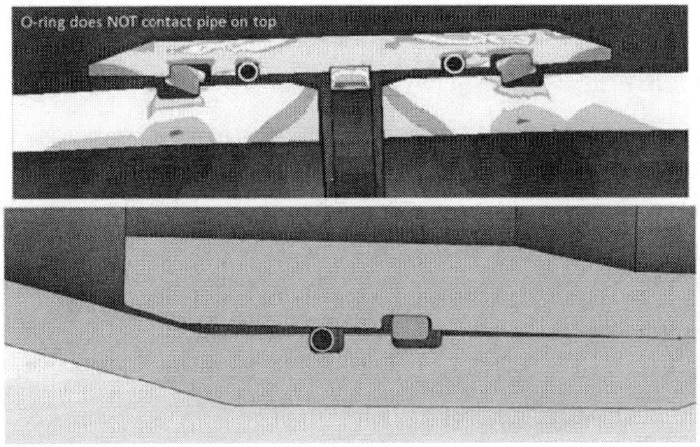

Engineering Improvements

- Improvements focused on maximizing sealing surface while maintaining key design parameters:
  - Minimized OD for minimized borehole size
  - Maximized tensile strength
  - Minimized assembly effort
- Considered Reiber style seal, but rejected due to weaker tensile, larger OD and less resilience to bend

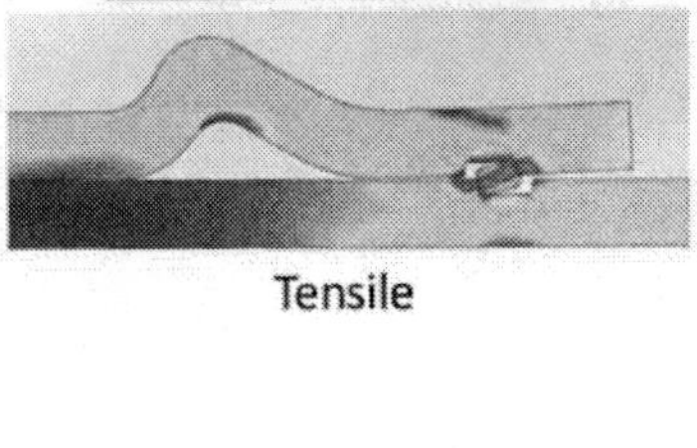

Tensile

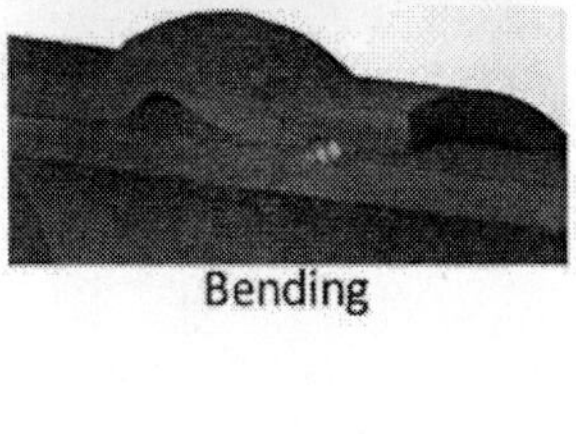

Bending

Engineering Improvements

- New profile gasket was designed that:
  - Provides over 2x the sealing surface as the comparable O-ring in normal conditions and 3.75x when system is deflected.
  - Improves groove retention/grip ratio eliminating seal dislocation.
  - Delivers similar spigot/bell insertion force.
  - Allows similar tensile strength.
  - Meets same Agency Listings/Certifications

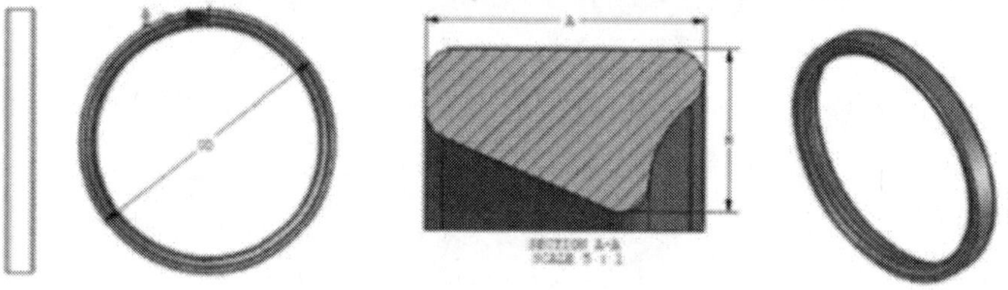

17

Engineering Improvements

- New seal design significantly improves systems' ability to not fail during over deflected conditions.
  - ***Only Uni-Bell guidance for deflection/bending is recommended***

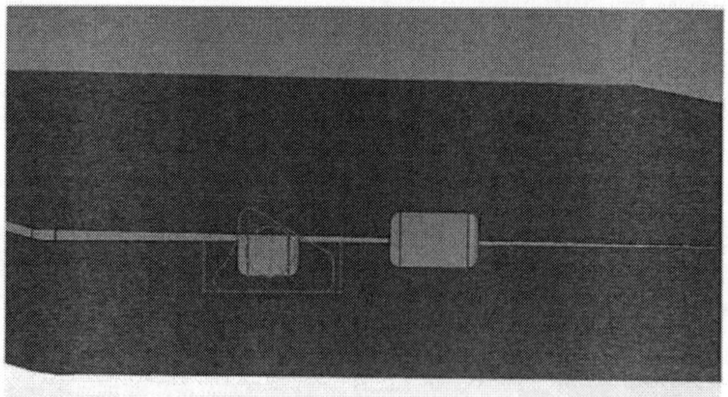

8" FEA model showing joint deflection due to one stick cantilevered

18

Formatting Issue - Best Available Paper/Graphic

Engineering Improvements

- New seal is being applied to 6"-12" RJIB and C905 Products

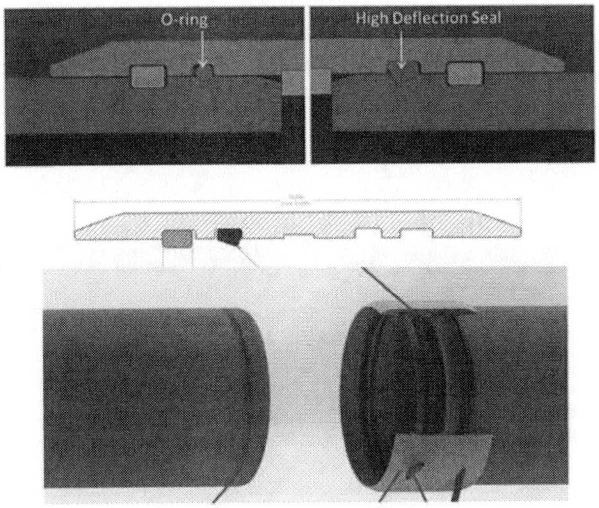

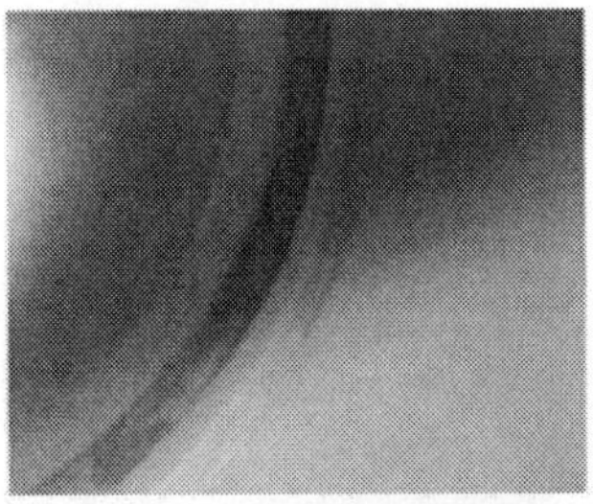

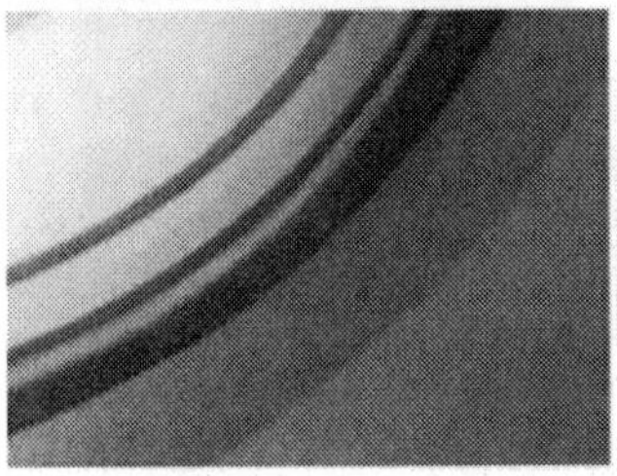

Product Validation

- Product testing validates system performance
  - To ensure Segmented Restrained Joint products meet job-site needs, compliance, regulatory and abuse testing is performed

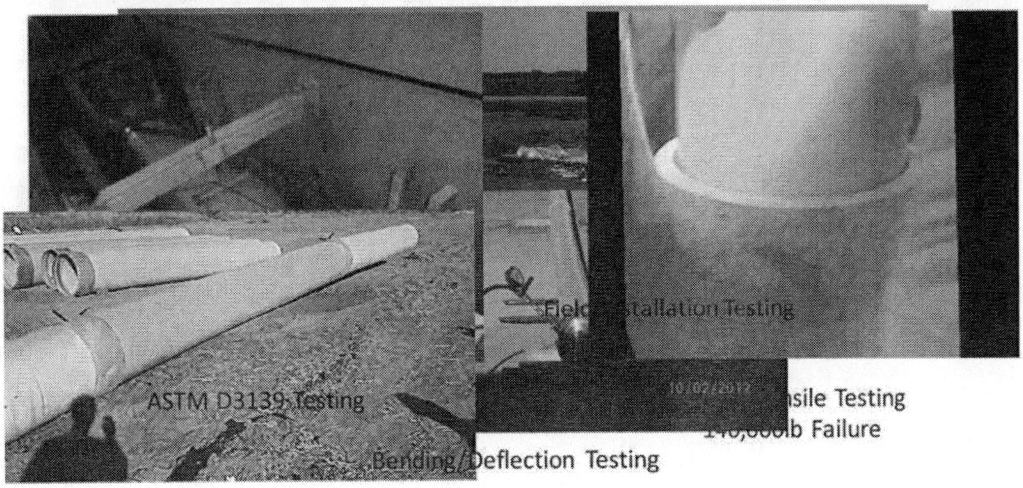

Formatting Issue - Best Available Paper/Graphic

Thank You!

- Questions?

Novel Pipe Cleaning System with Lining Integration – A Complete Rehab solution for Aging Water Pipes

Randall J. Cooper, P.Eng. M ASCE

OMAHAWK SYSTEM

While trenchless linings offer cost-effective <u>renewal</u> solutions....

Spray-in-place pipe (SIPP)

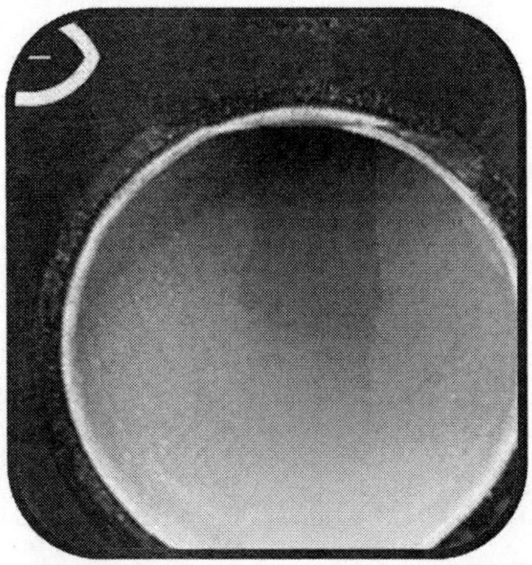

Cured-in-place pipe (CIPP)

Every pressure pipe lining project <u>starts</u> with pipe cleaning.

Current Methods: scraping, boring, flushing

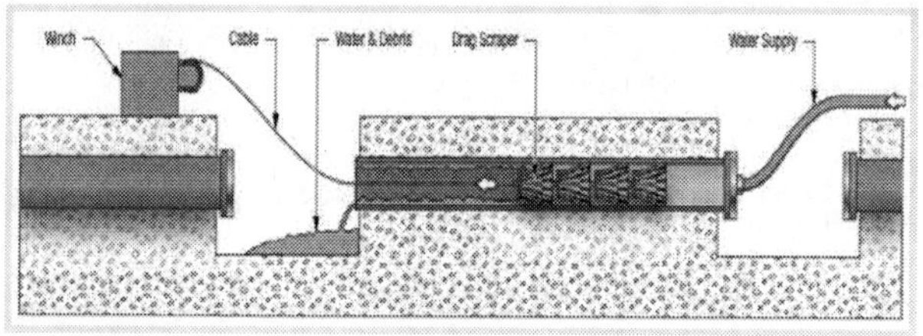

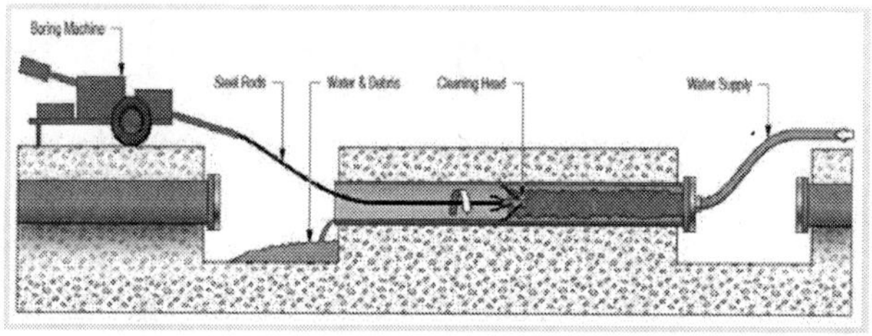

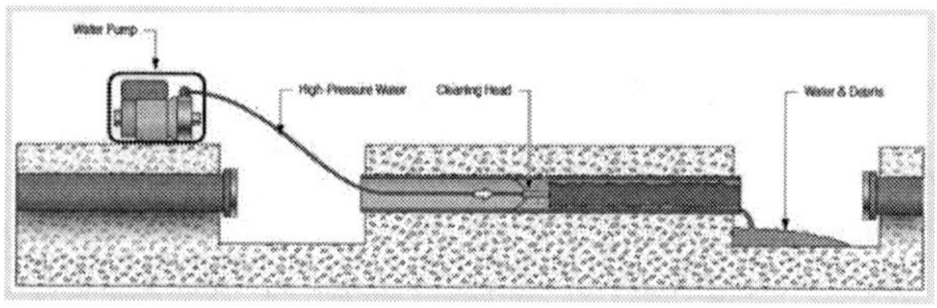

Challenges with Cleaning methods:

- ❖ Pipe is cleaned but not prepared for lining.
- ❖ Linings such as coal tar cannot be fully removed.
- ❖ Current methods can damage services and pipe wall.
- ❖ they do not dry the pipe for liner bond.
- ❖ They generate large quantities of wet waste.
- ❖ They use (and foul) a lot of fresh

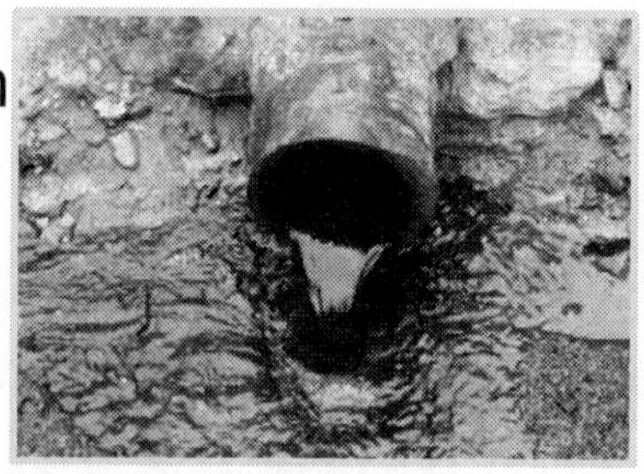

Technology Development Goals:

- Clean and inspect pipes 50% faster,
- Dry them for 100% superior liner bond,
- Not use a drop of water,
- Significantly reduce waste generation,
- NOT damage pipe walls or service connections,
- Capture all waste in a closed, filtered system; and
- Meet NACE 3 surface preparation standards.

Cleaning progression……

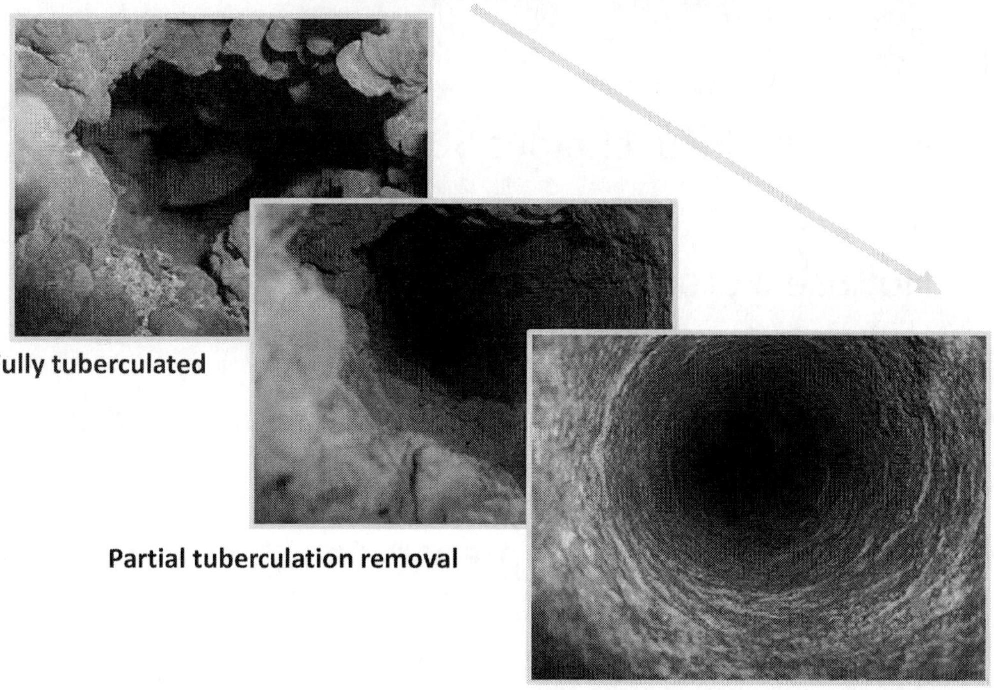

Fully tuberculated

Partial tuberculation removal

Tuberculation removed; bitumen liner visible

Cleaning progression……

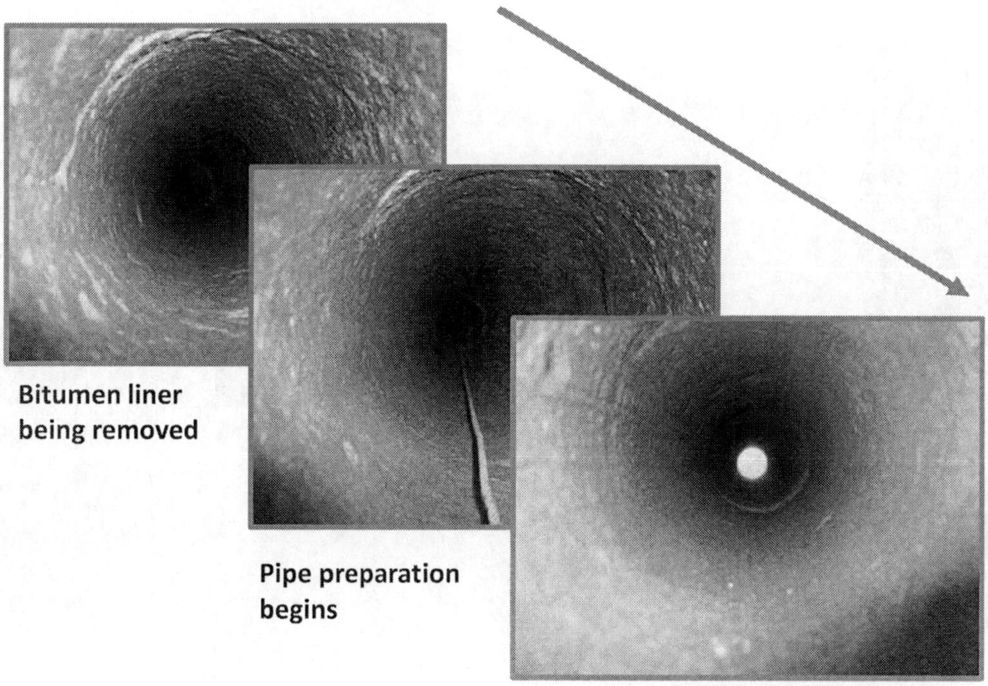

Bitumen liner being removed

Pipe preparation begins

Fully prepared for liner bonding

Development Outcomes With Airborne Stone:

- ✓ Cleans pipe to bare metal (4" – 12")
- ✓ Waterless, dust-free process
- ✓ Dries the pipe for best bond
- ✓ No damage to service connections
- ✓ Very fast (enables same-day return)
- ✓ Used on metallic, AC, & concrete pipes
- ✓ Reduced construction footprint

Why not develop a novel renewal solution…

….a trenchless solution that provides a far more economic solution, with <u>decades</u> of life extension, and the option for rapid return-to-service .

<u>70%</u> of aging pipes still have tangible life.

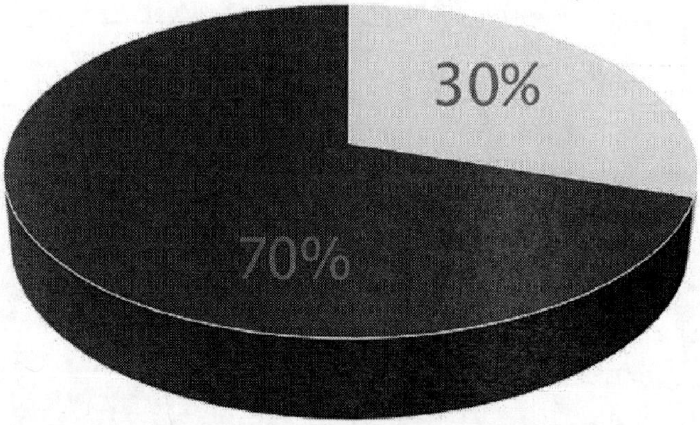

Source: Pipe Inspection data from Pica and Echologics

•

What is tangible life?

Minimal break history coupled with the retention of no less than 60% (SF 1.2) of design wall thickness.

If we have remaining wall thickness in our pipes.....

..why not line for that condition?

Airborne Lining Development Objectives:

- Develop a low cost, NSF61, non-structural polymeric barrier lining system for metallic pipes.
- Designed to prevent future interior corrosion, deposit build-up and water quality problems.
- Process is simplified and designed for preventative maintenance application.
- Process can be funded by utility operating & maintenance budgets.

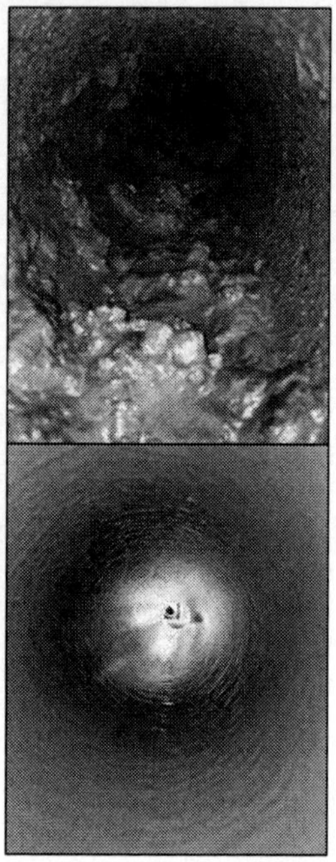

Tomahawk Airborne Lining Development

Video Presentation in
Original e-media

Waterloo - Albert St. Demo – July 2015

Purpose: To demonstrate Tomahawk cleaning and airborne lining process to potential applicators and municipal customers. To validate performance of cleaning and lining as part of joint R&D project.

City of Waterloo, Albert St. Demo – July 2015

Waterloo - Albert St. Demo – July 2015

✓ 53 meters (175 feet) lined in 10 minutes

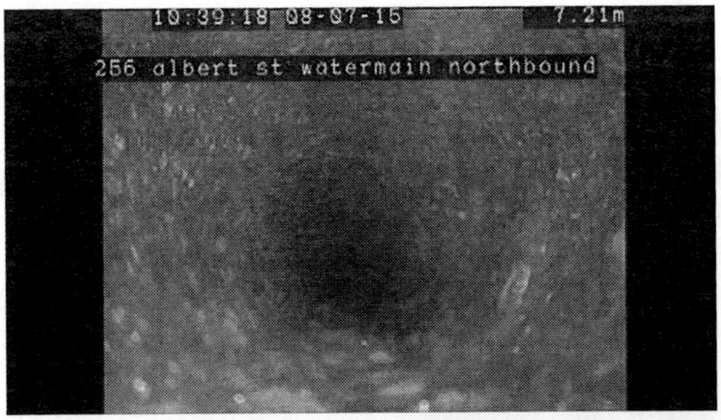

Waterloo water main before cleaning inspection video

Waterloo water main after stone cleaning and lining inspection video

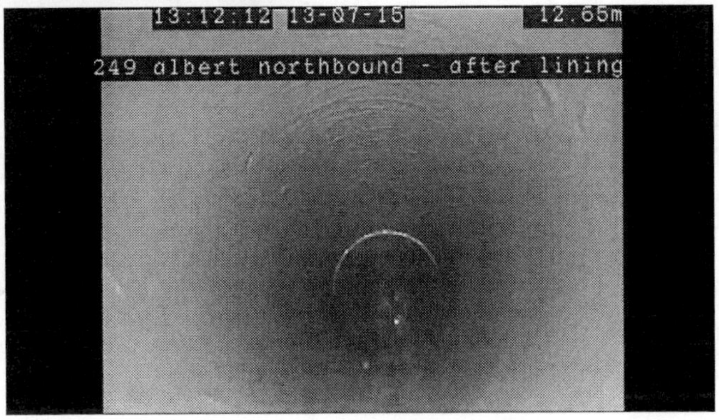

Waterloo - Albert St. Demo – July 2015

Service Connection Coating Coverage:

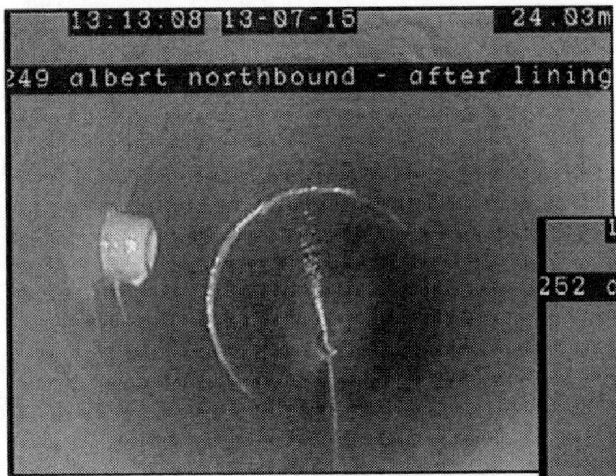

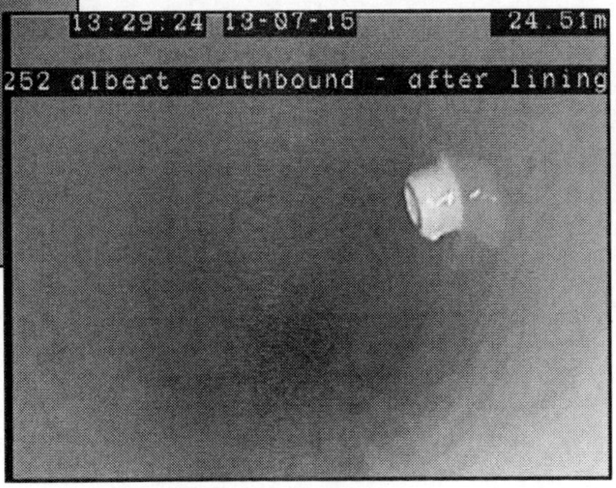

Waterloo - Albert St. Demo – July 2015

Pipe Joints Coating Coverage (NB 24.2m):

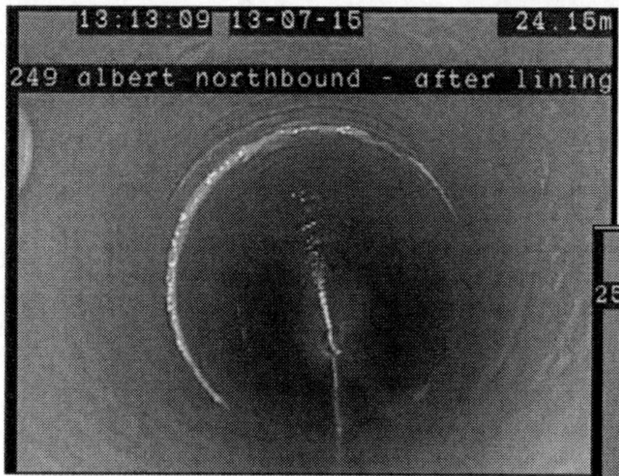

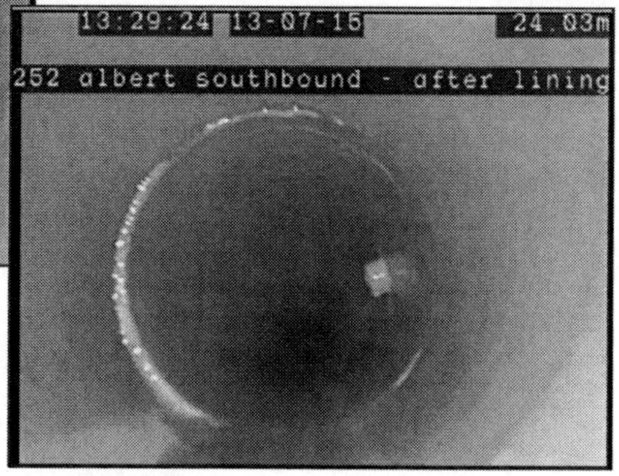

City of Waterloo: July 2015 Same-day Results

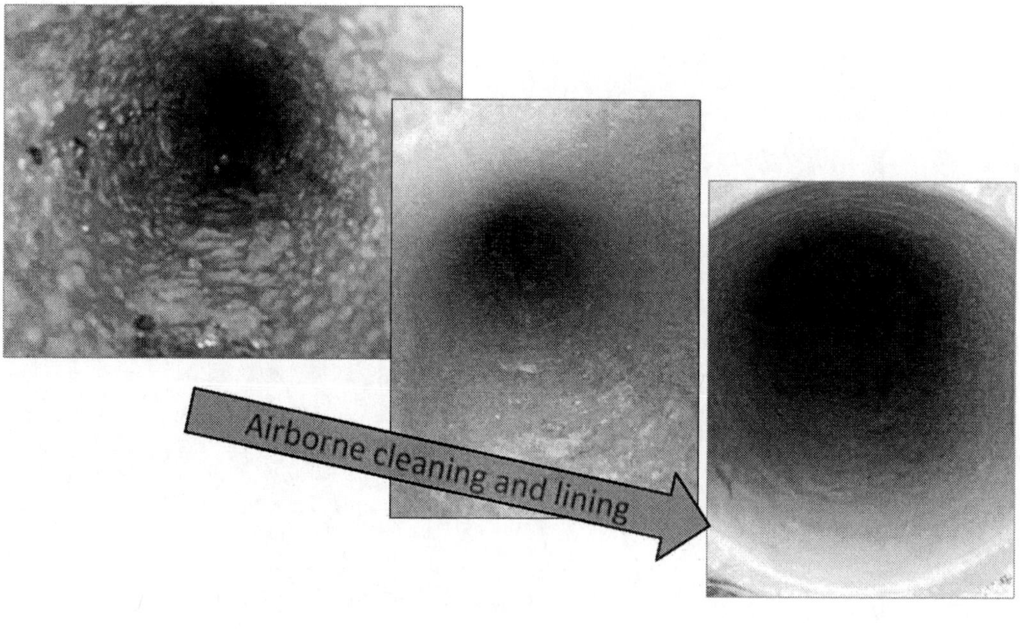

Waterloo - Albert St. Demo – July 2015

Liner Bond: This pipe section was purposely broken when removing from ground four days after it was lined on Albert St. The liner coating closely conforms with the pipe fracture line, indicating <u>excellent</u>, post-break bond.

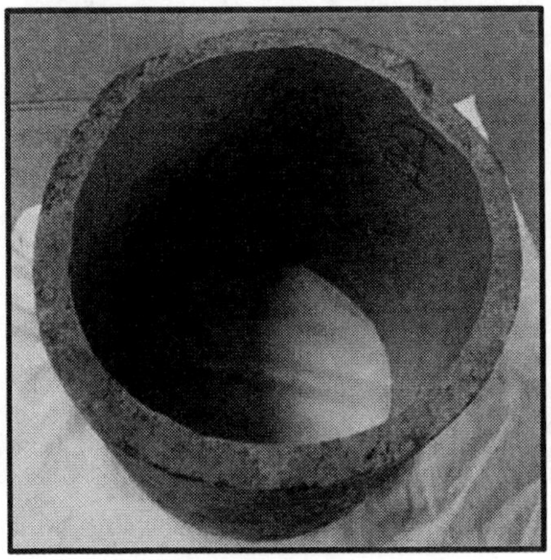

Waterloo - Albert St. Demo – July 2015

A service connection was then installed into the tapped pipe, producing a bonded connection. Below left: broken edge of pipe is visible with new service connection. Below right: close-up view of service connection.

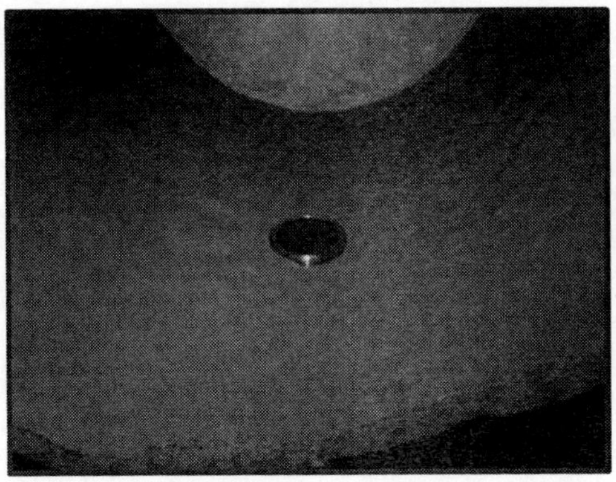

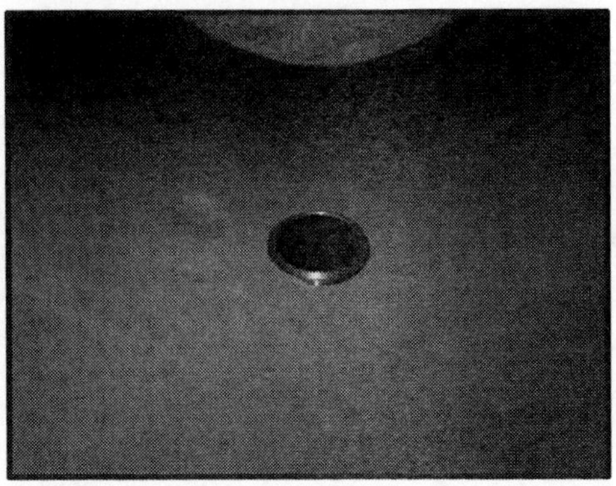

Tomahawk Airborne Lining

Clean then apply barrier coat in minutes

Low-cost, bonded barrier coat (AWWA Class I):

- Extends pipe life by decades
- Improves flow capacity & water quality
- Encapsulates lead found in pipe and joints

Summary:

- Cleaning aged water mains with airborne stone provides a dry cleaning process that is very fast, reduces the construction footprint, and provides a "best-bond" surface for all classes of pressure pipe liners.

- The process has now been further developed and piloted to include airborne NSF 61 barrier coatings, making it a complete, economical life-extension technology for pressure pipes.

- This novel cleaning and lining technology enables rapid return to service and can be deployed by water utility maintenance crews.

- Future research will focus on new composite polymeric linings that provide higher strength and total corrosion protection.

www.tomahawksystem.ca

THE TOTAL SOURCE

educational sessions

Roanoke Rapids Sanitary District Chooses Pipe Bursting to Replace Water Distribution Mains

Brent Johnson, P.E. – CDM Smith, Inc.
Mike Woodcock – Portland Utilities Construction Company, LLC
Gary Shepherd – Underground Solutions, Inc.

Underground Construction Technology
International Conference & Exhibition
February 3-4, 2016
Georgia World Congress Center
Downtown Atlanta, GA

Roanoke Rapids, NC

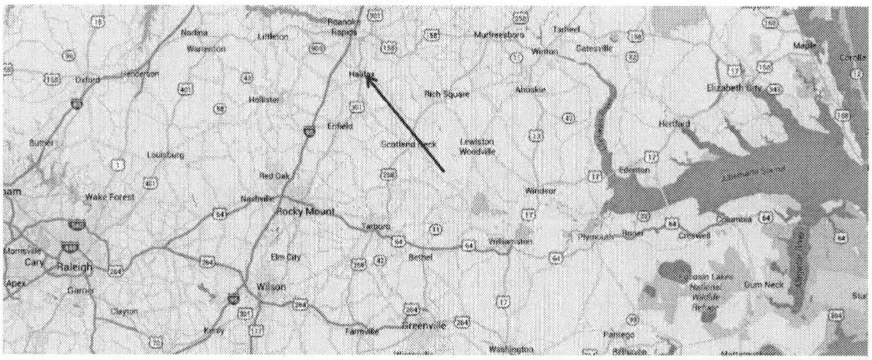

- Roanoke River runs through town on the way to Albemarle Sound and the Atlantic Ocean
- Textile Mills harvested power of the river in the 1800's

Underground Construction Technology
International Conference & Exhibition
February 3-4, 2016
Georgia World Congress Center
Downtown Atlanta, GA

RRSD Rehabilitation Program

- Old cast iron mains from 1950's
- Historically a clean & line agency (2012 and earlier)
- Question about life span of older mains if just cleaned (tuberculation returning) and even after lining (lead joints and leaking)
- New approach considered in 2012
  - Open cut
  - Pipe bursting

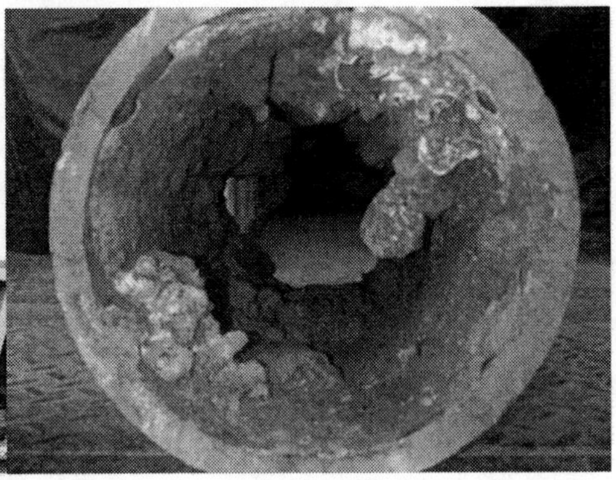

RRSD Rehabilitation Program

- Current Project
  - 10,200' of 6" Water Main
  - Small amount of 4",8",10" Water Main
  - 50 Gate Valves
  - 2 Hydrants
  - 117 Water Services
- Valve Assessment (Replace Broken Valves) – 4 as a part of the project, 7 to be replaced by RRSD prior to project
- How to best accomplish the work?

Underground Construction Technology
International Conference & Exhibition
February 3-4, 2016
Georgia World Congress Center
Downtown Atlanta, GA

Economic Analysis I – Open Cut vs. Pipe Bursting

- RRSD a water utility, not involved with roadways – Public Works / NCDOT governs roads
- Many roads have concrete base or surface; restoration costly

Underground Construction Technology
International Conference & Exhibition
February 3-4, 2016
Georgia World Congress Center
Downtown Atlanta, GA

Economic Analysis I – Open Cut vs. Pipe Bursting

- Pipe bursting becoming more commonplace in NC
- Contractor base expanding

Underground Construction Technology
International Conference & Exhibition
February 3-4, 2016
Georgia World Congress Center
Downtown Atlanta, GA

Pipe Bursting vs Open Cut

- Both can upsize main, upgrade service connections/appurtenances
- Open Cut – rehabilitate other infrastructure concurrently, good specs, multiple contractors
- Pipe Bursting Advantages
  - Less disruption to Public
  - Typically faster installation rates
  - Utilizes the same utility and easement corridor and location

Pipe Bursting Drawbacks

- Excavation required for access pits and appurtenances (valves, fittings, services, and repair clamps)
- Limited quality control of pipe bedding and side fill support
- Possible "pressure bulb" affects nearby utilities
- Possible heaving or slumping of roadway

Economic Analysis 2 – Ductile Iron vs. Fusible PVC Pipe

- Material cost favors Fusible PVC pipe in 6" through 12"

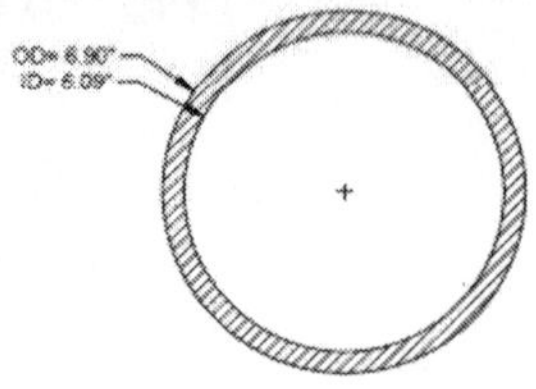

6" DR 18 Fusible C-900®

- DIP restrained bells require oversize expander head – adding to risk of heave

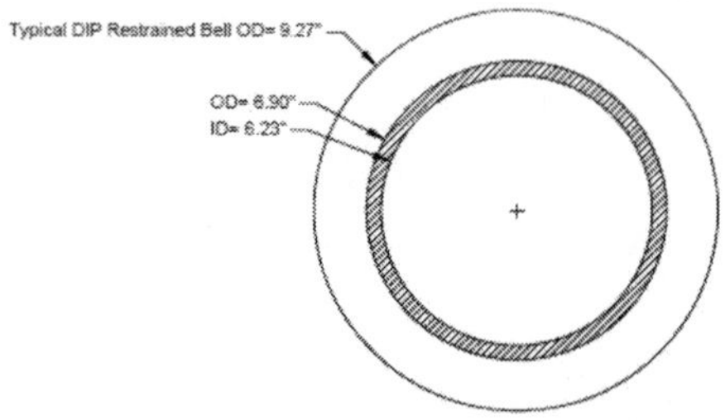

6" CL 350 Restrained Ductile Iron

Economic Analysis 2 – Ductile Iron vs. Fusible PVC Pipe

- Fittings and tapping the same
- DIP $10/lf higher cost
- HDPE not considered due to different skill set for utility owner

Road Block Ahead

- NCDEQ – Division of Water Resources claims to never have heard of Water Pipe Bursting
- Greensboro experience brought to bear
- Meetings and presentations
- Finally - Approval

Bids Accepted April 7, 2014

		Unit	Quantity	Unit Bid Price	Extended Total	Unit Bid Price	Extended Total	Unit Bid Price	Extended Total
1	Water Main								
	a. 6" Fusible PVC Waterline, DR18 (PC 235)	LF	9,730	$44.04	$428,509.20	$72.00	$700,560.00	$72.90	$709,317.00
	b. 4" Restrained Joint DIP (PC 350)	LF	40	$159.98	$6,399.20	$46.00	$1,840.00	$46.20	$1,848.00
	c. 6" Restrained Joint DIP (PC 350)	LF	650	$124.76	$81,094.00	$44.00	$28,600.00	$49.50	$32,175.00
	d. 8" Restrained Joint DIP (PC 350)	LF	140	$138.22	$19,350.80	$53.00	$7,420.00	$55.00	$7,700.00
	e. 10" Restrained Joint DIP (PC 350)	LF	40	$176.38	$7,055.20	$65.00	$2,600.00	$71.50	$2,860.00
2	Gate Valves								
	a. 4" Gate Valve	EA	3	$1,284.33	$3,852.99	$690.00	$2,070.00	$1,100.00	$3,300.00
	b. 6" Gate Valve	EA	32	$1,427.95	$45,694.40	$815.00	$26,080.00	$1,540.00	$49,280.00
	c. 6" Gate Valve (for Exist. FH)	EA	11	$1,427.95	$15,707.45	$960.00	$10,560.00	$1,595.00	$17,545.00
	d. 8" Gate Valve	EA	2	$2,123.37	$4,246.74	$1,200.00	$2,400.00	$1,650.00	$3,300.00
	e. 10" Gate Valve	EA	2	$2,679.89	$5,359.78	$1,715.00	$3,430.00	$2,200.00	$4,400.00
3	Restrained Joint Ductile Iron Fittings	LBS	10,000	$6.31	$63,100.00	$5.75	$57,500.00	$7.70	$77,000.00
4	Fire Hydrant Assemblies	EA	2	$4,980.44	$9,960.88	$4,130.00	$8,260.00	$3,850.00	$7,700.00
5	Water Service Reconnections	EA	117	$851.14	$99,583.38	$440.00	$51,480.00	$605.00	$70,785.00
6	2nd St. Creek Crossing Including Restoration	LS	1	$18,941.71	$18,941.71	$9,500.00	$9,500.00	$27,500.00	$27,500.00
7	Asphalt Pavement Removal and Replacement	SY	500	$101.86	$50,930.00	$120.00	$60,000.00	$110.00	$55,000.00
8	Concrete Pavement Removal and Replacement	SY	330	$133.38	$44,015.40	$132.00	$43,560.00	$71.50	$23,595.00
9	Gravel Replacement - 8" (Road/Alley/Driveway)	SY	250	$15.65	$3,912.50	$22.00	$5,500.00	$8.80	$2,200.00
10	Offsite Fill Material (Contingency)	CY	200	$20.51	$4,102.00	$24.00	$4,800.00	$16.50	$3,300.00
11	Gate Valve Replacement (Contingency)								
	a. 4" Gate Valve	EA	10	$1,815.20	$18,152.00	$2,800.00	$28,000.00	$1,500.00	$15,000.00
	b. 6" Gate Valve	EA	10	$2,157.22	$21,572.20	$3,000.00	$30,000.00	$1,700.00	$17,000.00
	c. 8" Gate Valve	EA	10	$2,824.01	$28,240.10	$3,720.00	$37,200.00	$2,000.00	$20,000.00
	d. 10" Gate Valve	EA	10	$3,495.93	$34,959.30	$4,800.00	$48,000.00	$2,500.00	$25,000.00
12	Mobilization	LS	1	$25,720.06	$25,720.06	$35,000.00	$35,000.00	$34,000.00	$34,000.00
Total Bid Price					$1,038,459.29		$1,204,360.00		$1,208,805.00

Specification Items of Note

- Temporary bypass
  - Requires installation of temporary piping
  - Disinfection and testing
  - Swap-over of service
  - Disassembly after work complete
  - Standard pipeline commissioning for rehabilitated section
- Static bursting equipment
- Certified Fusion Technicians
- Contractor Qualifications
  - Similar projects
  - Static equipment
  - Personnel
  - Pipe supplier training verification

Construction Notes, Pictures Thoughts

Pressure Pipe Bursting Basics

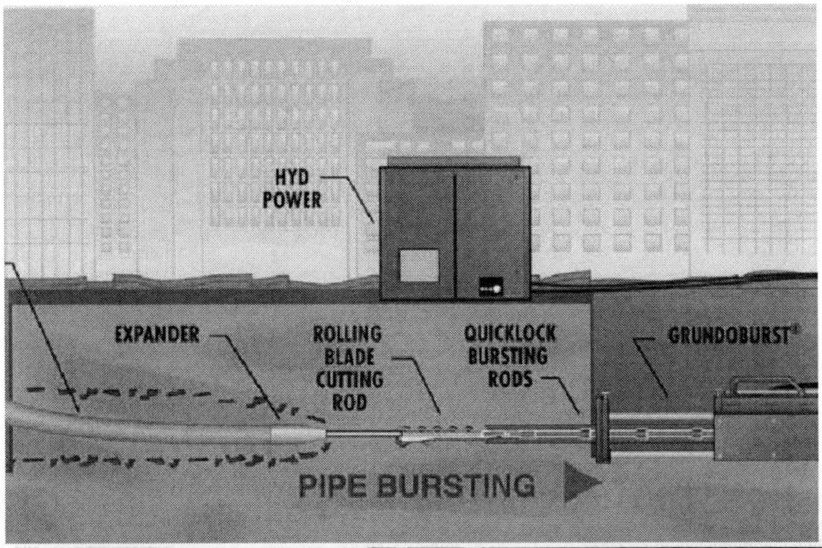

Underground Construction Technology
International Conference & Exhibition
February 3-4, 2016
Georgia World Congress Center
Downtown Atlanta, GA

How Static Pipe Bursting Tooling Actually Works (courtesy TT Technologies)

A Roller Blade Cutter utilizes a system of progressively larger diameter cutter wheels mounted to a cutter frame. This is a one piece with two cutter wheels and expander combination. The cutter first slices the existing pipe and couplings, or other appurtenances; the cutter is immediately followed with an expander that will fracture or open the split pipe and create an annular space sufficient to allow safe passage of the product pipe

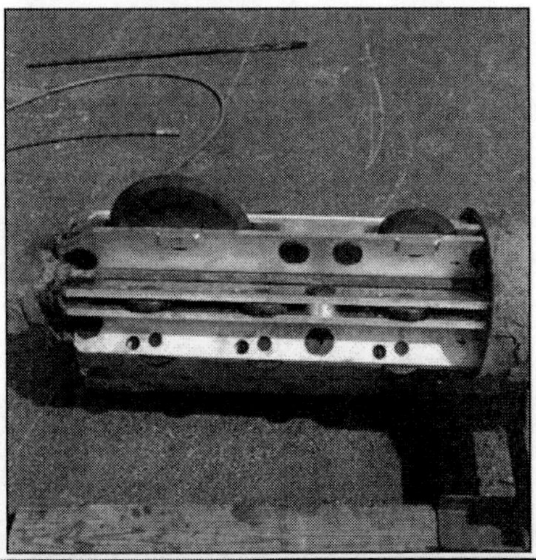

Underground Construction Technology
International Conference & Exhibition
February 3-4, 2016
Georgia World Congress Center
Downtown Atlanta, GA

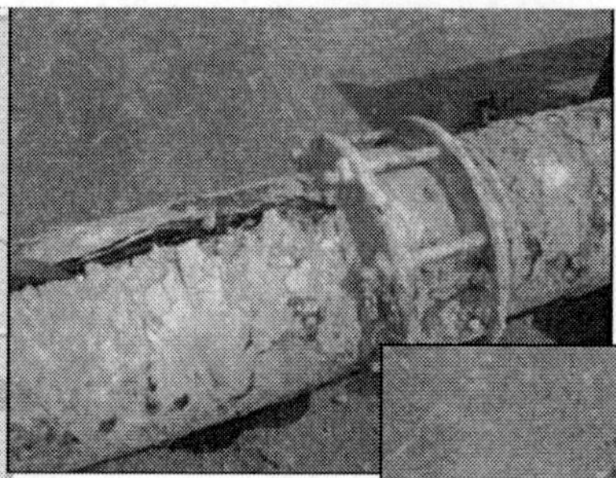

Pipe is Cut with First Blade

Coupling is Cut with Last Blade

Expander Opens up Coupling

Layout Room for Fused Pipe

- Central location for several strings best practice
- Pull pipe into place
- Or, create ramps for driveways for local stringing

Underground Construction Technology
International Conference & Exhibition
February 3-4, 2016
Georgia World Congress Center
Downtown Atlanta, GA

Temporary Bypass Required

Underground Construction Technology
International Conference & Exhibition
February 3-4, 2016
Georgia World Congress Center
Downtown Atlanta, GA

Temporary Bypass Required

Pipe Insertion Pit Calculator

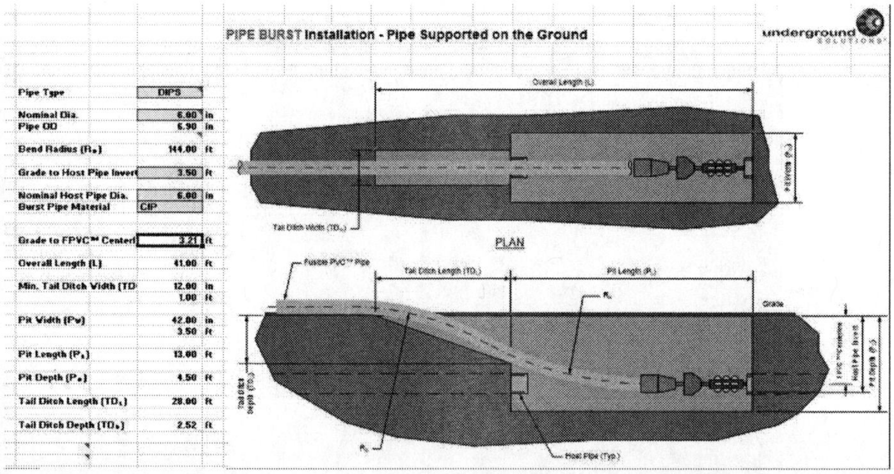

Underground Construction Technology
International Conference & Exhibition
February 3-4, 2016
Georgia World Congress Center
Downtown Atlanta, GA

Pipe Insertion and Static Machine Pits

Connections and Cleanup

UCT
Underground Construction Technology
International Conference & Exhibition
February 3-4, 2016
Georgia World Congress Center
Downtown Atlanta, GA

Lessons Learned

- Pavement Quantities
  - Inaccurate locates
  - No restriction on pit size
- Low bid quantities on open cut – therefore hard to make field changes
- Pipe removed for bursting operation was in good shape – more condition assessment up front

LESSONS LEARNED FROM EXISTING 20" DUCTILE IRON PIPE WITH 20" FUSIBLE PVC

JEFF GREENE, KRG UTILITY, INC.

Presenters

Jeffrey J. Wing, PE LEED AP

- Master of Applied Science, Environmental Policy and Management (Energy/Sustainability), University of Denver, 2011
- Bachelor of Science, Civil Engineering, University of Alaska at Anchorage, 1986
- Project Manager at Kimley-Horn and Associates (12 years)
- NASCO's Certified (Pipeline, Manhole, and CIPP)
- Have volunteered with Engineering Ministries over a 21 years

Jeff Greene

- Graduated from NC State with a Masters in Water Resource Engineering
- President of KRG Utility, which performs sewer and water rehabilitation by pipe bursting, pipe cleaning and inspection.

Project Overview

NEI Rehabilitation Phase 2 Project

- Goal is to decrease the amount of Sewer Spills
- Replacement of approximately 5,140 linear feet (LF) of a 20-inch-diameter ductile iron pipe (DIP) force main.
- **Largest 20-inch DIP bursting job** in the U.S. with the **longest single DIP bursting pull.**
- Replacement of three air and vacuum/pressure
- Five different location along 3 miles of route.
- Each segment varied from 400 LF to 2,200 LF

Project Overview

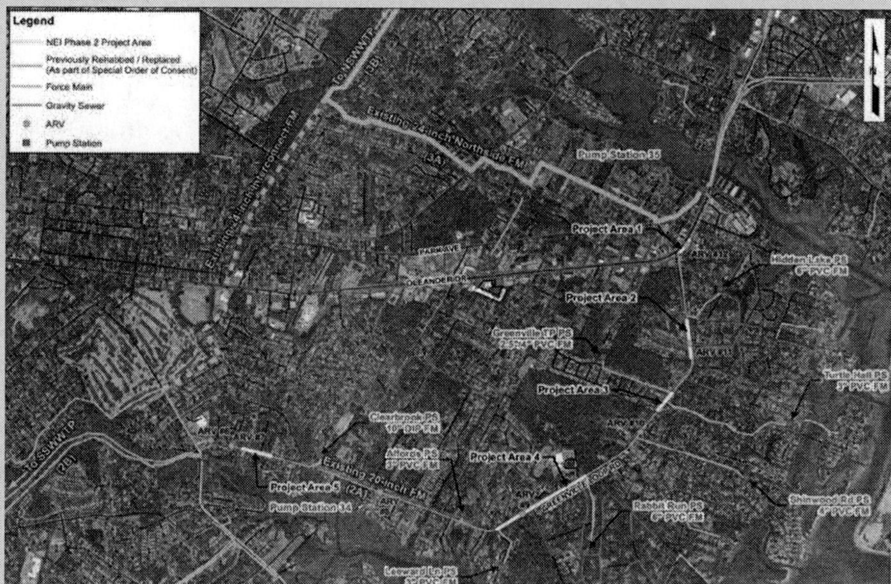

Project Overview

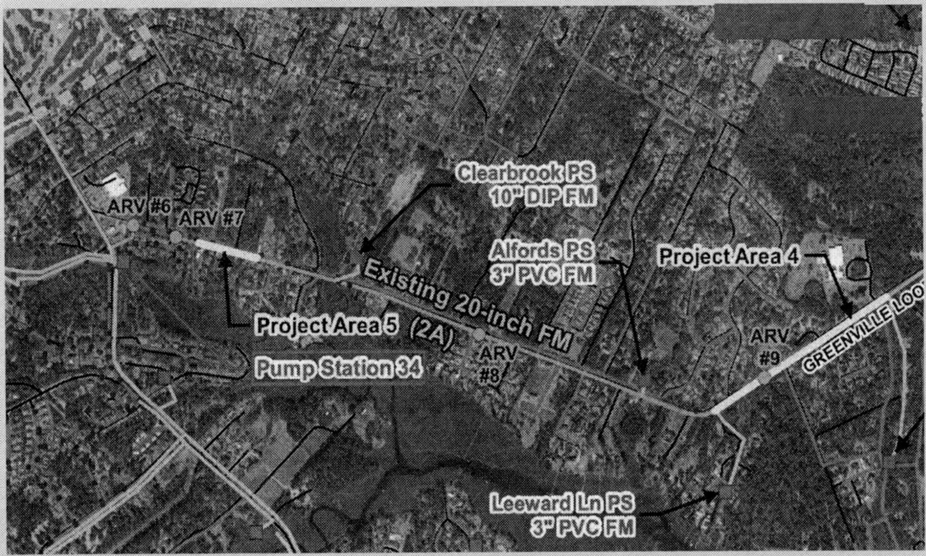

Project Overview

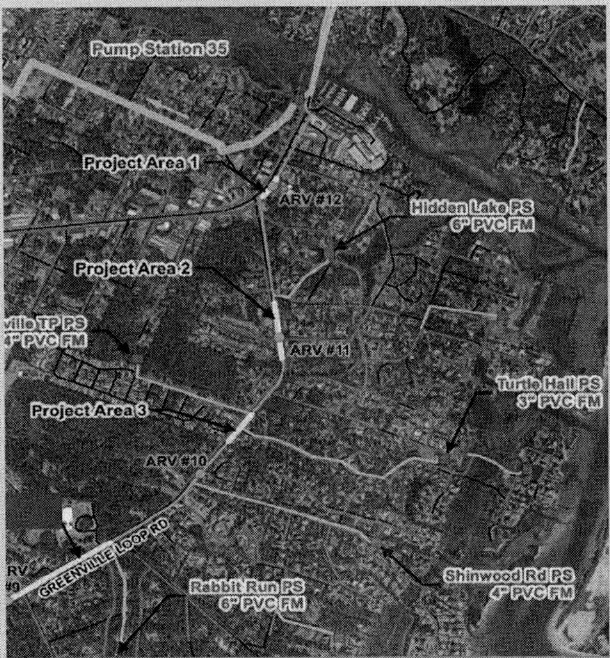

Project Team

- Client: Cape Fear Public Utility Authority, Wilmington NC
  Project Manger: Craig Wilson, PE
- Contractor: State Utility Contractors, Inc., Monroe, NC
  Superintendent: Dustin Wagner, PE
- Pipe Bursting Sub: KRG Utility
- Engineer: Kimley-Horn, Cary, NC
  Project Manager: ***Jeff Wing, PE LEED***
- Project Observer: ***Jack Smiley*** Construction Services
- Pipe Bursting Equipment: TT Technologies
- Underground Solutions: Fusible PVC Supplier

Award Winning Team

On January 30, 2016, Carolinas AGC (CAGC) bestowed the 2015 Pinnacle Awards—the most prestigious recognition in the Carolinas construction industry—**Best Utility Project Under 5 Million Dollars**

CAGC's goal: *Advancing the construction industry to enhance the quality of life and deliver a sustainable difference in the Carolinas.*

AGC represents more than 26,000 firms

Challenges

With narrow ROW and other utilities adjacent to the existing Interceptor, a trenchless solution was the most economical and least invasive project approach.

Collaboration was the Key!

Field adjustments included:

- Addition of line stops
- Revisions to pit locations
- Revisions to traffic control plans.
- Update of Erosion and Control Permit

A "No-Dig" Project Still Includes Land Disturbance!

This image cannot currently be displayed.

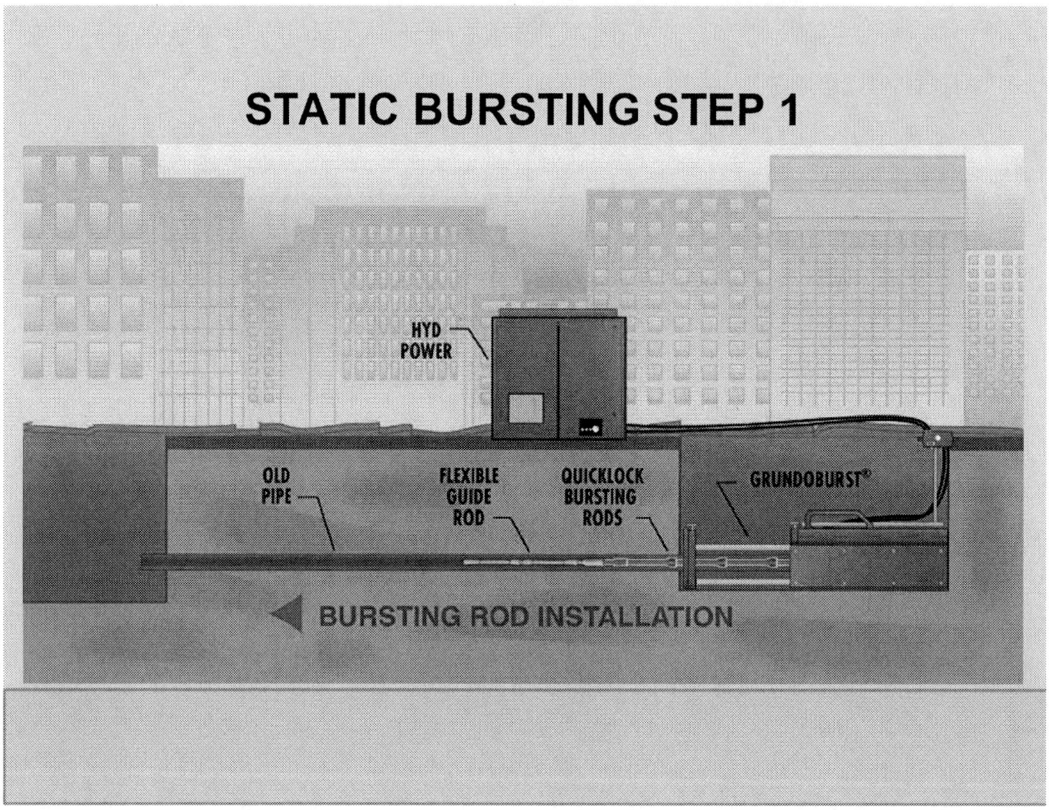
STATIC BURSTING STEP 1
HYD POWER
OLD PIPE
FLEXIBLE GUIDE ROD
QUICKLOCK BURSTING RODS
GRUNDOBURST®
BURSTING ROD INSTALLATION

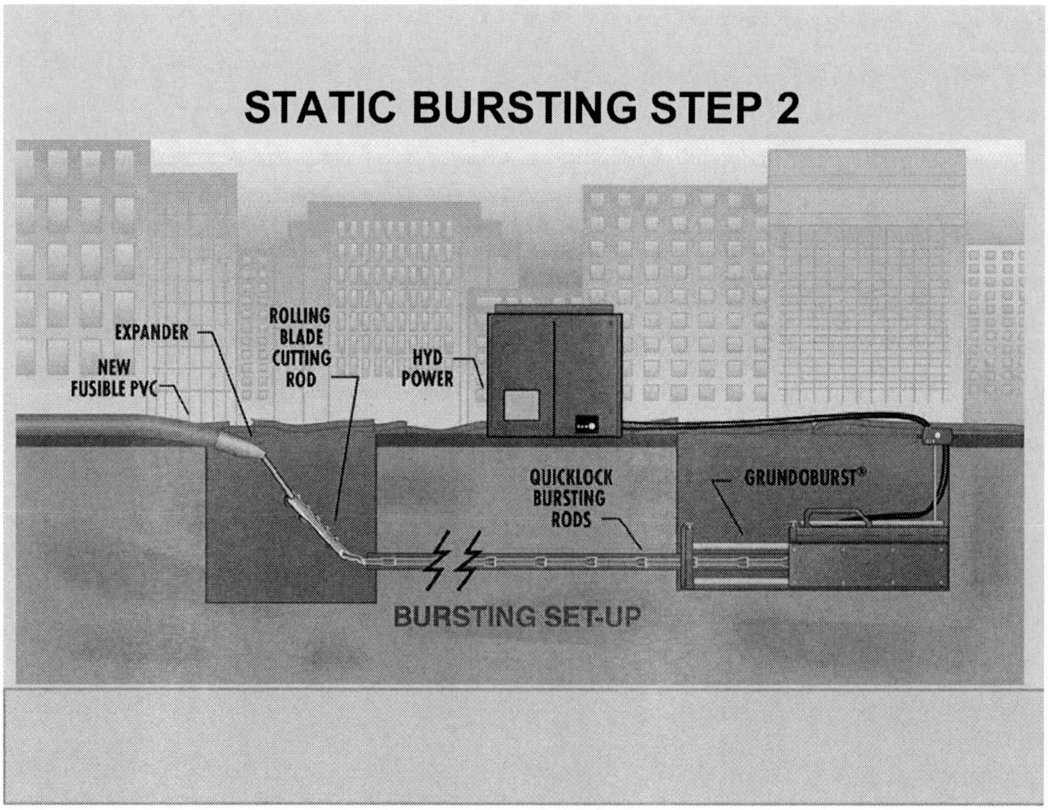
STATIC BURSTING STEP 2
EXPANDER
NEW FUSIBLE PVC
ROLLING BLADE CUTTING ROD
HYD POWER
QUICKLOCK BURSTING RODS
GRUNDOBURST®
BURSTING SET-UP

STATIC BURSTING STEP 3

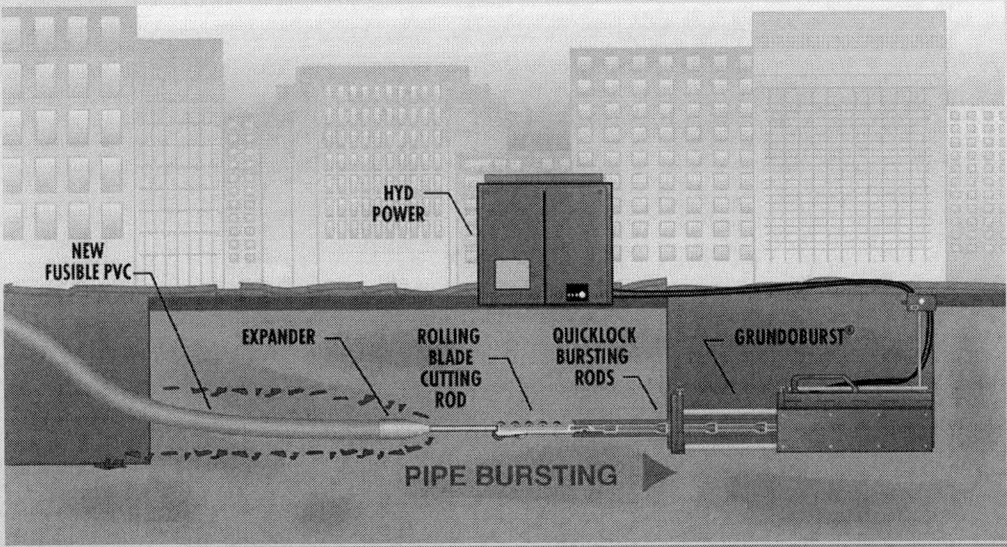

STATIC BURSTING

Installation: Static Pipe Bursting

20-inch DR18 Fusible PVC® pipe

LESSONS LEARNED

- Static versus Dynamic
- Pull Length
- Soil Conditions
- Split fittings/pipe "bunch-up"
- Field Changes and Collaboration
- Pulling PVC compared to HDPE

Challenges

Revisions

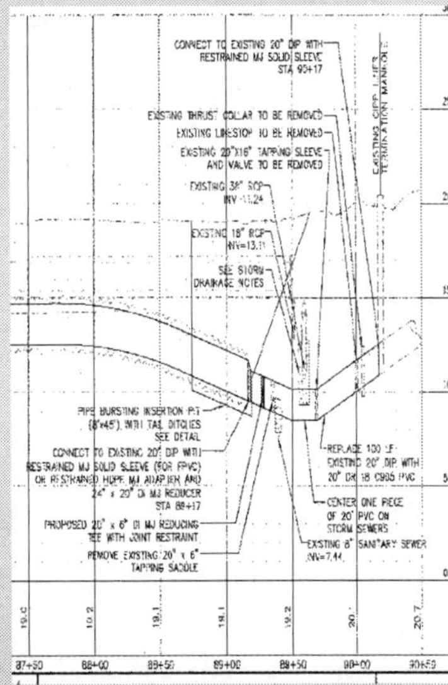

MORE LESSONS LEARNED

- More Pre-Design Subsurface Utility Engineering (SUE)
- Ground Heave
- Bypass Cost
- Multiple Partial Substantial Completion was Beneficial.
- Collaboration is Essential for Field Changes.

CONCLUSION- A Successful Project

- The team successfully accomplished a 20-inch-diameter DIP bursting project that has the longest single DIP bursting pull in the United States.
- There were no sewer spills.
- There were no reported traffic accidents.
- The team worked together for expedited submittal approval and took a collaborative approach to addressing issues in the field to meet the final date.
- The project won the Pinnacle award for the best Utility Project in North and South Carolina.

Questions?

Thank You

THE TOTAL SOURCE
educational sessions

Pre-chlorinated Pipe Bursting at Arlington National Cemetery

MPC Murphy Pipeline Contractors, Inc.

U.S. Army Corps of Engineers Norfolk District

2015 Trenchless Project of the Year

Arlington National Cemetery Background

- U.S. government purchased 1,100-acre property in the 1860s (owned at the time by the family of Mary Anna Randolph Custis, wife of Robert E. Lee)
- Founding in 1864 when first military burial took place during the Civil War
- Landscape serves as a tribute to the service and sacrifice of every individual (over 400,000) laid to rest within the hallowed grounds

Arlington National Cemetery Background

- Ideal location due to aesthetically pleasing view of Washington, D.C.
- Winding pathways conform to natural topography of site, which is naturally landscaped with flower gardens and trees
- 11,000 daily visitors with 25 to 30 daily funerals, changing of the guard at the Tomb of the Unknowns and other ceremonies

Water Main Replacement Program Background

- 60 years ago, series of cast iron water mains installed for fire suppression and potable water
- Need of replacement due to leaks/breaks causing severe disruption
- US Army Corp of Engineers Norfolk District required solution to:

- Reduce environmental impact to landscaping
- Reduce risk of unknowns with digging due to history and age of site
- Reduce social impact to visitors/funerals
- Provide 100 year plus new design life

Program Overview

- Pre-chlorinated Pipe Bursting technology selected
- Murphy Pipelines awarded work through US Army Corp of Engineers and Patriot Construction 8(A), SDVOSB general contractor
- 5 construction phases replacing 44,500 feet of 6-in, 8-in and 10-in cast iron water main with 6-in, 8-in and 10-in HDPE pipe
- Sensitive areas included: Welcome Center by park entrance, former President John F. Kennedy's grave, the Memorial Amphitheater and Tomb of the Unknowns with the changing of the guard occurring every hour

Fusing & Pre-chlorination of HDPE

- Staging area located away from burst locations to minimize site impact
- Butt fusion is used to connect pipe sections to corresponding burst lengths
- Pressure testing
- Lines capped and disinfected with hypochlorite solution of at least 25 mg/L
- Two consecutive days of samples, taken 24 hours apart

Pipe Bursting Operations 8:00 AM

- Entry and exit pits excavated 4’ x 12’
- Service connections pits 3’ x 3’
- 8:00 AM – Decommission Main
- Bursting equipment is set in pit and rods are shuttled through host pipe

Pipe Bursting Operations 10:30 AM

- 10:30 AM – Bursting head (or ductile slitter) and expander are attached to the rods which have reached the entry pit
- Pre-chlorinated and sealed HDPE pipe attached to the expander
- Pull back begins

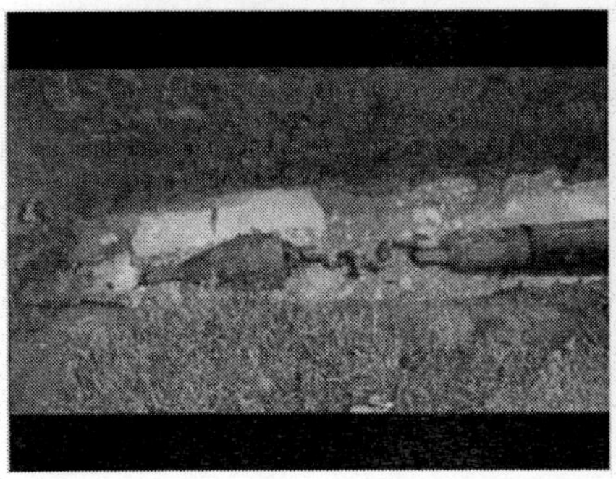

Pipe Bursting Operations Noon

- Rods are removed from the exit pit as pipe is pulled into place
- Noon – New pipe is installed

Pipe Bursting Operations Noon to 3:00 PM

- Noon to 3:00 PM – Connections made
- Ductile Irion fittings, mega lugs, stainless steel inserts, mechanical service saddles

Pipe Bursting Operations 3:00 PM

- 3:00 PM – Post Chlorination Process
- Hypo-chlorus solution sprayed on all connections
- Super-chlorinate new pipeline with slug of chlorine to 300ppm
- 4:00 PM – New main back live

Pipe Bursting Results

- Engineering costs reduced by 78.4% due to following existing utility path
- HDPE pipe provides 100 year plus new design life
- Production rate of 300 – 600 feet per day
- Crew visible to each area for one day
- Surgical excavations along project with a reduction in restoration of 87% vs. open cut
- Compact equipment
- Sensitive areas of cemetery not impacted
- Zero change orders
- Social costs reduced as all landmarks remained open, no streets were closed and visitors had full access to cemetery

Challenges & Successes

- Working in an operational national landmark that would not be closing to tourists and visitors (11,000 per day)
- Shutting down construction activity for funeral processions (up to 30 per day)
- Designing burst program to reduce excavation throughout sensitive areas

- Project progressed without incident to tourists, visitors and funerals
- Largest pre-chlorinated pipe bursting program in North America
- 2015 Trenchless Project of the Year Award Winner

THE TOTAL SOURCE

educational sessions

Pre-chlorinated Pipe Bursting at Arlington National Cemetery

MPC Murphy Pipeline Contractors, Inc

U.S. Army Corps of Engineers
Norfolk District

2015 Trenchless Project of the Year

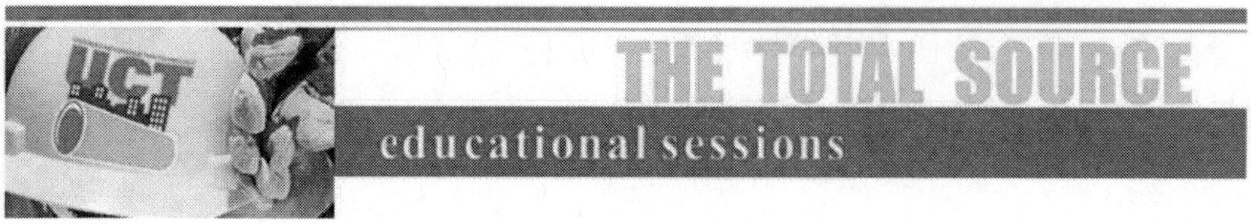

PCCP Transmission Main Inspection

A Cobb County-Marietta Water Authority Case Study

Cobb County-Marietta Water Authority

- System Overview
- History of PCCP
- Capitol Replacement Program
- Condition Assessment Program
  - Case Study
- Cost Savings

CCMWA System Overview

- Drinking water supplier since 1951
- 13 wholesale customers
- 800,000 clients
- 158 MGD
- 200 miles of large diameter pipe
  - 16- through 64-inch
  - 73 miles PCCP

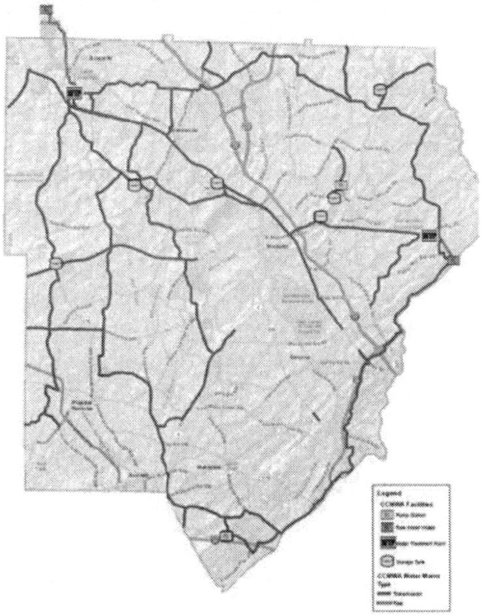

History of PCCP at CCMWA

- Installed over 160 miles of PCCP since 1951
- Experienced failures due to:
  - Construction damage
  - Mortar coating failure
  - Poor quality prestressing wires
  - Corrosive environment
  - Capacity
- Began full-scale replacement of all PCCP lines

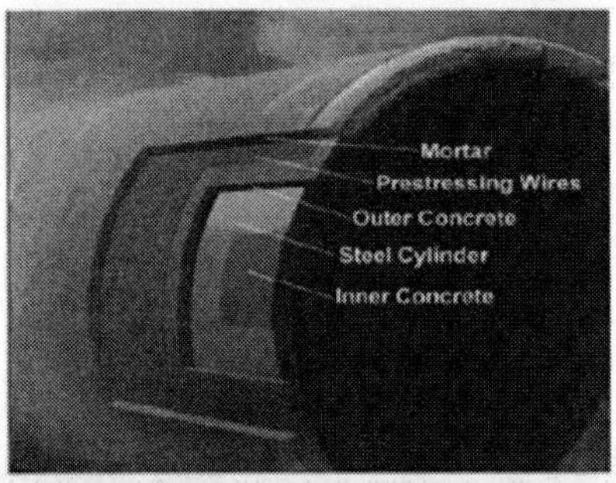

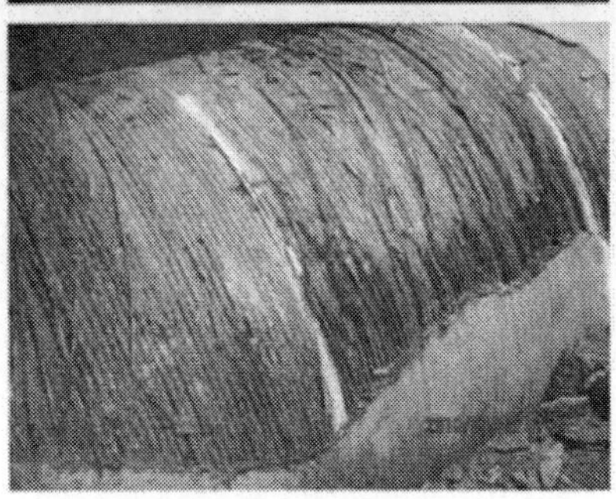

CCMWA PCCP Capital Replacement Program

- 46% of PCCP inventory has been replaced
- 73 miles remain in service
- Capitol replacement became infeasible both financially and logically once the "Class IV" pipelines were replaced
- Oldest pipeline in the system remains serviceable since 1958

Condition Assessment of PCCP

- Condition Assessment Program implemented in 2013
- Inspections include leak and gas pocket detection and electromagnetics
- 9.7 miles inspected

Case Study – Raw Water Line

- Lake intake to Wyckoff Water Treatment Plant
- 30-, 36-, and 42-inch diameter PCCP
- Manufactured in 1964
- Approximately 4.25 miles
- Past failure on 30-inch section

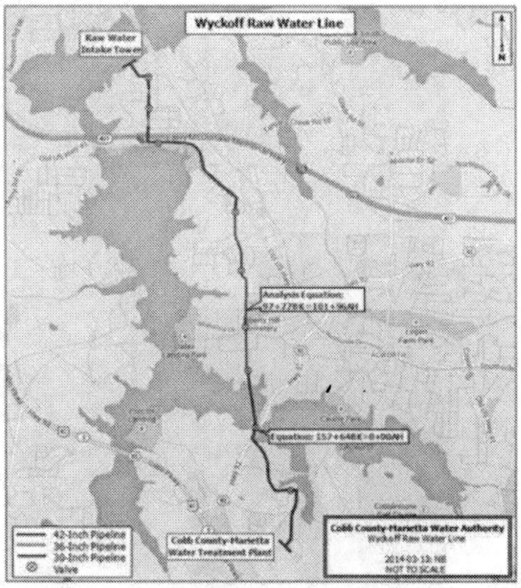

Case Study – Scope of Services

- Leak and Gas Pocket Detection
- Electromagnetic Inspection
- Structural Analysis
  - AWWA C301 Analysis
  - AWWA C304 Analysis
  - Finite Element Analysis
- Remaining Useful Life and Re-Inspection Analysis

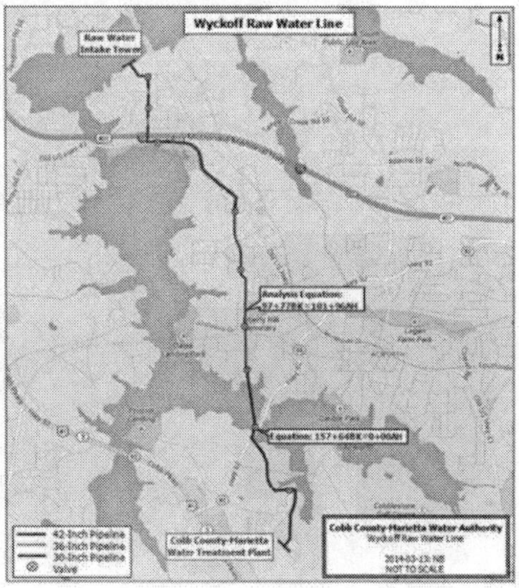

Case Study – Leak Detection

Case Study – Leak Detection

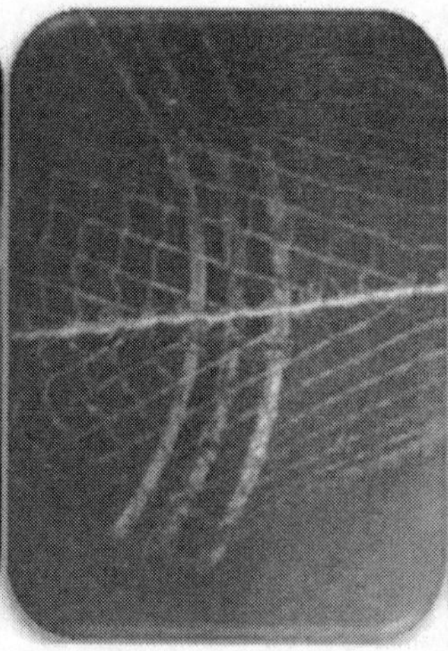

Case Study – Electromagnetics

Case Study – Electromagnetics

Case Study – Results

Distressed Pipes in the Wyckoff Raw Water Line							
Rank	Pure Reference Number	Low Station	Pipe Length (feet)	Break Position (feet)	Number of Broken Wire Wraps by Region	Total Number of Broken Wire Wraps	Notes
30-Inch Pipeline							
1	824	123+58	16	3.0	10	10	1
2	533	73+53	16	3.5	5	5	1
42-Inch Pipeline							
1	30	~472	16	3.5	35	35	2
2	187	15+13	16	9.5;12.0	15;15	30	1
3	189	15+45	16	10.0	15	15	1
4	14	~216	16	13.0	10	10	2
5	188	15+29	16	7.0	10	10	1
6	454	61+32	16	11.5	10	10	1
7	1121	154+52	16	5.0	5	5	1
8	1140	157+24	16	12.5	5	5	1

Case Study – Results

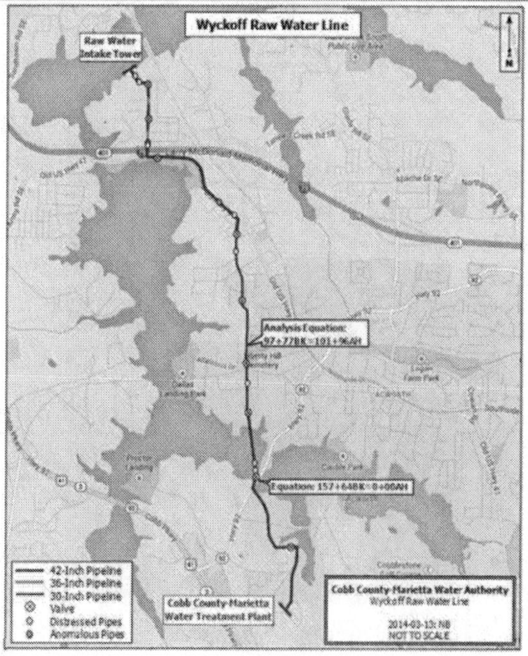

Case Study – Structural Analysis

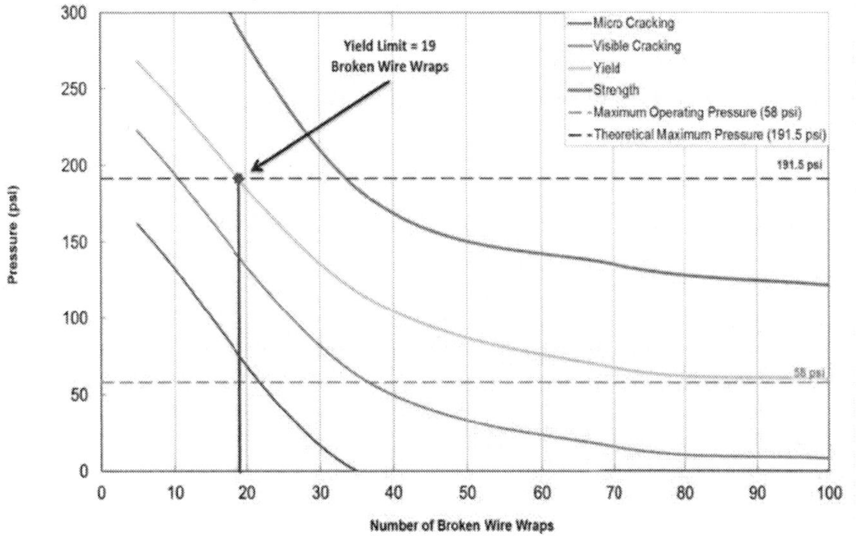

Case Study – Statistical Analysis

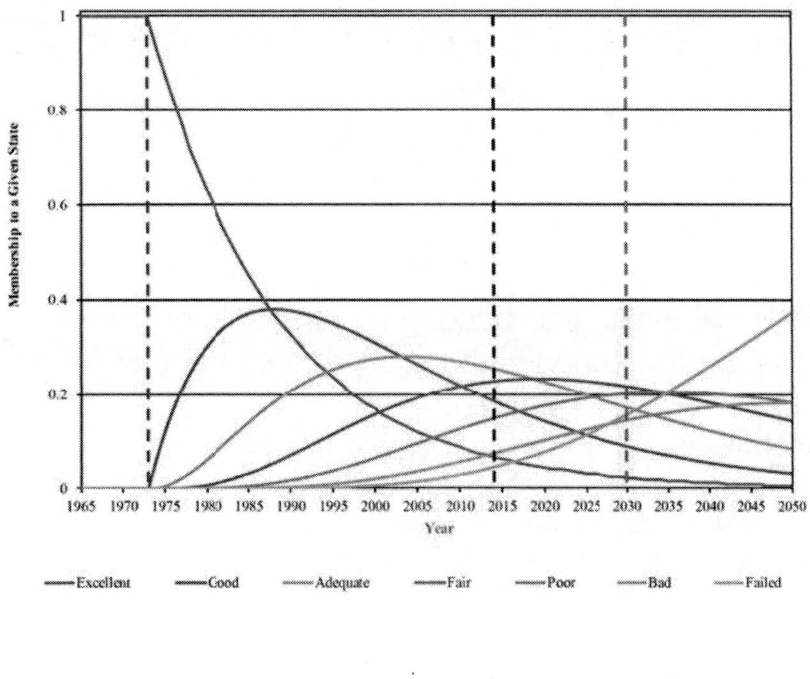

Case Study – Conclusions

- No leaks or gas pockets
- 10 pipe sections indicating wire break damage
  - 0.72% of total pipe sections
- Two pipe sections indicating high level of distress based on structural analysis
- Statistical analysis suggests serviceability of 16 years
- Re-inspect in 5 to 8 years

Cost Benefit Analysis

Financial Comparison of Management Option Costs		
Management Option	**Net Present Value of Cost Unti1 2038 ($M, USD)**	**Percent of Capital Replacement Cost**
Capital Replacement	$82.0	N/A
Assess & Address	$11.3	13.7%

- Financial evaluation based on cost of capital replacements compared to periodic condition assessment inspections
- Assumes:
  - Average $615/foot +15% for replacement
  - 2.5% escalation rate
  - 7-year inspection cycle with 1% repair rate

Cost Benefit Analysis

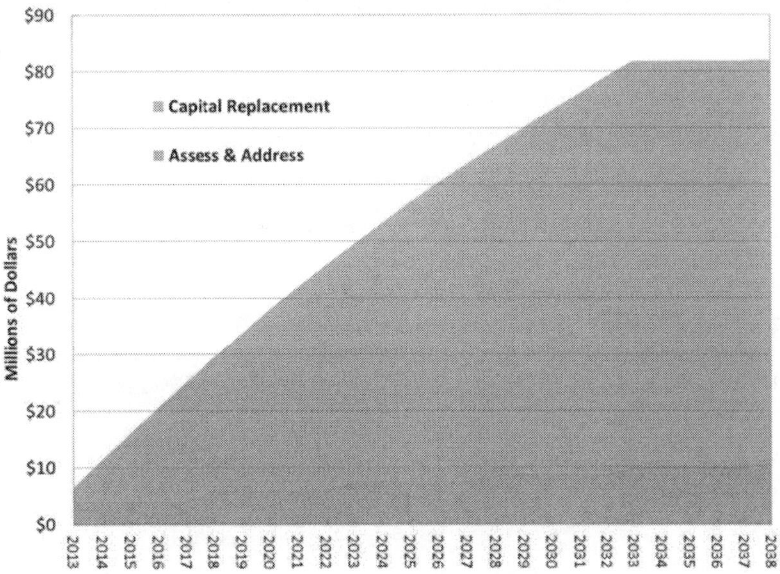

Next Steps

- Continued inspections of PCCP inventory
- Identify limitations within system to allow for intelligent capitol planning
- Implement long-term pipeline management

Questions?

Jennifer Steffens, PE
Pure Technologies US Inc.
jennifer.steffens@puretechltd.com
404-751-5060

THE TOTAL SOURCE

educational sessions

Leak Technology

What is New and Being Used Around the World

What exactly is 'technology'

Application of knowledge to the practical aims.

Changing and manipulating the way we think.

Includes the use of materials, tools and techniques to make life easier, more pleasant, increase productivity and to ensure that the problem that was impossible becomes possible.

Technology began to influence human endeavor as soon as people began using tools.

Technology advancement is part of the dream that any ideas should never be dismissed – somewhere there is the solution

Technologies Available

Satellite Earth Mapping

- Using technology that is used to look for water on other planets, this product analyses satellite imagery to detect leaks and non-revenue-water.
- The result? Leak detection that covers thousands of square miles at once, and that can pinpoint a leak in a 6 meter radius area, saving significant labor associated with finding leaks with current tools.

Satellite Earth Mapping

Taken from satellite mounted sensors – to spot leakage in underground distribution pipes. The raw imagery is then overlaid on GIS systems and is processed by unique algorithms.

This algorithm detects drinking water, by looking for a particular spectral "signature" typical to drinking water.

The user is presented with a pinpointed leakage graphic report overlaid on a map with streets, pipes and size information.

GPIR – Water Pipe

Ground Penetrating Infra-Red

- It detects water pipes made of **all types of materials**;
- It detects **leaks** and **damages** in water pipelines of **all types of materials;**
- It has a penetration capability of up to **2 meters** into the ground;
- The **image resolution** capabilities of the damaged pipe will be less than **5 cm;**
- The **survey time** (detection and imaging) will be **10 sec/meter** along the pipe axis;

2005 GPR Technology

GPR-2015

Visual	IR Data
Anomaly at the fire hydrant.	
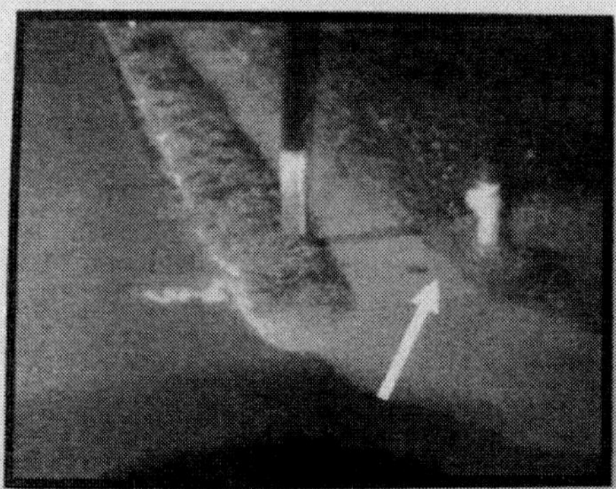	
Originally marked at sidewalk; repositioned to street.	

GPR
Crack in the road, South bound lane.

about us
project results
objectives
contact
EUROPEAN COMMISSION
GROWTH Programme
Leaking
news
innovative concepts
decision support system for leakage control
he Problem
't.EAKI • (GROI-2001- 0284) is an EU co-ft.med iro· aiming a deveklpmen of irnlva · platform fir the detec mof n unden;,-ol.nd water p· .LEAKING sesl<s the soution to this severe ial p..blk:health with effcien rardestroc · tes · tectv,· .T platform deveklpment i'lleg"ates mi:rowave radiometry and g--ol.nd penetration radar eq.Jpment with anced Dgital Sigial ProcesshJ atgcrithms.
Global w mhJ may lead to a sg,ifi: t str· age of the g-cx.n:lwabar reserves. At the same time,
SEWERIN
United Utilities
CTI
ICCS NTUA

Internal Pipe Repair

- Internal repair clamps
- Epoxy coating
- Internal spray lining (Spin Casting) over leaks
- Platelets
- 'Magic Pea'

Rehabilitation of Service Lines

- Epoxy coating of service pipes
- Cost saving
- Time period to reline service
- 2016 will be standard approach in UK
- Projects in the USA moving forward are a university epoxy coating of lead services and City of Flint, Michigan

Noise Loggers

- Getting better
- Offer many functions now to try to dismiss false positives
- Can offer acoustic footprints only or can auto correlate
- Can be mapped into google mapping
- Only record noise

Noise logging

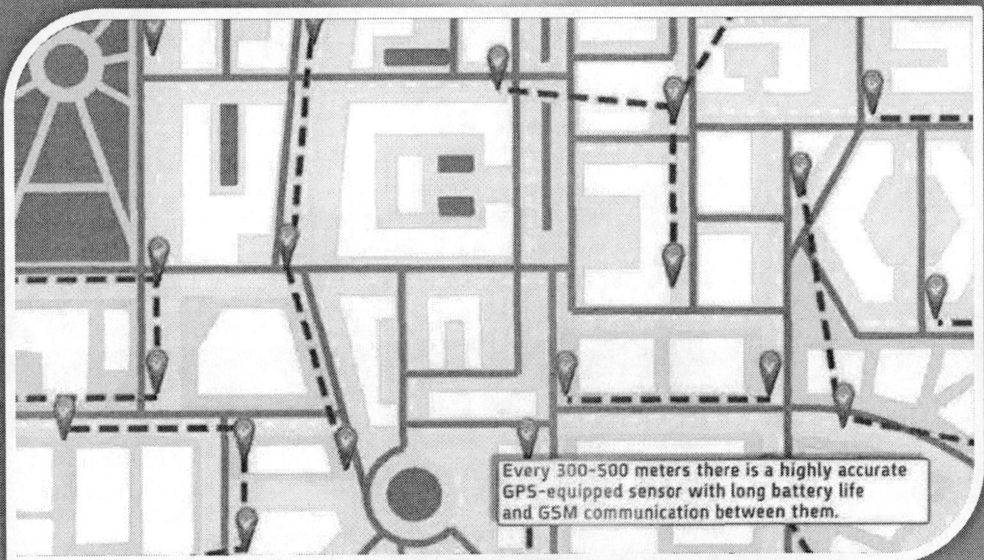

Electronic Condition Assessment

Average pipe wall thickness vs Actual pipe wall thickness

As different as an *X-ray* vs *MRI*

- Velocity of sound
- Ping of a transient
- Magnetic flux
- Multiple loggers measuring average pipe wall thickness – noise loggers but not only leaks located
- Visual (internal camera)
- Ultrasonic Measurement

Checking your assets

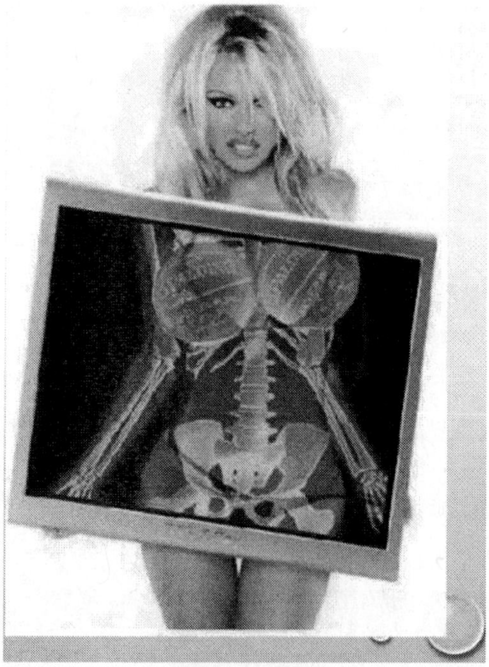

ü Some results may be embarrassing

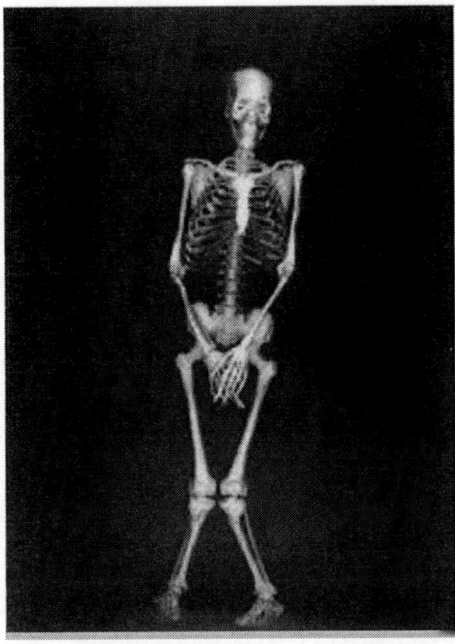

Some results you like

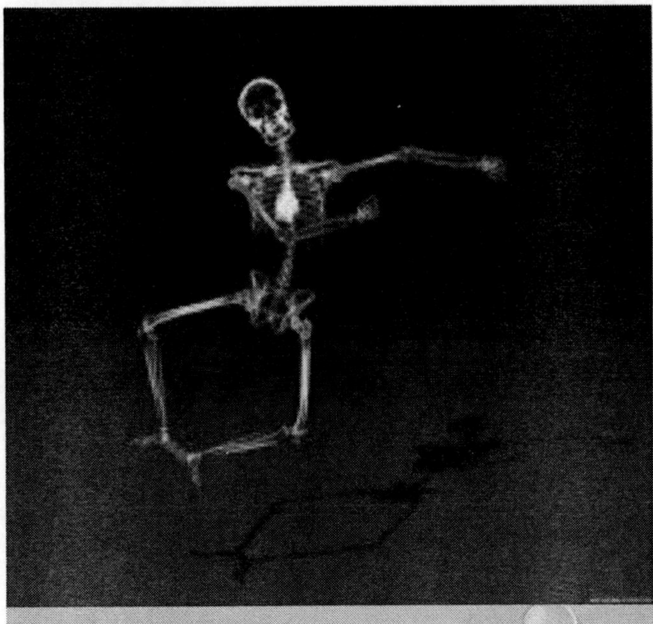

MRI scan of a pipe

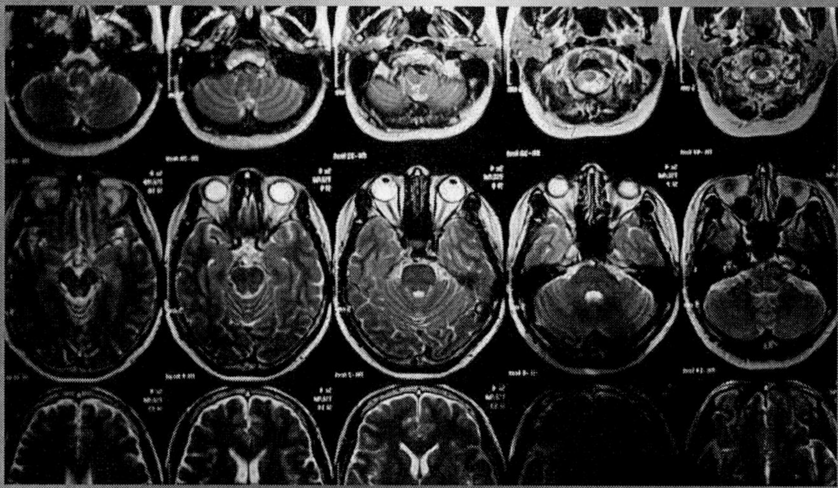

Detail and Accuracy

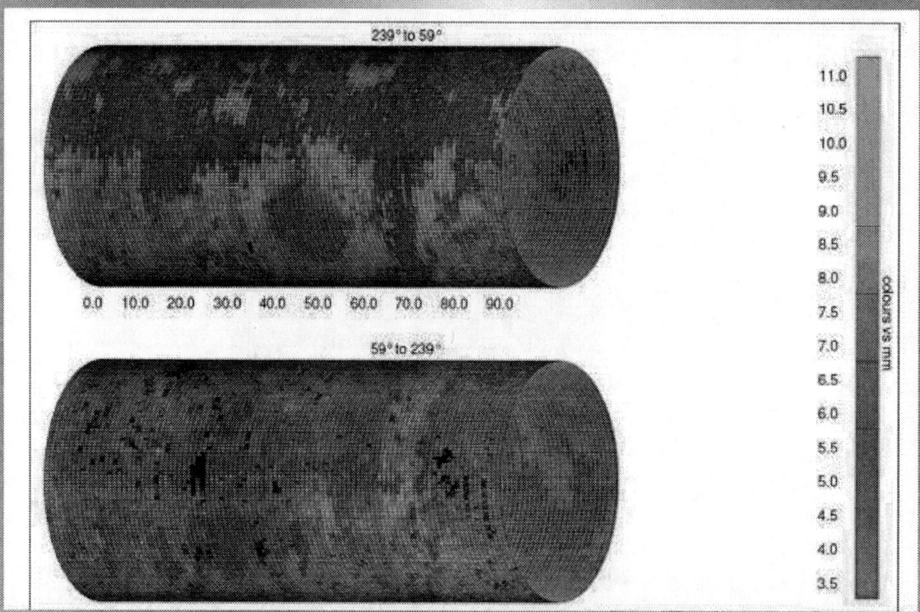

Exhumed CI pipe

Other Technologies for Leak Detection

- Gas insertion leak location
- Generic algorithms from past reported leaks
- Predictive leak location – (new)
- Correlators – distribution & transmission
- Transient monitors
- Temp difference for leakage ?
- Insertion technology – pushrod, tethered and free flowing systems
- Step testing
- Internal/external electronic voltage leak detection
- Traditional ground microphones and listening sticks

Takeaways

- So now we know what is available; what do you do?
- No, you don't jump in and buy every thing you see that is new
- Any technology above what you currently have is new, so this is where you set your goals

Conclusion On Technologies Available

- We have to climb the ladder slowly and securely – its not an escalator where you stand still and get to the top
- Many don't even use what is available today as they are waiting for what is coming tomorrow, causing NRW and catastrophic failures
- Use what is best to deliver your solution no matter how simple the technology is – every thing has its place in the world of water loss

Hamilton Says

- Don't try to be too clever, as you may end up looking like you know nothing
- Whereas if you start off 'knowing nothing', you might end up looking clever

Start Somewhere!

Thank You!

Mark.Ames@jd7usa.com

THE TOTAL SOURCE

educational sessions

Steel Slip Lining of 90 inch PCCP

David Markert

Cole Kratochvil

Kiewit

Val Vista Water Main History

- Constructed from 1972 to 1974
  - 5 separate contracts
  - 72", 90", 96", and 108"
- Pipeline Capacity 220MGD
  - Raw water from CAP Canal
  - 30% of Phoenix drinking water
  - 40% of Mesa drinking water
  - Gravity pipeline (85psi max.)

Water Main Alignment

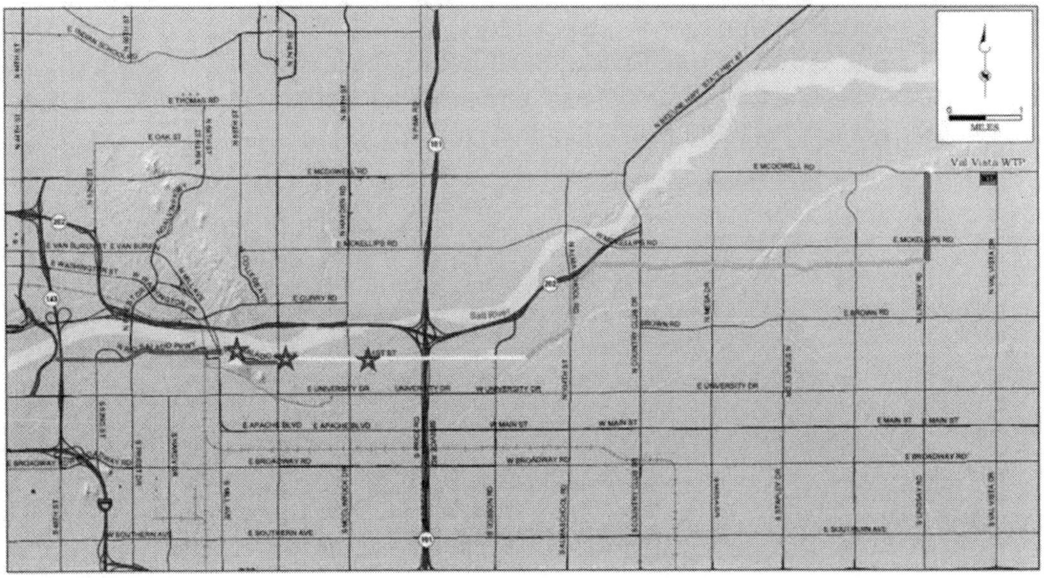

Project Timeline

- 2001-04-Major and Minor Assessments
- 2005-06 Pilot and Extended Pilot Program
  - 9,700LF 72"
- 2007-08 Full Design/Prioritization
- 2008-09 Major Rehabilitation
  - 18,500LF 72" – 96"
- 2010 Rehabilitation
  - 5,600LF 72"
- 2014 Rehabilitation
  - 6,262 LF 84" and 87"

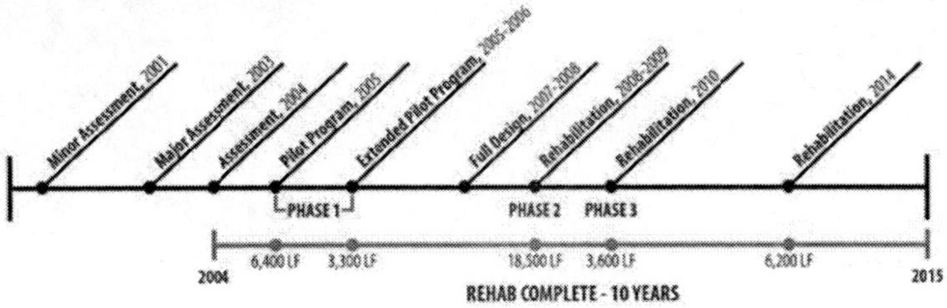

Project Overview

- Construction Manager at Risk (CMAR) Project with the City of Phoenix
- Guaranteed Maximum Price (GMP)-$13,670,000
- 6,262 LF of 84" & 87" collapsed and solid can liners
- Access consists of 3 Portal locations

Slip Lining Overview

- Dewatering/Isolation
- Excavation and installation of slide rail shoring
- 84" solid and 87" collapsed can liners
- Circumferential and longitudinal welding required
- Annular space grout
- Cement mortar lining (CML)
- Fill/Disinfect
- Surface Restoration

Installation of Slide Rail Shoring

Comparison: Split Can v. Solid Can

ASPECT	SPLIT	SOLID
Procurement	Plate steel = labor intensive	Rolled steel = machine welds
Schedule (fit-up & welding)	10 to 20 cans / day	25 to 35 cans / day
Quality	Longitudinal – Full Pen Circumferential – Fillet Grout Plug - Seal	Circumferential Grout Plugs
Capacity Reduction	~3 to 4-inches in past	~6 to 8-inches
Section Lengths (Geometry and existing condition dependent)	4-20LF	4-24LF

Installation of 87" Split Can Steel Liner

- 87" split liner collapsed to 83"
- Liner was banded with two 1" steel bands and secured with three welded tabs

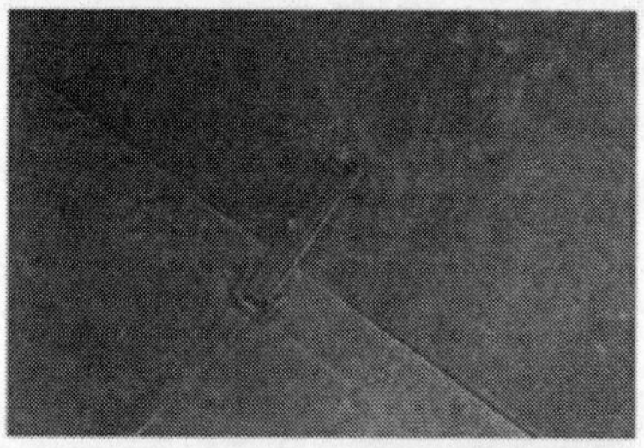

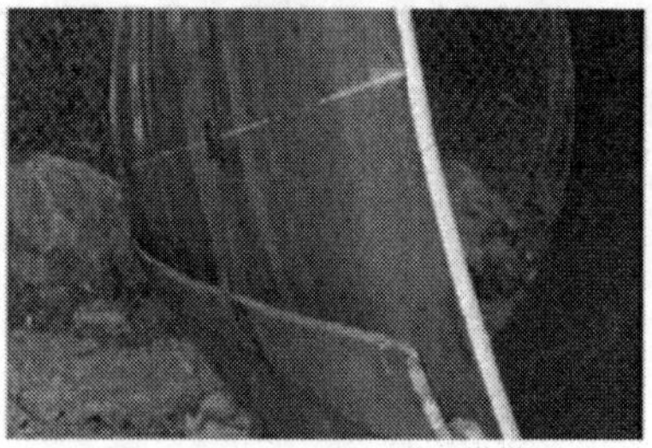

Installation of 87" Split Can Steel Liner

- Bell and spigot connection
- Tolerance on bell and spigot – ¼"
- Tolerance on longitudinal ¼" to ¾"
- Fabricated guides to maintain tolerance on longitudinal seam

Expanding Collapsed 87" Liner

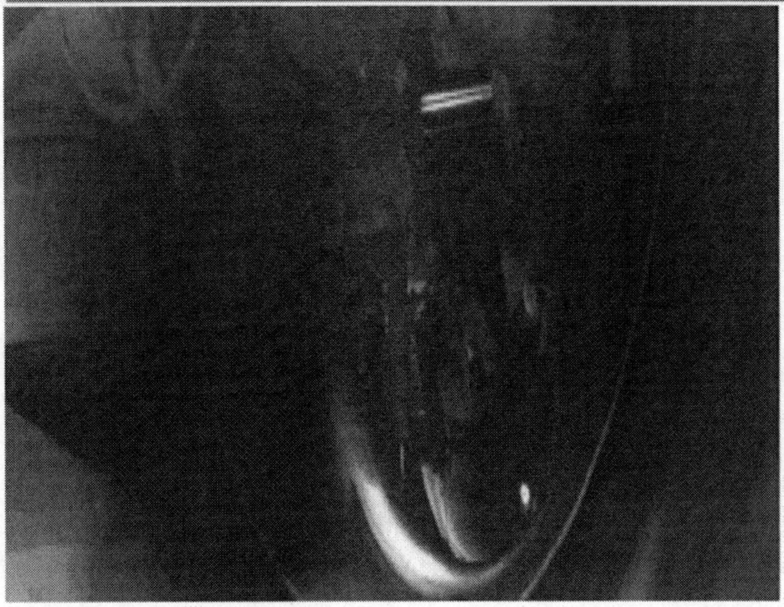

Installation of 84" Solid Can Liner

- Bell and spigot connection
- Tolerance – ¼"
- Eliminates longitudinal weld
- Increased production

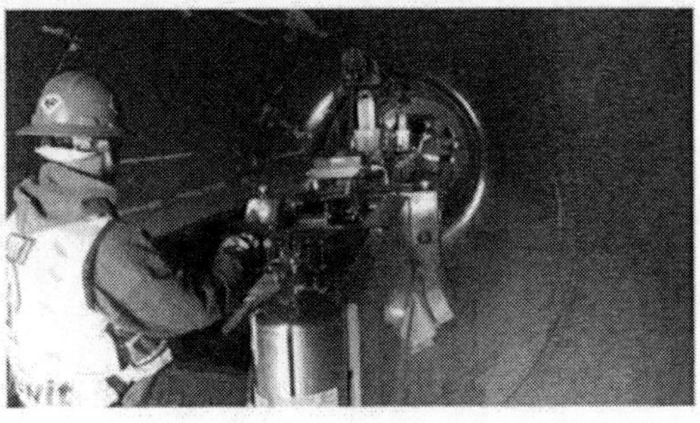

Annular Space Grouting

- 100 PSI grout for annular space
- Installed at maximum of 5 PSI
- Grout plugs installed and seal welded

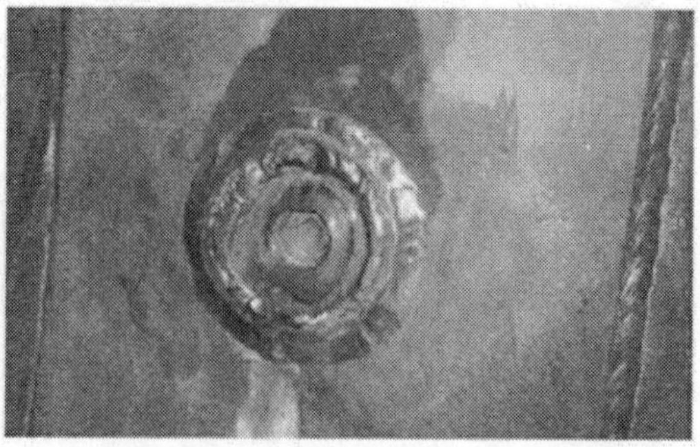

Cement Mortar Lining (CML)

- ½" field applied cement mortar lining (CML)
- Hand applied
- Steam curing until main is filled

Portal Closures

- Mortar lined steel closures with manways
- Reinforced encasement

Surface Restoration

Project Successes

- Schedule
  - Two weeks ahead of schedule
- Budget
  - $1.5M under budget
- Quality
  - 5009 Individual Liner inspections (.08% failures)
- Safety
  - 0 Recordable/Lost time injuries

Questions?

Kiewit

THE TOTAL SOURCE

educational sessions

Water Main Replacement in an Urban Environment

City of Atlanta
Mayor Kasim Reed

Department of Watershed Management

Water Main Replacement in an Urban Environment

Jo Ann J. Macrina P.E.
Commissioner

February 3-4, 2016
Underground Construction Technology
International Conference & Exhibition
Atlanta, GA

Presenters

John Tolbert – Watershed Manager II, Atlanta DWM
Jonathan Walker – Construction Project Manager II, Atlanta DWM
Hari Panthayi, P.E. – Project Manager, Atlanta DWM
Raymond J. Wilke, P.E. – Senior Watershed Director, Atlanta DWM

2013 3rd Quarter 3

Agenda

- **Managing Atlanta's Water Distribution System**
- **Recent Water Main Replacement Projects**
- **Criteria for Water Main Replacement**
- **DWM Linear Asset Management Program**

2013 3rd Quarter 4

Managing Atlanta's Water Distribution System

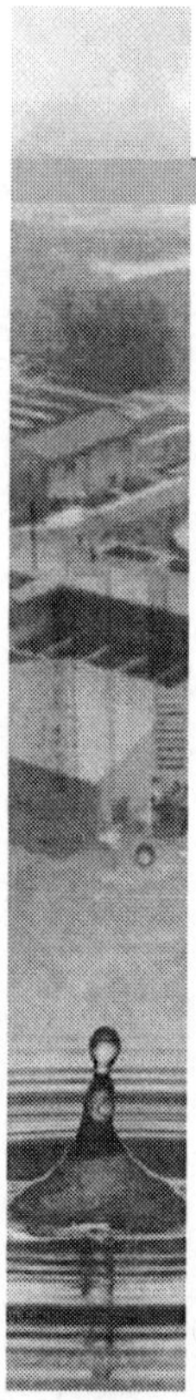

Managing Atlanta's Water Distribution System

➢ History

- Water distribution system dates back to 1875
- First treatment plant – South River headwaters
  - Feeding downtown Atlanta with 3 million gallons each day
  - The pipe from this plant to downtown was the beginning of the City's water distribution system

➢ Today's Approximations

- Distribution system – 2,700 miles
- Valves – 61,000
- Fire hydrants – 24,000
- Median age – exceeds 50 years
- Assessed value of the system – in excess of $2 billion
- Replacement value – in excess of $3 billion

Managing Atlanta's Water Distribution System

City of Atlanta Service Area

Managing Atlanta's Water Distribution System

Office of Linear Infrastructure Operations

- System Management: Department of Watershed Management (DWM)
  - Office of Linear Infrastructure Operations (OLIO)
    - Distribution system repairs and maintenance
    - Raw water supply
    - Water pipe integrity
    - Water quality
    - Main break repairs
    - Water distribution facilities installation
    - Leak detection
    - Exercising water valves
    - Testing fire hydrants
    - Installing water meter services

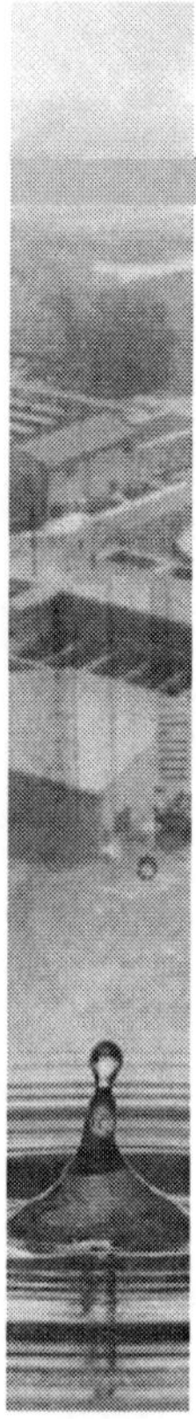

Managing Atlanta's Water Distribution System

- OLIO
  - 20 water distribution work crews
  - Service area – approximately 379 square miles

- Contract Operations
  - Water facilities replacement
  - Valve and hydrant installation
  - Small and large water meter installation and repair

Recent Water Main Replacement Projects

Recent Water Main Replacement Projects

- Atlanta Streetcar
- Northside Drive Water Main Replacement
- Southeast Atlanta Water Distribution System Improvements (WDSI)
- East Atlanta WDSI
- Southwest Phase II Atlanta WDSI

2013 3rd Quarter 11

Recent Water Main Replacement Projects

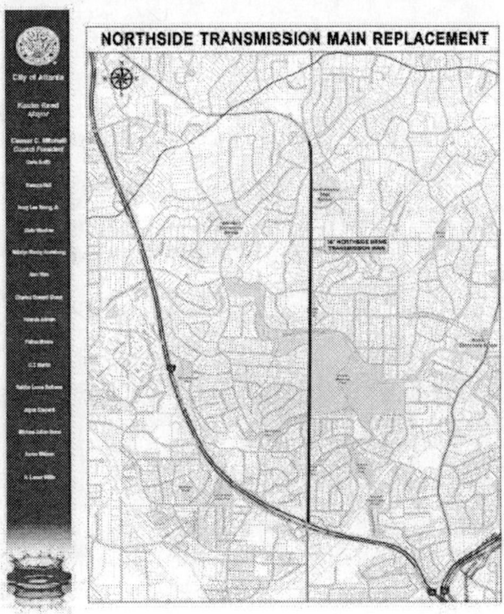

2013 3rd Quarter 12

Recent Water Main Replacement Projects

Pipe Replacement Project

2013 3rd Quarter

13

Recent Water Main Replacement Projects

Water Main Break

2013 3rd Quarter

14

Recent Water Main Replacement Projects

Pipe Replacement Project

2013 3rd Quarter 1

Criteria for Water Main Replacement

Criteria for Water Main Replacement

Possible Replacement Criteria

- Age of Pipe
  - Some water mains that exceed 100 years
- Interior degradation
  - Scaling and corrosion
  - Leads to pipe failure
  - Impacts fire flow capability
- Material
  - Ductile iron, cast iron, steel, etc.
  - Some materials have shorter lifespan

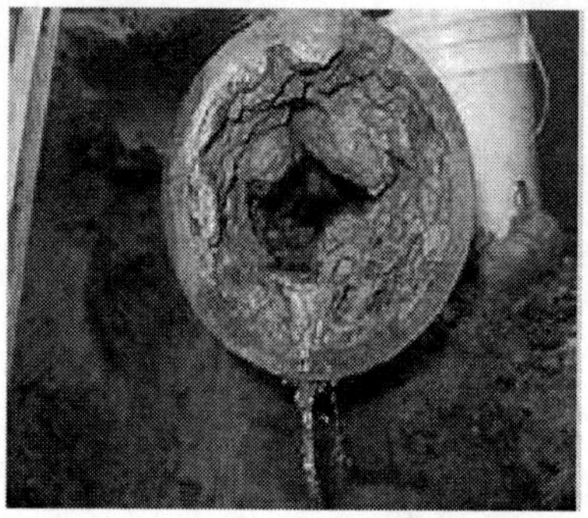

Criteria for Water Main Replacement

- Water Quality Issues
  - Impacts older neighborhoods
  - Colored or cloudy water
  - Unpleasant taste or smell

- Increased demand
  - Rapid growth over the last 2 decades
  - Requiring system upgrades
- Continual main breaks
  - Indicator of pipe degradation
  - Replacement may be in order
- Continual Leak Assessment
  - Valuable part of an urban water condition assessment and replacement program

DWM Linear Asset Management Program

DWM Linear Asset Management Program

Criticality Model

- Used to identify high-risk pipes. Model uses various parameters such as:
  - Age
  - Material
  - Repair history
  - Service area, etc.
- Data is used to determine:
  - Main's likelihood of failure
  - Main's consequence of failure

- The model helps identify and prioritize pipes that may require replacement.

2013 3rd Quarter

6

DWM Linear Asset Management Program

Risk Profiles Identify Assets that Require More Detailed Strategies to Mitigate Failure

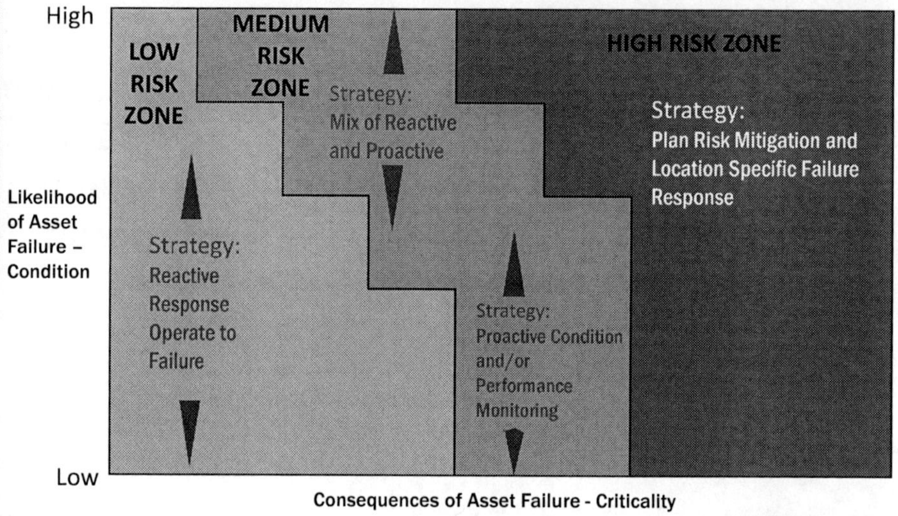

2013 3rd Quarter

7

DWM Linear Asset Management Program

Criticality Scoring Matrix

Likelihood of Failure			
Criteria	**Description**	**Scoring**	**Weighting**
Material	Reinforced Concrete, Steel pipe	1	30%
	High Density polyethylene, Polyvinyl Chloride, Ductile Iron Pipe.	2	
	Cast Iron Pipe, Copper Pipe	4	
	Vitrified Clay Pipe, Wrought Iron, Corrugated Metal Pipe.	10	
	Material Unknown	7	
Age/Remaining Useful Life	Age between 81 and 100% remaining useful life	1	30%
	Age between 61 and 80% remaining useful life	2	
	Age between 40 and 60% remaining useful life	4	
	Age between 20 and 40% remaining useful life	7	
	Age less than 20% remaining useful life	10	
Work Order Maintenance	0 /Point Repairs	1	20%
	≥1/ Joint/ Point Repair	10	
	>1/ Pipe Repair	4	
Operational Parameter	Operating Pressure 90-150 psi	7	10%
	Operating Pressures 40-90 psi	4	
	Operating Pressures < 40 psi	1	
	Operating Pressures > 150 psi and/ or Flow Velocity > 5ft./s	10	
Type of Bury/Cover Condition	Aerial	7	10%
	Non Aerial	4	

2013 3rd Quarter

8

DWM Linear Asset Management Program

Criticality Scoring Matrix

Consequence of Failure			
Criteria	**Description**	**Scoring**	**Weighting**
Pipe Diameter	≤ 6-inch	1	30%
	8 to 10-inch (Include 8 and 10 inch pipes)	2	
	12 to 20-inch (Includes 12, 16 and 20 inch pipes)	4	
	24 to 36-inch (Includes 24, 30 and 36 inch pipes)	8	
	>40 inch (Includes anything >42 inch pipes)	10	
	Diameter Unknown	7	
Transportation Impact/ Cover Condition	>25' from any road	1	20%
	Within 25' of any road (local, subdivision)	4	
	Within 26' to 100' of a railroad, planned rail road track station, bridge, or major road (US, CR, SR, Interstates)	6	
	Crosses or within 25' of any major road (US, CR, SR, identified City Road or Interstate), railroad, planned railroad track, or bridge.	10	
Level of Service	Operating Pressures > 150 psi	1	15%
	Operating Pressures between 40-150 psi with a minimal impact on fire protection	4	
	Operating Pressures < 40 psi with significant effect on fire protection	10	
Critical Infrastructure	Proximity (0-100 ft.) to Tier 4 Critical Facilities (Government Offices)	1	25%
	Proximity (0-100 ft.) to Tier 2 Critical Facilities (Schools, Retirement Communities/housing, Personal Care Homes, Other Health Services)	4	
	Proximity (0-100ft.) to Tier 2 Critical Facilities (Libraries, prisons, senior/community center)	7	
	Proximity (0-100ft.) to Tier 1 Critical Facilities (Airport, Major hospitals, Dialysis Centers)	10	
Population Impacts	Proximity outside Downtown limits	4	10%
	Wholesale Users / Top 100 users	8	
	Proximity with-in Downtown limits	8	
	Airport	10	

2013 3rd Quarter 9

DWM Linear Asset Management Program

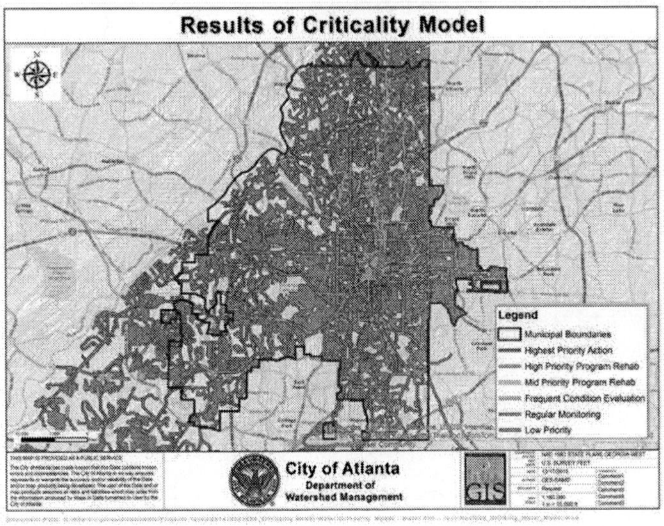

2013 3rd Quarter 10

DWM Linear Asset Management Program

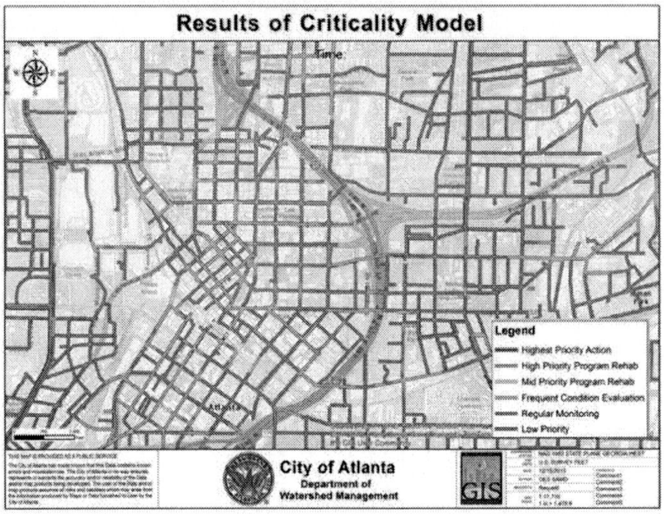

DWM Linear Asset Management Program

Pipe Condition Assessment

- Uses various technologies to determine pipe condition acoustics
  - CCTV
  - Visual
  - Electromagnetic
  - Other

- These technologies provide data on
  - Main leak locations
  - Average pipe thickness
  - Other structural defects

- The results of these inspections help prioritize pipes for replacement

DWM Linear Asset Management Program

Sensors and data transmitters used for pipe inspection

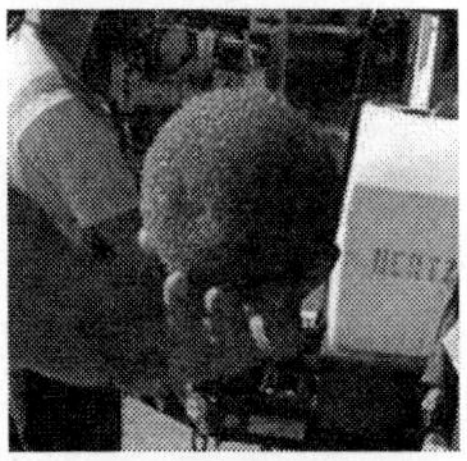

2013 3rd Quarter

13

DWM Linear Asset Management Program

OLIO Coordination

Hansen CMMS data from OLIO's operations feeds Asset Management analysis and thus updates to the Criticality Model.

- These activities include
  - Main break/repair history
  - Leak survey and assessment
  - Valve and hydrant exercise program
  - Pipe replacement during construction activities

2013 3rd Quarter

14

THE TOTAL SOURCE
educational sessions

Fused PVC Conduit and Casings for High Voltage Cables

Tom Marti
Underground Solutions, Inc.
Warrendale, PA

Header Placement

Fused PVC Conduit Background
Trenchless Technics Used for Installation
Fused PVC Conduit Properties
Case Studies

Topics Included

- Fused PVC Conduit Background
- Trenchless Techniques Used for Installation
- Fused PVC Conduit Properties
- Case Studies

Underground HV/EHV(High Voltage/Extra High Voltage Conduit Drivers

- Increasing numbers of high and extra high voltage power cables are being routed underground
- Non-metallic conduit and casings are now being used to efficiently provide the alignment to bury these cables
- Open cut direct bury conduit banks
- River crossings/highway crossings, environmentally sensitive areas, and other obstacles require trenchless methods
- Fusible PVC® pipe provides a known and proven material for these conduits and casings

HDD and Sliplining

- HDD
  - Casing
    - 16" to 36" diameter typically
    - Non-metallic
    - Internally debeaded
  - No casing
    - "Slick Bore"
    - Conduits pulled-in loose
    - Behave individually
- Sliplining
  - Conduits are fused and assembled into a bundle
  - Spacers are used to hold conduits in alignment inside the casing
  - Slide type/ roller type spacer supports
  - Typically will include smaller conduit for communication/grounding/ grout delivery

PVC Casing and Conduit Configuration

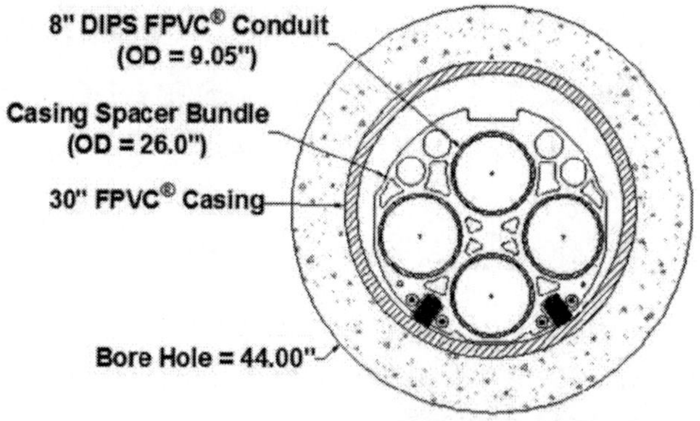

FPVC® Conduit and Casing

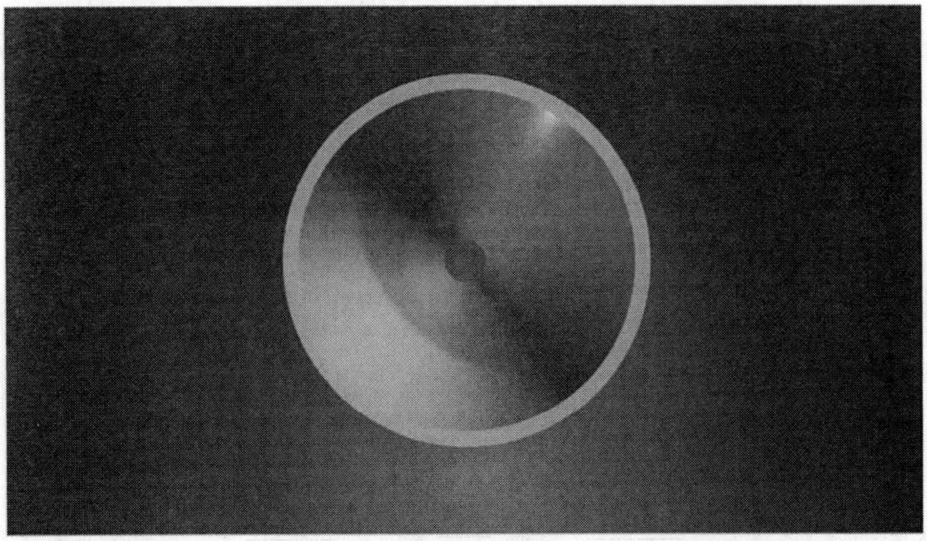

“Slick Bore”(No Casing)

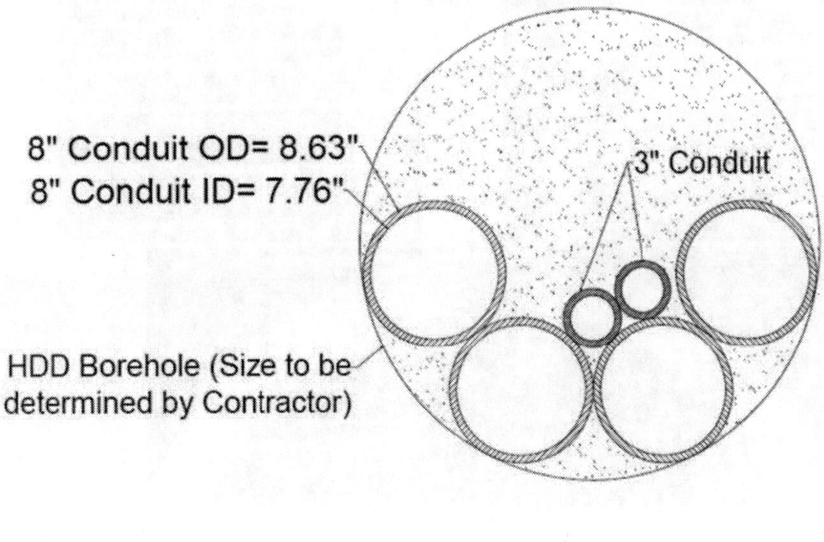

De-beading operation is designed with a robust QA/QC process to ensure successful cable installation

- Pneumatic or manual (depending on pipe size) de-beading equipment sent down pipe (40' – 45' sticks) to locate and remove fusion joint
- Shavings removed with squeegee plugs as de-beading equipment is retrieved
- Mandrel check for clearance can be followed by video check with GoPro® or track/push camera
- Longer distance remote de-beading available for intermediate fusions or repairs

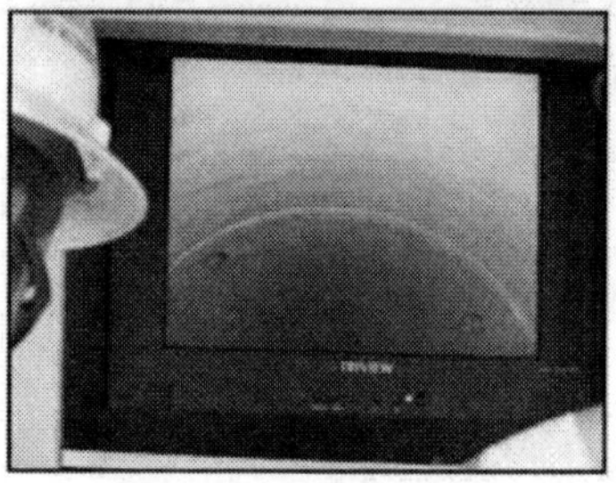

9

Internal Debeading

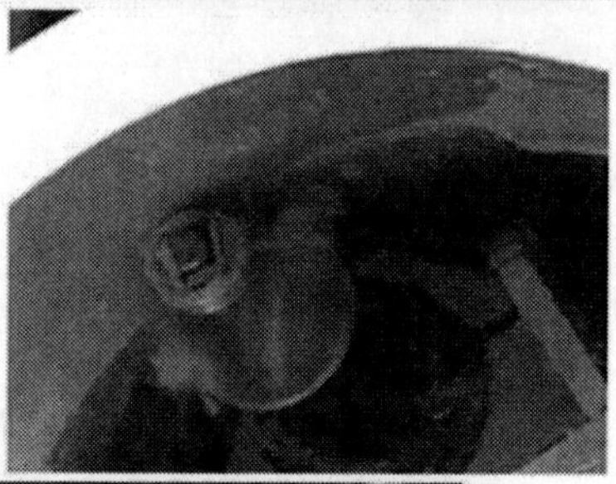

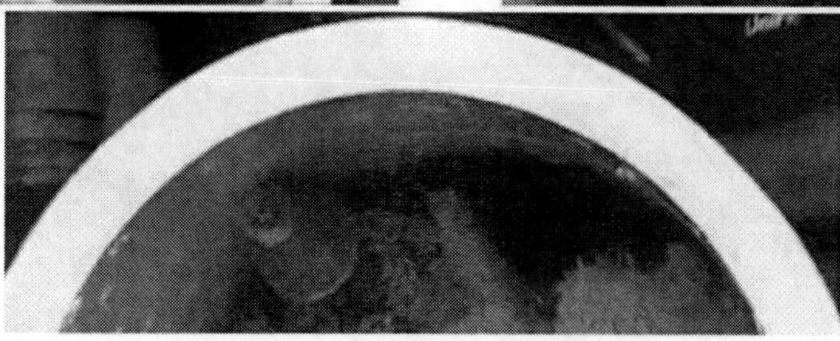

PVC Casing and Conduit Properties

Property	Specification	Fused PVC
Tensile Strength psi	ASTM D638	7,000
Specific Gravity	ASTM D1505	1.40
Hydrostatic Design Basis At 73° F, psi	ASTM D2837	4,000
Modulus of Elasticity psi	ASTM D638	400,000
Hardness (Rockwell R)	ASTM D785	117
Coefficient of Linear Expansion In./In. deg F	ASTM D696	0.3×10^{-4} .36"/100'/10°F
Hydrocarbon Permeation[6]		Highly Resistant
Thermal Resistivity	IEEE-442	565 °C-cm/W
Internal Coefficient of Friction		0.04-0.07

PVC Casing and Conduit Properties

Property	Specification	Fused PVC
Tensile Strength psi	ASTM D638	7,000
Specific Gravity	ASTM D1505	1.40
Hydrostatic Design Basis At 73° F, psi	ASTM D2837	4,000
Modulus of Elasticity psi	ASTM D638	400,000
Hardness (Rockwell R)	ASTM D785	117
Coefficient of Linear Expansion In./In. deg F	ASTM D696	0.3×10^{-4} .36"/100'/10°F
Hydrocarbon Permeation[6]		Highly Resistant
Thermal Resistivity	IEEE-442	565 °C-cm/W
Internal Coefficient of Friction		0.04-0.07

Thermal Resistivity and Cable Jacket COF Test Results

Geotherm USA

6354 Clark Ave.
Dublin, CA 94568

September 25, 2013

Underground Solutions
230 Executive Drive, Suite 110
Cranberry Township, PA 16066
Attn: Tom Marti

Re: Thermal Resistivity/Conductivity of Fusible PVC Samples

The following are the results of thermal resistivity measurements conducted on 3 block samples of Fusible PVC sent to our laboratory.

The tests were conducted in accordance with the IEEE standard (IEEE-442) using laboratory type thermal probes and Thermal Property Analyzer Model TPA2000. The tests were conducted at room temperature of 22 °C (72 °F).

The average density of the 3 samples was 89 lb/ft³.

The average thermal resistivity was 565 °C-cm/W.

The average thermal conductivity was 0.177 W/°C-m.

Please contact us if you have any questions or require additional information.

Geotherm USA, LLC

Deepak Parmar

565°C-cm/W

Thermal Resistivity testing conducted by Geotherm USA to IEEE standard 442

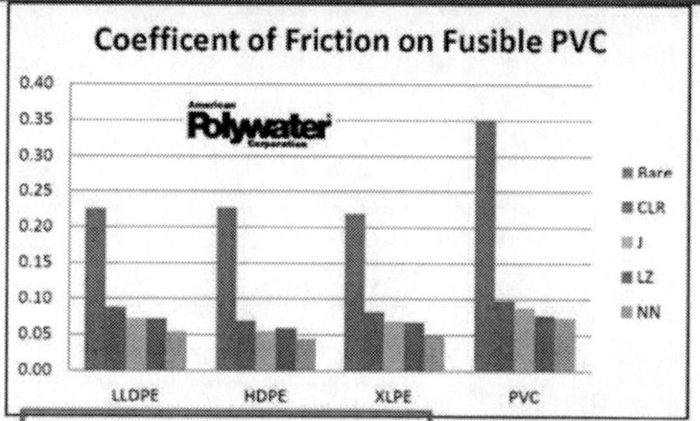

PVC Internal COF 0.04-0.07

13

Ampacity

- Measure of electrical current flow through a conductor
- Metal near/around a conductor can reduce a cable's ampacity by as much as 30%
- Magnetic fields created and transient currents from metal
- Non-metallic PVC essentially eliminates the interference and ampacity loss

HDD Installation of FPVC® Conduit – Denver, CO

Clients: Cherokee Power Station, Pinnacle Development

- 480 LF HDD installation of (7) bundle of 6" DR18 FPVC® conduits
- Direct pull-in with loose bundle (no banding)
- Conduits internally de-beaded
- Project completed May 2014

15

Slipline Installation of FPVC® Conduits - Hackensack, NJ

Clients: PSEG New Jersey and J. Fletcher Creamer & Son, Inc.

- 15 bundle: (10) 6" and (5) 5" FPVC® conduits
- Installed in 1,500 LF, 36" steel casing (22,500 LF of conduit)
- Fuse, de-bead, bundle and pull sequential operation with limited layout space
- 6" conduits carry 69 kV high voltage power cables; 5" conduits carry 26/13/4 kV power cables (future capacity)

16

HDD Installation of FPVC® Conduits - Lexington Park, MD

Clients: (SMECO) Southern Maryland Electric Cooperative
Mears Group, Inc.

- (2) 4,400 LF HDD crossings with total of (7) 8" DR14 FPVC® conduits
- Direct HDD pull-in (no casing/slick bore) with conduits pulled in loose (no banding or bundling)
- Full length of conduits fused and de-beaded prior to pull-in
- Conduits carry 230 kV XLPE high voltage power cables

17

HDD Installation of FPVC® Conduit – Corpus Christi, TX

Clients: Padre Island Water Line Improvements Project
Bridges Specialties; Mears Group, Inc.

- 5,535 LF HDD of 4" DR 14 FPVC® conduit (single pull, uncased)
- Full length fused and de-beaded prior to pull-in (no intermediate fusions)
- Parallel 5,545' HDD installation of 18" DR 18 Fusible PVC® waterline

18

HDD Install of FPVC® 4" Conduit Bundle– Pantego Creek, NC

Clients: Lee Electric, Inc., Tidewater Electric Co-op

- 17,280' of 4" DR 14 conduit to carry 25KV cable
- (4) bundle 4" FPVC® conduits installed via HDD and open cut
  - 4320' of conduit fused in 4 lengths
  - 3800' installed by HDD
  - 520' to remain in the open cut
- Conduit internally de-beaded
- Conduit installation complete in September, 2014

19

HDD Install of FPVC® Casings & Conduits – Woodbridge, NJ

Clients: Woodbridge Energy Center
Carson Corporation

- 11,000 LF of 30" DR25 FPVC® conduit casings (6) HDDs: (2) 2,500 LF, (2) 1,600 LF, (2) 1,400 LF
- (4) bundle of 8" DR21 IPS FPVC® conduits sliplined into each casing (45,000 LF total)
- Wheeled spacer bundles located every 5 LF
- Casings de-beaded and thermally grouted

20

FPVC® Casing Layout Woodbridge, NJ

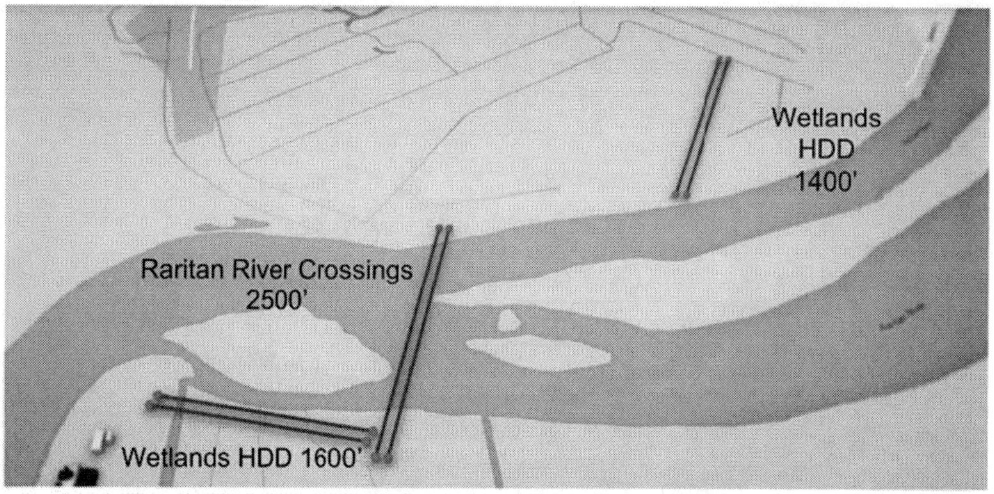

CAT

CAT

Connections with FPVC® Pipe can be Easily Accomplished with Industry Standard Couplings and Fittings

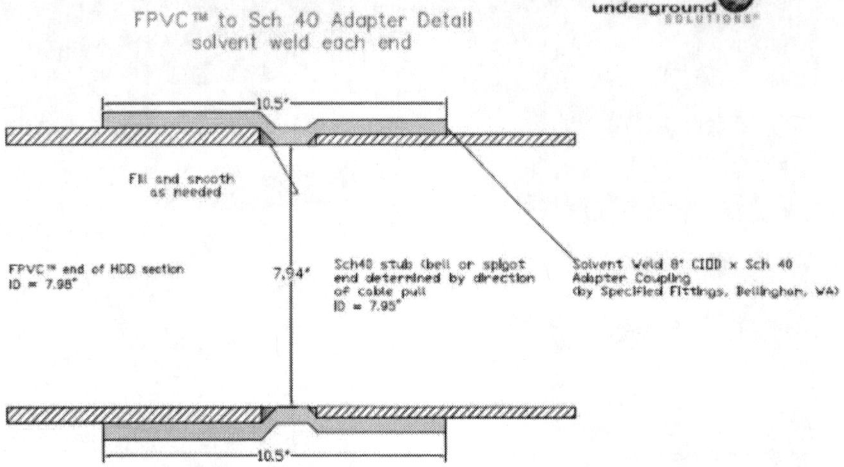

Conclusions

- Fused PVC provides a trenchless capability for high voltage electrical conduits and casings with the same non-conductive plastic material commonly used for duct banks
- Fused PVC provides excellent thermal resistivity and internal COF
- Fused PVC is extruded and fused to the same high quality standards as pressure water pipe
- Fused PVC allows lengths for non-metallic casings and conduits that previously could not be considered

Questions?

UNDERGROUND SOLUTIONS, INC.
POWAY, CA. 858-679-9551
WWW.UNDERGROUNDSOLUTIONS.COM

THE TOTAL SOURCE

educational sessions

Horizontal Directional Drilling (HDD) Technology Advancements to Mitigate Environmental Issues and Reduce Carbon Footprints

Seth Matthesen
Ditch Witch®
Senior Product Manager, HDD

Agenda

- Pre-Planning
- Recycling Drilling
  - Recycling HDD Drilling Fluids
  - Micro Drilling-Fluid Recyclers
  - Benefits of Micro Systems
  - Savings
  - Counting Gallons and Dollars
  - Environmental Impact
  - Reducing Risks
- Case Study
- Summary

Abstract

- Drilling fluids are essential in any HDD installation.
- Fluid carries cuttings out of the bore hole.
- Disposing of used fluid can be time consuming and costly.
- Topics: Drilling-fluid returns and the environmental perception

Recycling HDD Drilling Fluids

Recycling of drilling fluids is standard procedure for large drilling rigs.

Recycling is a multistep process:

1. Fluid passes through shaker screens
2. Desander cones remove sand
3. Shaker screen removes additional solids

Clean fluids are reused.

Recycling HDD Drilling Fluids

Micro Drilling-Fluid Recyclers

Recycling equipment is usually designed for large drilling rigs.

Small recycling systems have not been fully utilized.

Micro recycling systems:
- Compact to work in urban areas
- Can be quickly moved

Micro Recycling Systems

Recycling drilling fluids:

- Reduces the volume
- Improves drilling productivity
- Impacts carbon footprint
- Limits disturbance to the environment

Benefits

Primary benefits of micro recycling:

- Lower volumes of water
- Production improved
- Fluid disposal cost reduced
- Reduced continual use of a vacuum system
- Moved around while containing fluid

Limitations

- May not be effective for all soil conditions
  - Clay-based soil
- Change in methodology
- Training effort

Counting Gallons and Dollars

To forecast potential savings, estimate:

- Amount of fluid needed
- Disposal costs
- Labor costs
- Transportation cost
- Miscellaneous costs
- Project with and without recycling

Potential Savings

	3,000 gallons		1,000 gallons
	Cost Per Day		Cost Per Day
Mixing system	$134.76	Micro recycling system	$277.63
Water	$15.00	Water	$5.00
Fuel used on vac/truck for mud recovery	$84.50		
Drilling additive cost for the day	$99.00	Drilling additive cost for the day	$33.00
Fuel used on mixing system	$50.00	Fuel used on micro recycling system	$66.60
Cost of dumping mud (determined from volume pumped/day	$249.75	Cost of dumping mud (determined from volume pumped/day	$64.80
Fuel used hauling mud off	$133.20	Fuel used hauling mud off	$44.40
Hourly charge of vacs being pulled (wear and tear)	$300.00	Cost of screens	$55.00
Hourly charge on trucks hauling mud off (wear and tear)	$150.00		
Total cost for the day of drilling	$1,216.21	Total cost for the day of drilling	$546.43

TOTAL SAVINGS IN ONE DAY	$669.78
TOTAL SAVINGS FOR MONTH OF EQUAL USAGE	$13,395.60

Environmental Impact

Reduces carbon footprint through:

- Decreased emissions
- Water conservation
- Less fluid cleanup

Safety

Fluid recycling reduces risks:

- Less labor for spoils handling
- Reduced amount of fluid returns escaping
- Reduced transport
  - Disposal of returns
  - Water-hauling trips

Case Study: Google-MasTec Austin, TX

Case Study: Google-MasTec Austin, TX

Case Study: Google-MasTec Austin, TX

Case Study: Google-MasTec Austin, TX

Case Study: Google-MasTec Austin, TX

Case Study: Google-MasTec Austin, TX

Case Study: Google-MasTec Austin, TX

Case Study: Google-MasTec Austin, TX

Case Study: Google-MasTec Austin, TX

Case Study Summary

- Environmentally-sensitive region
- Limited disposal sites
- Limited CDL labor force
- Additional revenue of $100,000 per month
  - Savings of fluid and disposal
  - Increased production

Summary

- **Micro fluid-recycling systems:**
  - Conserve water
  - Lower fluid volumes
  - Reduce work area cleanup
  - Reduce fuel and labor costs
  - Reduce carbon footprint
  - Make the project more environmentally friendly

Reducing Costs and Risks in HDD with Innovative Tooling Generation

Jeremy Stenslie, Mechanical Engineer, Herrenknecht USA
Underground Construction Technology
Atlanta, GA
Feburary 3, 2016

Reducing the Cost & Risk of HDD using FFHO and DHJP

Full Face Hole Opener (FFHO)

Down Hole Jet Pump (DHJP)

Horizontal Directional Drilling Steps

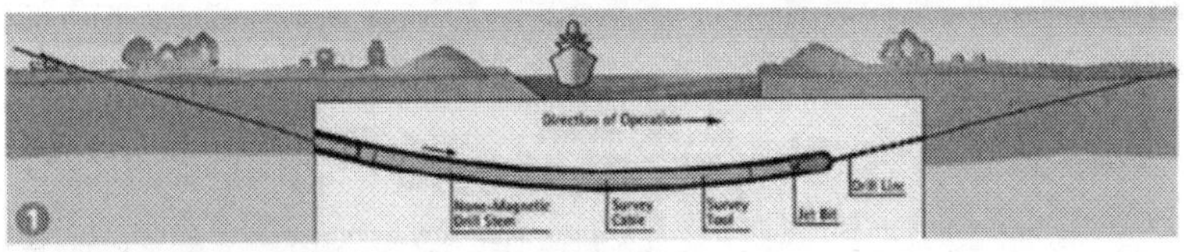

Pilot Drill

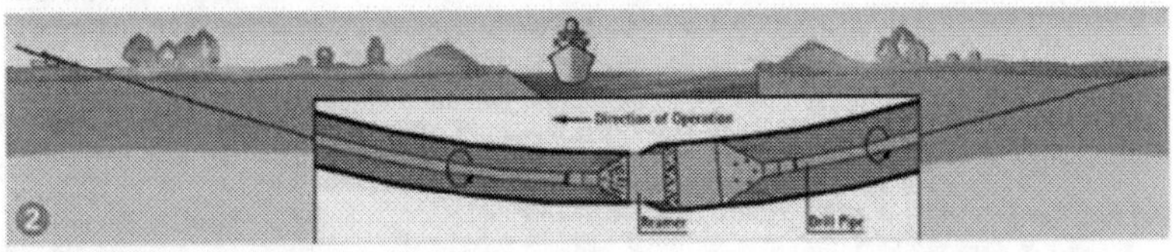

Reaming

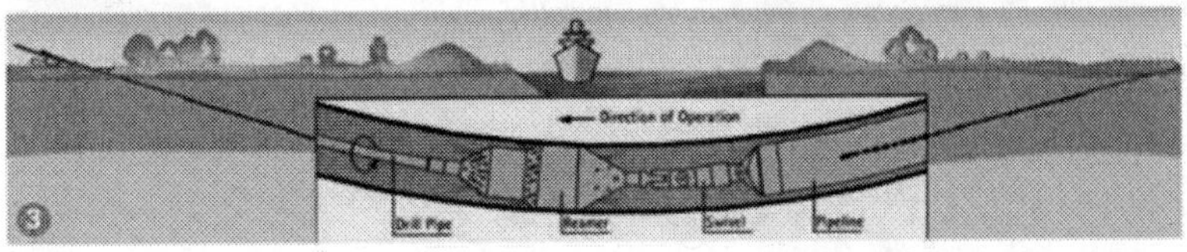

Pipeline Pullback

Fullface vs. Conventional Reaming Steps

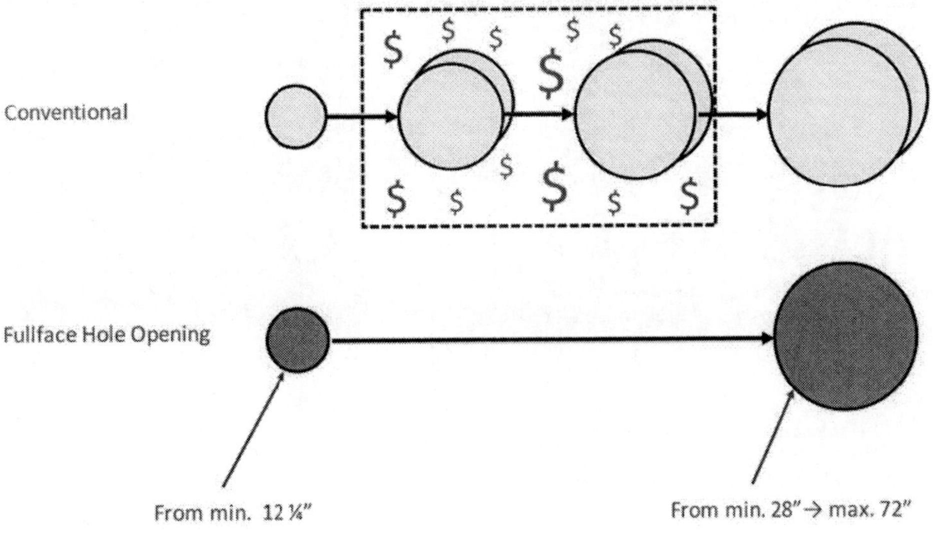

Full Face Hole Opener Design

- Replaceable Front Centralizer
- Tool-Face Cleaning
- Cutting Scraper

Replacable and Rebuildable Tooling

- Heavy duty proven bolting system
- Replaceable parts: bearings, seals
- Pressure-compensated

Efficiency through Cost Reduction

- Less drilling tools needed
- Drilling tools with replaceable cutters
- Cutters with longer bearing and hub lifetime
- Effective wear protected body for a longer lifetime

→ **Saving investment costs**

- Less tripping operations
- Less joint changes
- Less tool changes
- Less cleaning runs
- Increased productivity

→ **Saving construction costs**

→ **Less Costs-per-foot**

Challenges for the (Fullface) Hole Opening

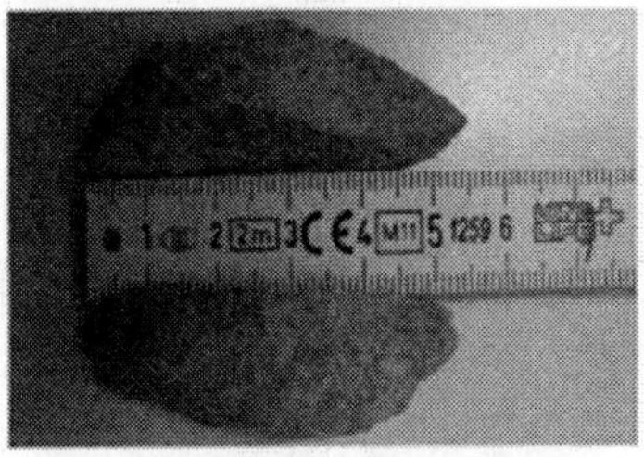

- Large cutting transport
  - Risk of settlement in the borehole
- Frac-out and fluid losses
  - Natural fractures & fissures in the formation
  - Over pressurizing of the borehole
    - Friction losses during pumping thru borehole
      - Fluid viscosity, density, flow rate
      - Bore hole wall character, length, cross section

Mud & Cutting Flow

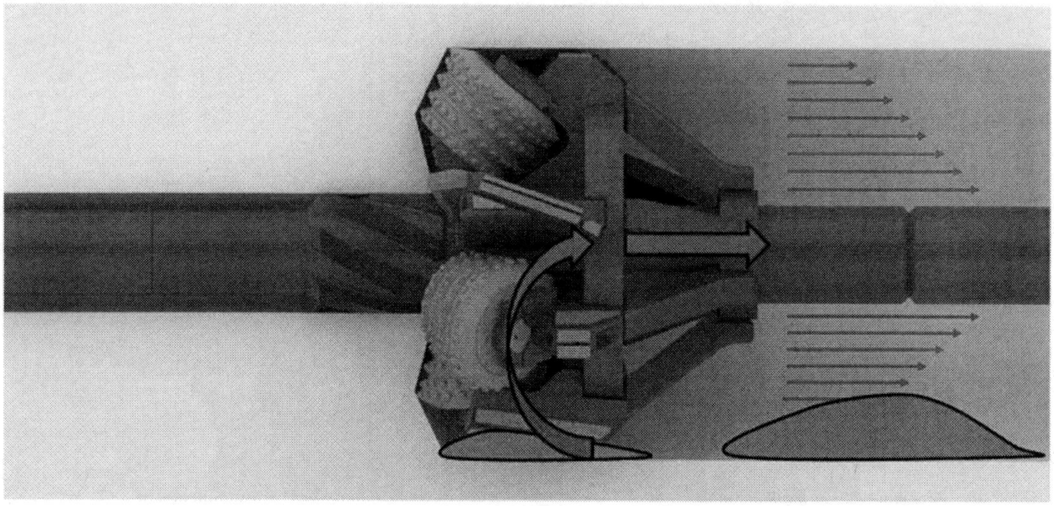

Drill Pipe used as Discharge Line

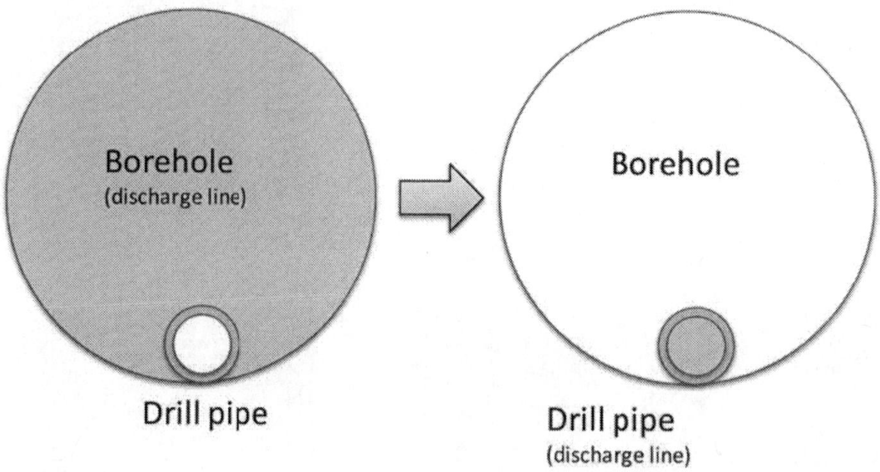

Down Hole Jet Pump (DHJP)

Down Hole Jet Pump (DHJP)

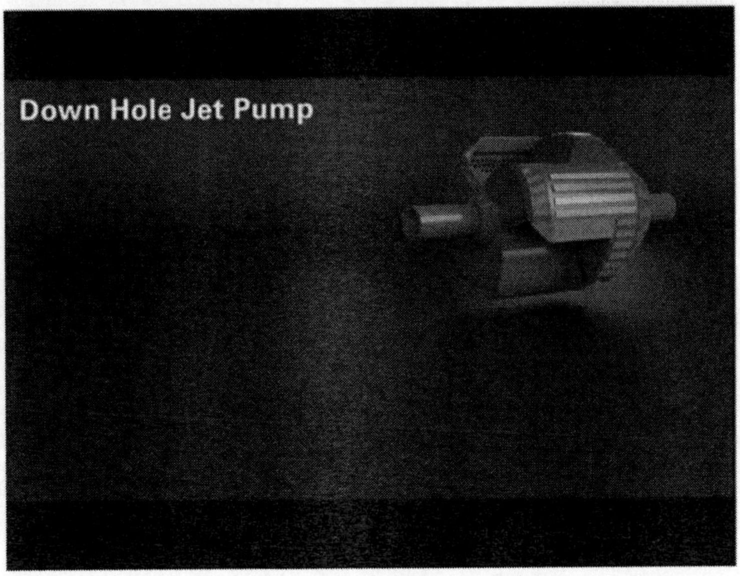

Down Hole Jet Pump (DHJP)

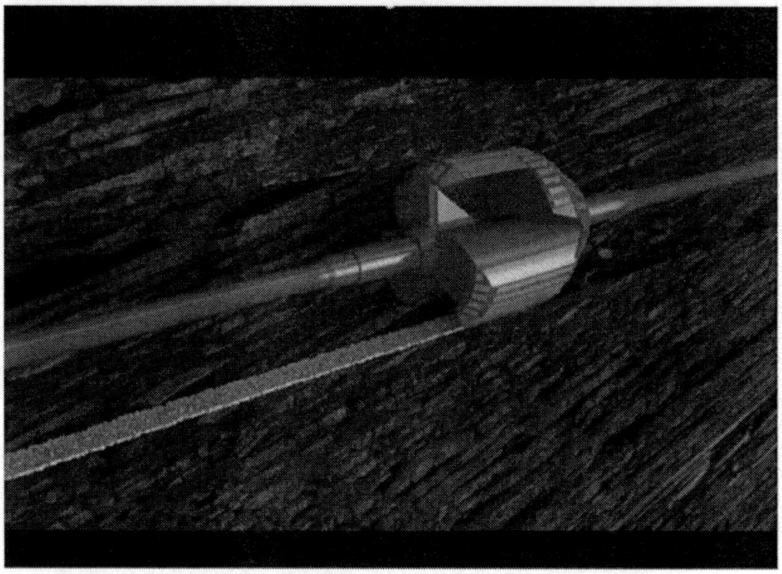

DHJP Testing Facility

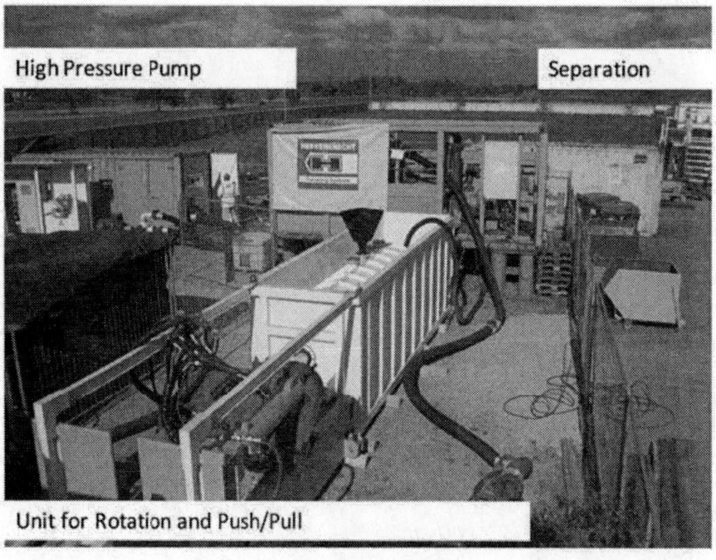

DHJP Testing Facility

1000m of discharge line circuit

DHJP Testing Facility

- DHJP in artificial borehole loaded with cuttings (dry)

- DHJP in artificial borehole after the testing (dry)

DHJP Testing Facility

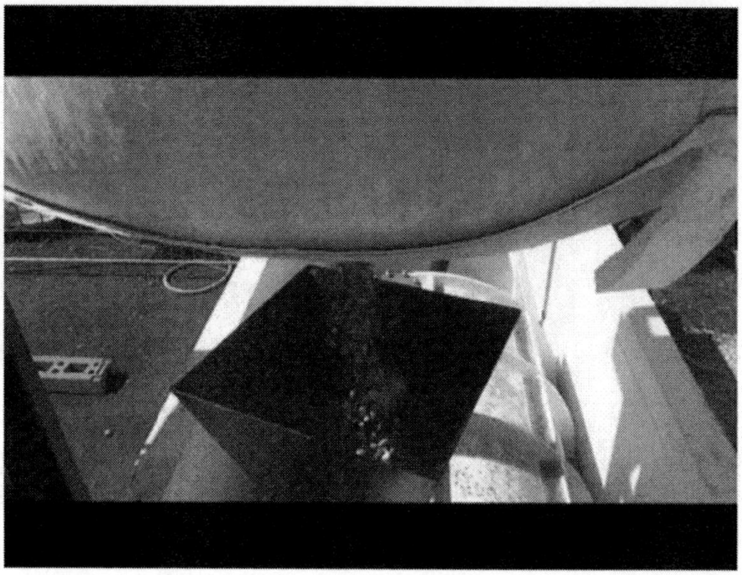

Requirements to operate an DHJP

- Min. Borehole diameter: 20" = 508 mm
- HP-Motive Flow:2000-2500 l/min
- HP-Motive Pressure: 50-60 bar
- Separation capacity: 3000-3500 l/min
- Recommended Drill Pipes: 6 5/8"
- Pipesite-swivel
- Mudreturnline (Pipesite-Rigsite)

DHJP Advantages at a Glance

- Minimum frac-out risk during the reaming operation
- Possibility to use a simple and cost saving mud program
- It is possible to create a 98% clean borehole
- Immediate formation feedback on the separation plant (1000m = 7 min)
- Possibility to transport larger cutting sizes
- Possibility for full-face reaming
- Works with non or partially filled boreholes
- Defined return flow direction
- Flow amounts (in & out the borehole) can be simply monitored and logged
- Direct connection of the mud flow to the recycling unit (No mud pit pump necessary)

Full Face Hole Opener & DHJP

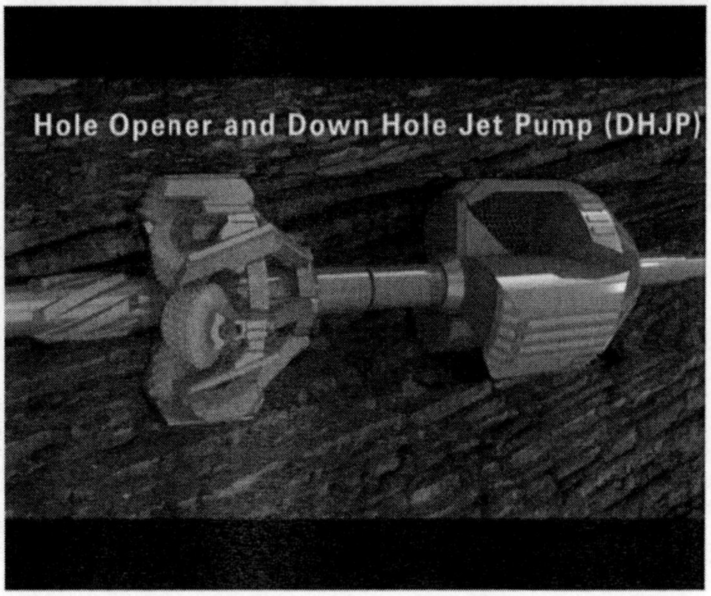

Thank you for your attention!
HERRENKNECHT
Tunnelling Systems

THE TOTAL SOURCE

educational sessions

Track VI-B = HDD: PRODUCTIVITY & FLUID MANAGEMENT

Tricone Pilot Bit Dull Grading

Objectives

To make participants familiar and comfortable

- Understand basic tricone designs
- Working with a tricone bits and understanding limitations and preventive maintenance necessary.
- Understanding dull grading tricone bits.
- Understanding the importance or dull grading and maintaining well documented bit performance.

Rotary Rock Bit Elements

Basic tricone bit construction

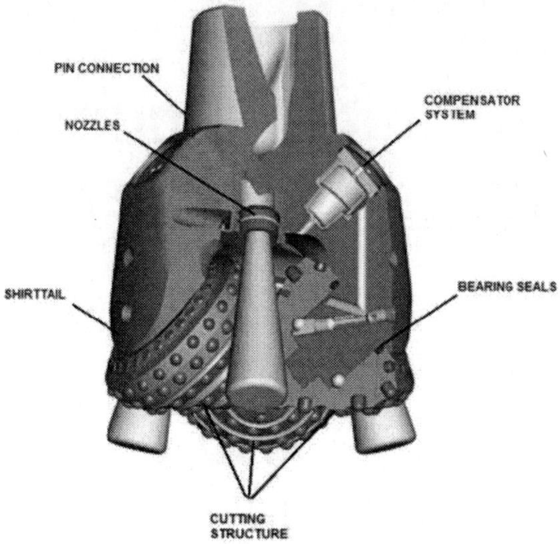

Rotary Rock Bit Cutting Elements

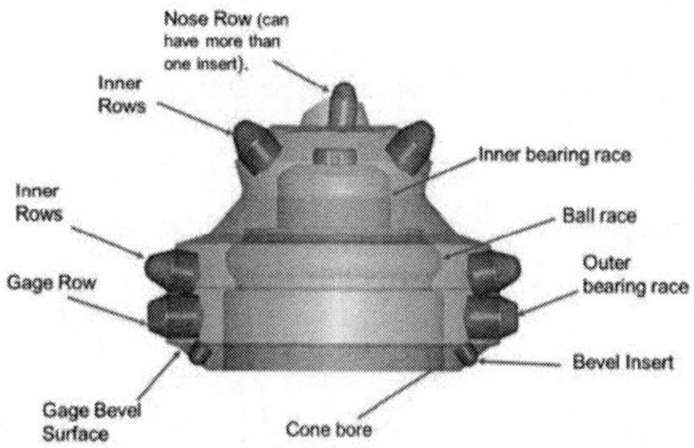

Rotary Rock Bit Elements

Bit Body

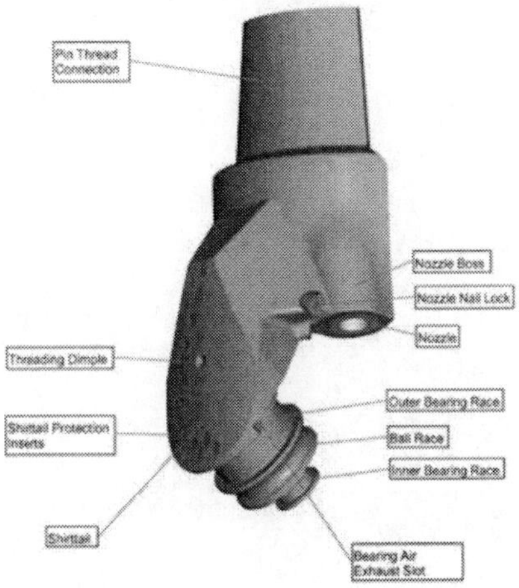

Sealed Bearing Jet Circulation

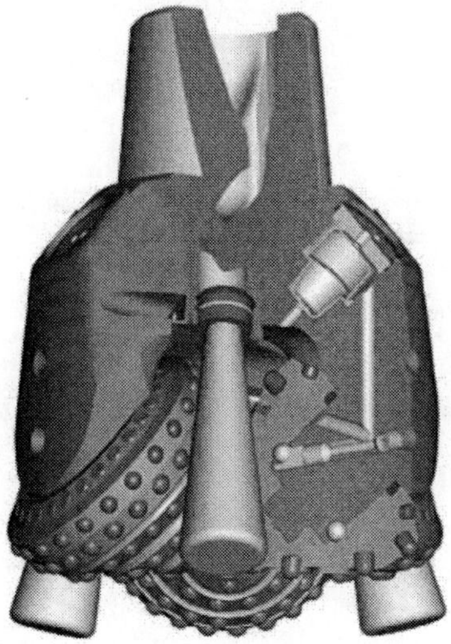

Applications:

- Directional Drilling
- Blast hole TCI (Tungsten Carbide Inserts)
- Water well (TCI & ST)
- Raise boring Pilot

Tricone Rock Bit Identification

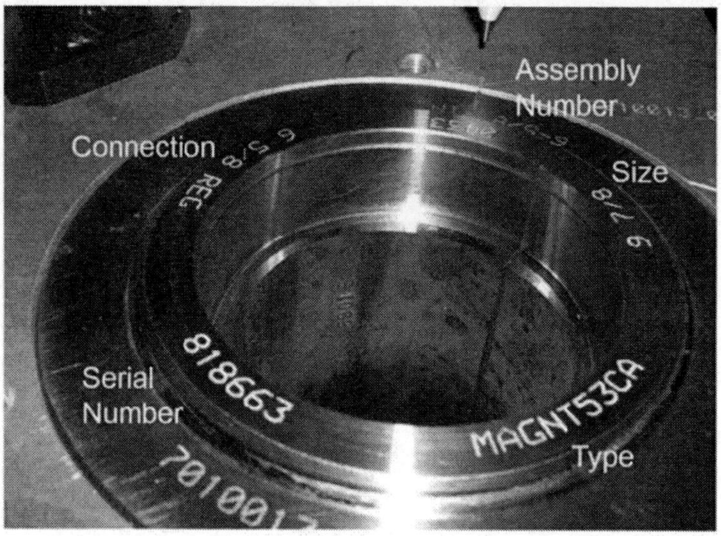

Preventive Care of Bits

- Do no drop bits
  - Dropping bits will cause internal damages to bearings and other components that are not apparent, but can affect the structure and performance of the bit.
- Clean bits after each use
  - Clean bits after each use and prevent leaving them in the hole overnight without any circulation. This will prevent the bits from generating to much corrosion externally and internally.

Preventive Care of Bits

Storage

- Store bits in a controlled environment to minimize the affect of exterior weather damage to metal components. Painting the bit will help protect the bit's body from environmental damages.
- If the bit is an open bearing bit, it helpful to clean the bit thoroughly and submerge it in a bucket of oil to extent its storage life.

- Corrosion
  - Properly clean and maintain bits in a controlled environment to prevent corrosion to the pin connections and other critical areas.

Preventive Care of Bits

- Application of heat
  - Do not apply heat to the bits with out consulting with bit manufacturer.
  - Bits have internal components that are affected by the heat application, most of the sealing components are rated to a certain temperature before changing geometry and properties.
- Do not re-greasing bits
  - Bits are greased using a special procedure, do not attempt to grease bits without consulting the manufacturer first.

Care During installation

- Cross threading the bits
  - Exercise care during the installation to prevent cross threading, most HDD bits and other tools use API treads that are very fines and easily damaged.
- Apply a good amount of thread grease
  - Insure that a sufficient amount of thread grease is applied to pin connections to prevent damage during installation and removal of the bits.

Good Practices During Operation

- Do no drill into a smaller hole (Pinched bit)
  - Do not drill with a bit into a smaller hole, it will cause cones to pinch in and cone will start to interfere.
- Do not drilled in a junked Hole
  - If possible do not drill into a hole full of junk (Metal, another bit, or another tool).

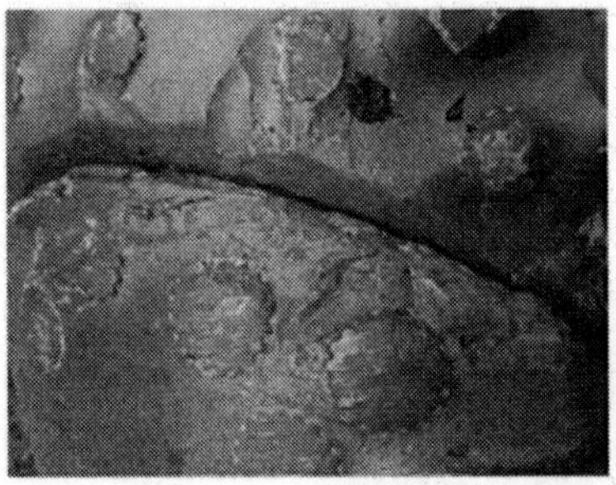

12

Good Practices During Operation

- Adequate thrust loading and RPM
  - Apply the correct amount of thrust and RPM to match the ground formation
  - Too much load will cause the bit to over penetrated damaging the cones, not enough will result in premature insert wear.
  - Too much RPM will wear inserts prematurely, not enough RPM will result in slow rate of penetration

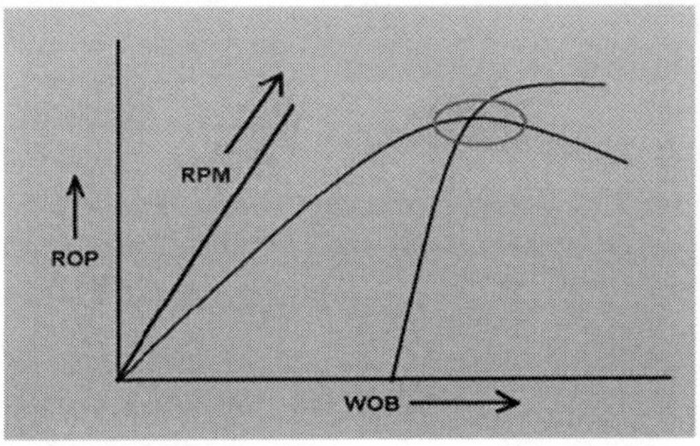

13

Good Practices During Operation

- Installation of the correct size of nozzles
  - Installing the correct size nozzles for the flow system will insure proper cleaning of the tools, and good flow velocities for evacuating the cuttings.
- Proper bottom hole cleaning or circulation
  - One of the most important parameters during operations is the fluid flow
  - Improper fluid flow will damage grease sealing components, bit body, bit cutting structure, and ultimately performance and production.

14

Bit Dull Grading System

Dull Bit Grading System

- Originally developed by the IADC* as a simple, universal system for describing the condition of dull bits used in oilfield drilling.
- Dull Grading is ***descriptive only***! It is NOT intended as a "failure analysis" tool.

*International Association of Drilling Contractors

- Bit Performance Data Base
- Value Added Service
- Recommended Bit Selection
- Information for Design Engineering or manufacturers

Bit Dull Grading System

Cutting Structure				B	G	Remarks	
Inner Rows	Outer Rows	Dulling Characteristics	Location	Bearing/Seal	Gage	Other Dull	Reason Pulled
(I)	(O)	(D)	(L)	(B)	(G)	(O)	(R)

Column 1 (I-Inner) – Used to report the condition of the cutting elements not touching the wall of the hole

0- No loss of cutting Structure
8-Total loss of cutting Structure

Bit Dull Grading System

Cutting Structure				B	G	Remarks	
Inner Rows	Outer Rows	Dulling Characteristics	Location	Bearing/Seal	Gage	Other Dull	Reason Pulled
(I)	(O)	(D)	(L)	(B)	(G)	(O)	(R)

Column 2 (O-Outer) – Used to report the condition of the cutting elements that touch the wall of the hole (Gage row).

0- No loss of cutting Structure
8-Total loss of cutting Structure

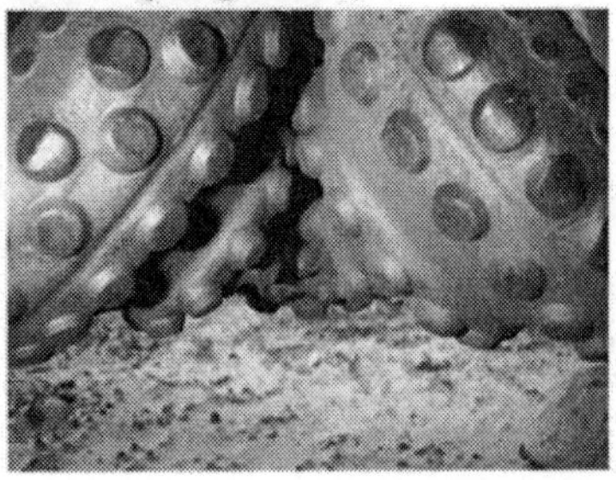

Bit Dull Grading System

Cutting Structure				B	G	Remarks	
Inner Rows	Outer Rows	Dulling Characteristics	Location	Bearing/Seal	Gage	Other Dull	Reason Pulled
(I)	(O)	(D)	(L)	(B)	(G)	(O)	(R)

Column 3 – Dulling characteristics – Cutting Structure) Uses 2 letter code to indicate the major dull characteristic of the cutting structure (see following codes)

BT =	Broken Teeth	HC =	Heat Checking	SD =	Shirttail Damage
BU =	Balled Up	JD =	Junk Damage	SS =	Self-sharpening
*CC =	Cracked Cone	*LC =	Lost Cone	ST =	Shirttail
*CD =	Cone Dragged	LN =	Lost Nozzle	TR =	Tracking
CI =	Cone Interference	LT =	Lost Teeth	WO =	Wash Out On Bit
CR =	Cored	PB =	Pinched Bit	WT =	Worn Teeth
CT =	Chipped Teeth	PL =	Plugged Bearings	NO =	No Other Wear
ER =	Erosion	PN =	Plugged Nozzle		

N =	Nose Rows	H =	Heel Rows	1 =	Cone # or #'s
M =	Middle Rows	A =	All Rows/Cones	2 =	Cone # or #'s
* Show Cone Number (s) under Location, column 4				3 =	Cone # or #'s

Bit Dull Grading System

Cutting Structure				B	G	Remarks	
Inner Rows	Outer Rows	Dulling Characteristics	Location	Bearing/Seal	Gage	Other Dull	Reason Pulled
(I)	(O)	(D)	(L)	(B)	(G)	(O)	(R)

Column 4 – (L- Location) Uses a letter or number code to indicate the location on the face of the bit where the cutting structure dulling characteristics occur.

N-Nose Row
M-Middle Row
G-Gage Row
A- All Rows
Cone # 1, 2, or 3

Bit Dull Grading System

Cutting Structure				B	G	Remarks	
Inner Rows	Outer Rows	Dulling Characteristics	Location	Bearing/Seal	Gage	Other Dull	Reason Pulled
(I)	(O)	(D)	(L)	(B)	(G)	(O)	(R)

Column 5- (B - Bearing) Uses a letter or number code, depending on the bearing types, to indicate bearing condition of the roller cones.

Non-Sealed Bearings
0-New Bearing
8-All bearing life has been used

Sealed Bearings
E- Seals effective
F- Seals Failed
N- Not Able to Grade
X –Fixed Cutter Bit (Bearingless)

Bit Dull Grading System

Cutting Structure				B	G	Remarks	
Inner Rows	Outer Rows	Dulling Characteristics	Location	Bearing/Seal	Gage	Other Dull	Reason Pulled
(I)	(O)	(D)	(L)	(B)	(G)	(O)	(R)

Column 6- (G- Gage) Used to indicate the Gage of the bit, letter "I" indicate no gage reduction. If the bit does have a reduction in gage, it is to be recorded in 1/16th of an inch.

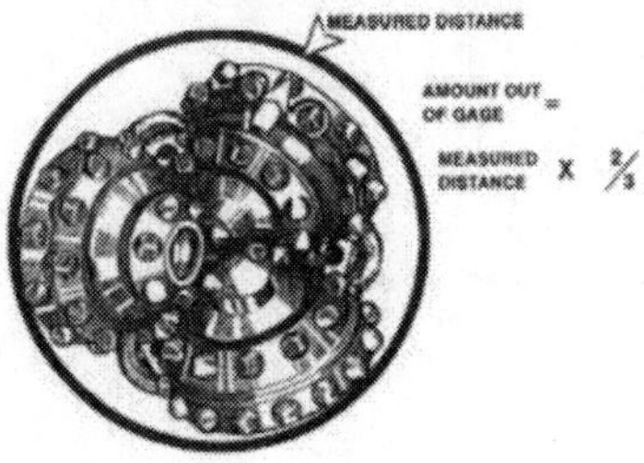

1- In Gage
1/16- 1/16" out of gage
2/16–1/8" out of gage
4/16- ¼" out of gage

The "Two Third's Rule" is correct for three cone bits

Bit Dull Grading System

Cutting Structure				B	G	Remarks	
Inner Rows	Outer Rows	Dulling Characteristics	Location	Bearing/Seal	Gage	Other Dull	Reason Pulled
(I)	(O)	(D)	(L)	(B)	(G)	(O)	(R)

Column 7- (O- Other Dull Characteristics) is used to report any dulling characteristics of the bit, in additions to the cutting structure dulling characteristics listed in column 3.

BT =	Broken Teeth	HC =	Heat Checking	SD =	Shirttail Damage
BU =	Balled Up	JD =	Junk Damage	SS =	Self-sharpening
*CC =	Cracked Cone	*LC =	Lost Cone	ST =	Shirttail
*CD =	Cone Dragged	LN =	Lost Nozzle	TR =	Tracking
CI =	Cone Interference	LT =	Lost Teeth	WO =	Wash Out On Bit
CR =	Cored	PB =	Pinched Bit	WT =	Worn Teeth
CT =	Chipped Teeth	PL =	Plugged Bearings	NO =	No Other Wear
ER =	Erosion	PN =	Plugged Nozzle		

N =	Nose Rows	H =	Heel Rows	1 =	Cone # or #'s
M =	Middle Rows	A =	All Rows/Cones	2 =	Cone # or #'s
* Show Cone Number (s) under Location, column 4				3 =	Cone # or #'s

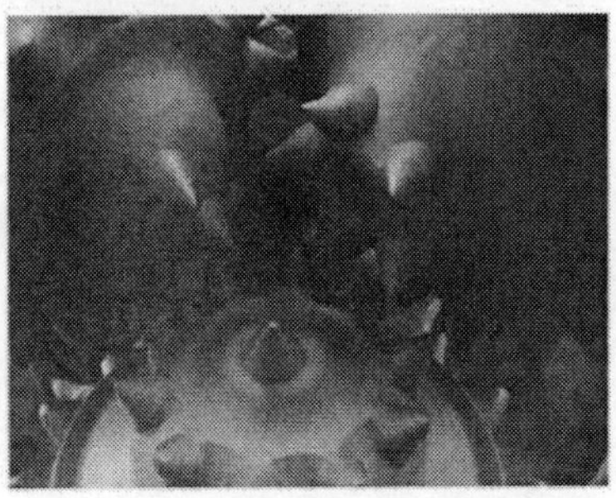

Bit Dull Grading System

Cutting Structure				B	G	Remarks	
Inner Rows	Outer Rows	Dulling Characteristics	Location	Bearing/Seal	Gage	Other Dull	Reason Pulled
(I)	(O)	(D)	(L)	(B)	(G)	(O)	(R)

Column 8 - (R- Reason Pulled) is used to report the reason for terminating the bit run.

BHA- Change Bottom Hole Assembly
CM- Conditional Mud
CP- Core Point
CMF- Down Hole Motor Failed
DP- Drill Plug
DSF- Drill String Failure
DST- Drill Stem Test
FM- Formation Change
HP- Hole Problems

HR- Hrs. on Bit
LIH- Left in Hole
LOG- Run Logs
PP- Pressure Pump
PR-Penetration Rate
RIG-Rig Repair
TD- Total Case Depth
TW- Twist off
TQ- Torque
WC- Weather Conditions

Bit Dull Grading System

Importance of maintaining a good bit dull grading system.

- Bit Performance Data Base
- Value Added Service
- Recommended Bit Selection
- Information for Design Engineering or manufacturers

Questions?

Header

Text

THE TOTAL SOURCE

educational sessions

ENHANCED DRILL BIT LIFE

Improved housing design featuring an alternative bolt pattern and geometry

Diego N. DELL'ERBA, PhD

PREMIER DRILL PIPE LTD.
A Tejas Tubular Products Company

CURRENT BIT and ADAPTOR PROBLEMS/FAILURES

1. Bit loose due to bolt break (shear failure of bolts)

2. Bit crack:
   - Starting between bolt holes
   - Starting at bend and progressing at 45 degrees toward bit tip

3. Dirt accumulates on bolts' holes

CURRENT DESIGN

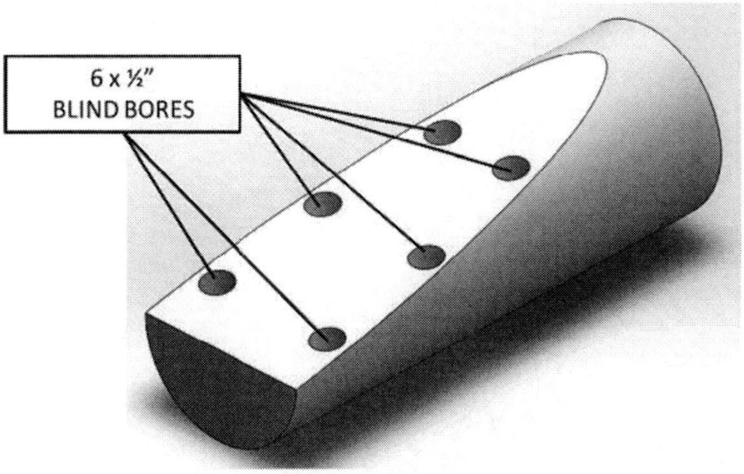

CURRENT DESIGN

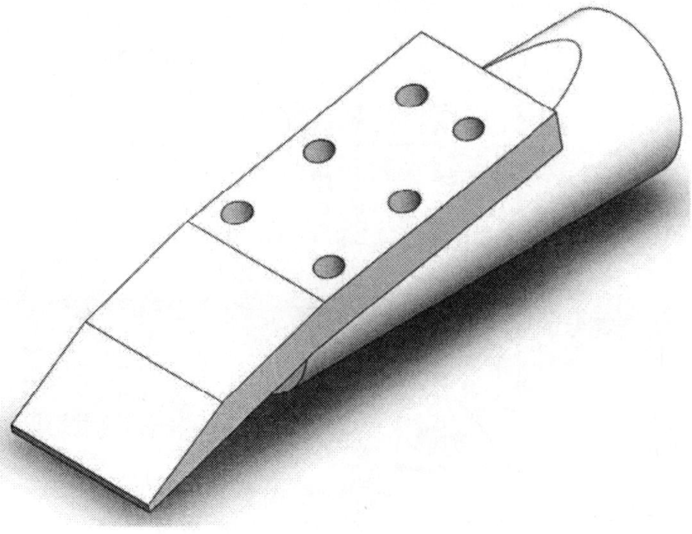

CURRENT DESIGN

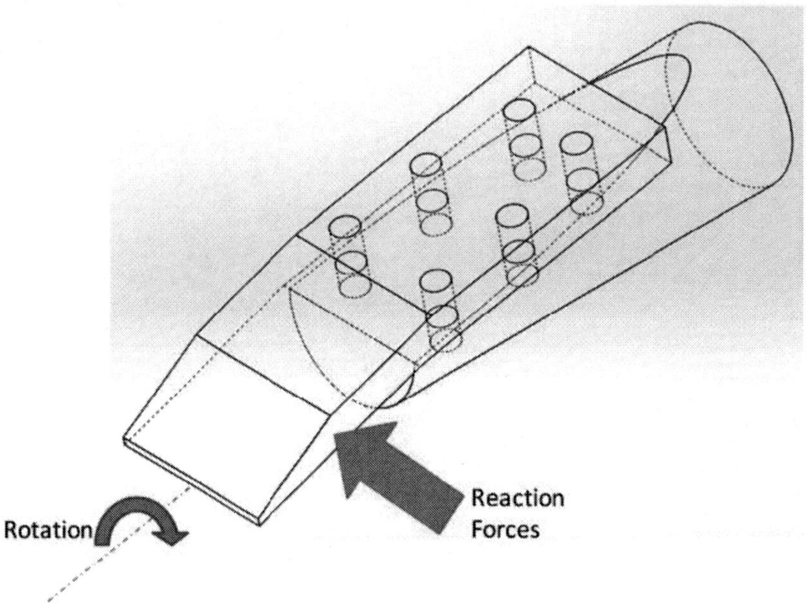

CURRENT DESIGN

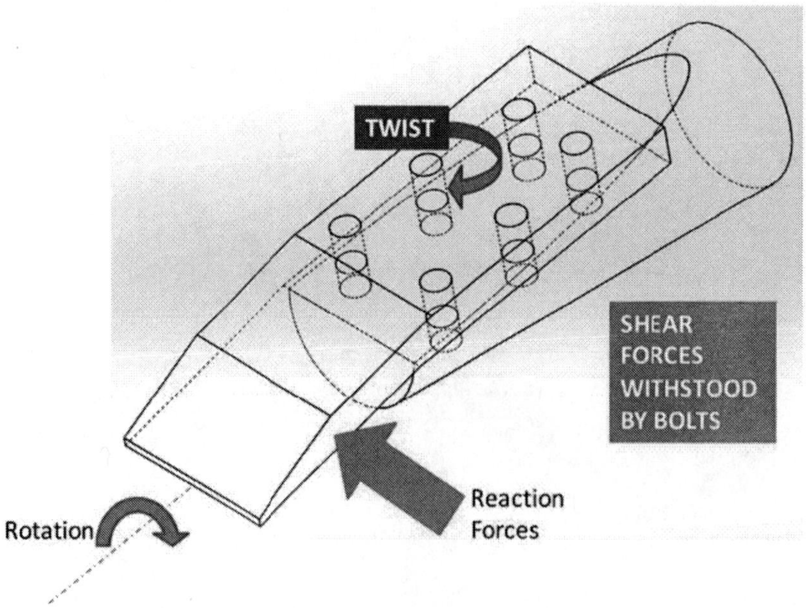

IMPROVED DESIGN

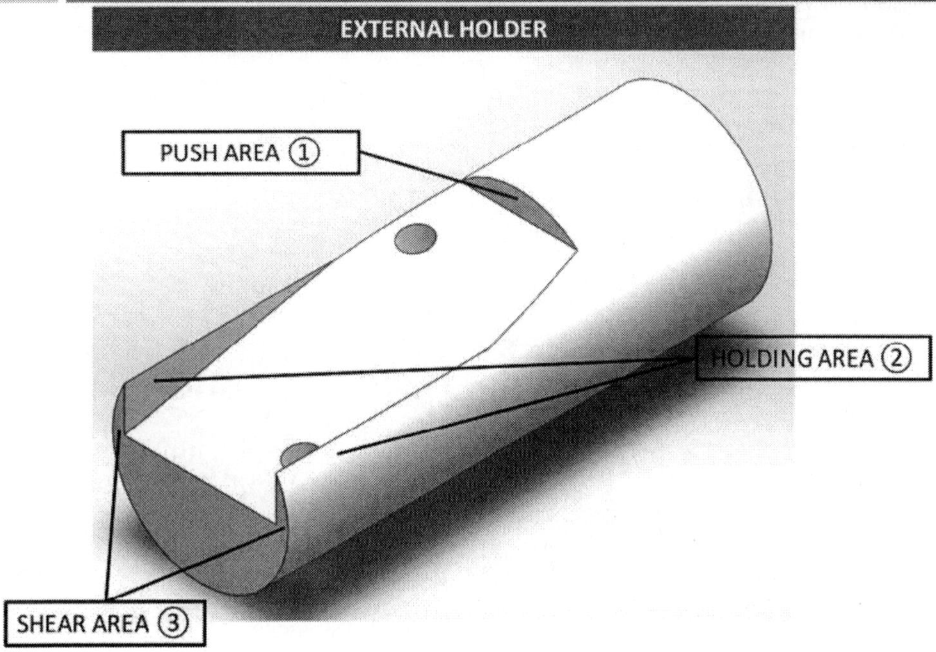

IMPROVED DESIGN

PUSH AREA ①
Provides the reaction force for pushing the bit (provided by bolts on current design)

HOLDING AREA ②
Holds the bit in place.
Provides the reactive force pair that oppose to the pair on bit due to drilling reaction (provided by bolts on current design)

SHEAR AREA ③
Provides the reactive force pair that oppose to the pair on bit due to drilling reaction (provided by bolts on current design)

IMPROVED DESIGN

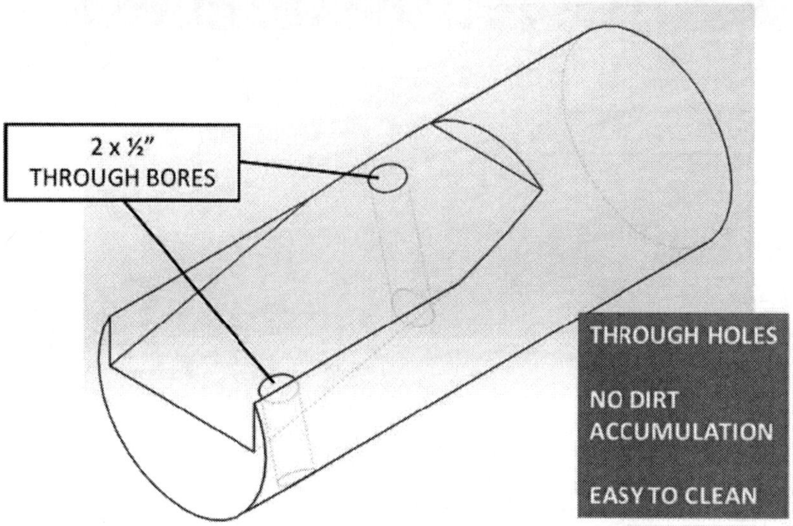

IMPROVED DESIGN

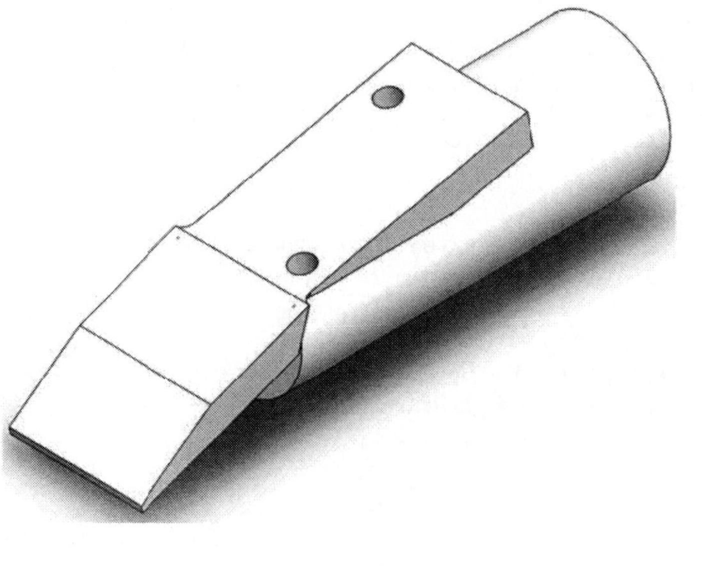

IMPROVED DESIGN

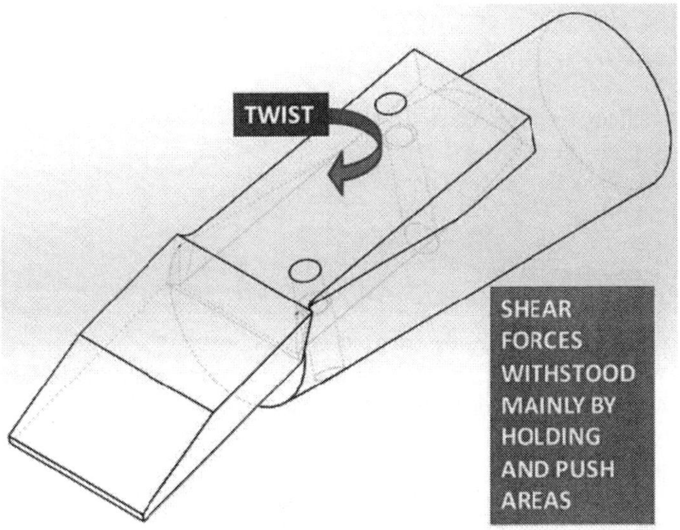

IMPROVED DESIGN

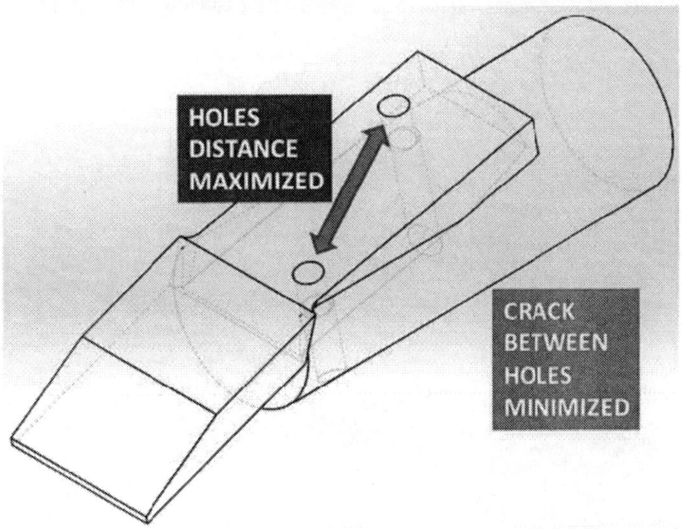

IMPROVED DESIGN - ADVANTAGES

OPERATIVE

Less bolts to be fitted on field
Perfect drill bit alignment
Easy cleaning due to thru holes

PERFORMANCE

Shear forces acting mainly on adaptor, NOT on BOLTS.
Maximized bolt distance as 2 bolts instead of 6 are needed in this new design
New PUSH area for pushing bit during drilling
Increased shear area

IMPROVED DESIGN

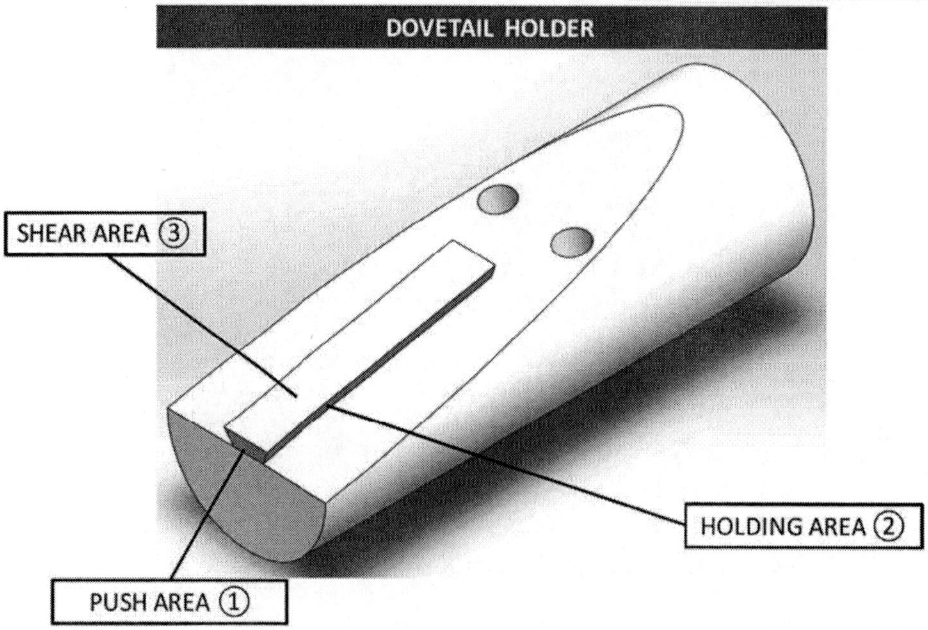

IMPROVED DESIGN

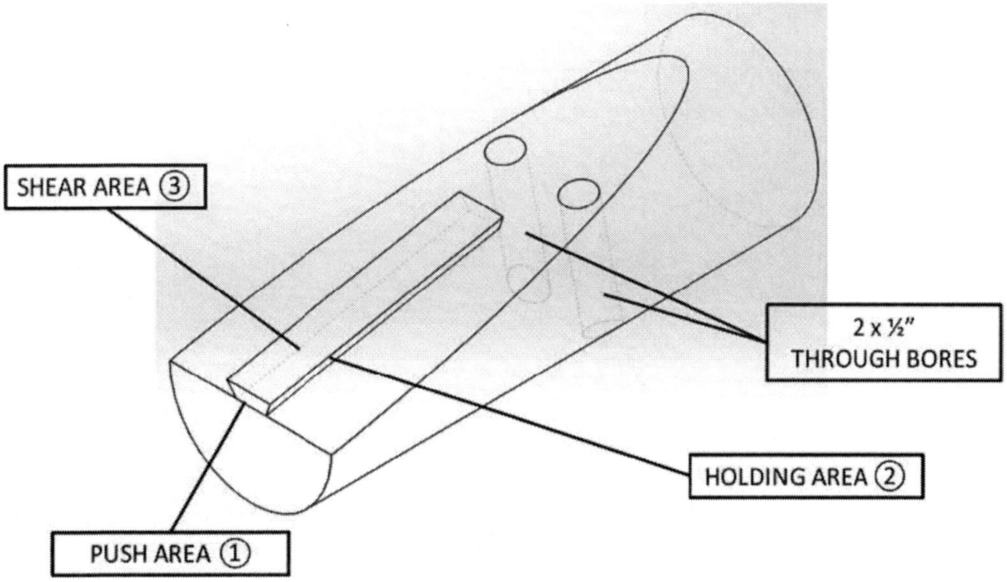

IMPROVED DESIGN

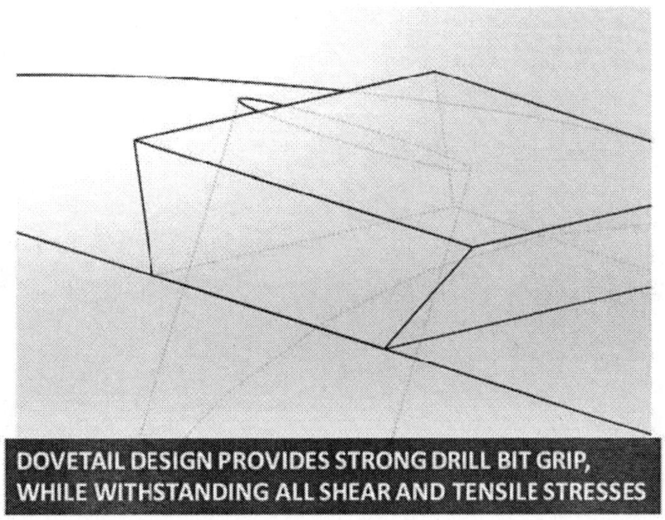

IMPROVED DESIGN

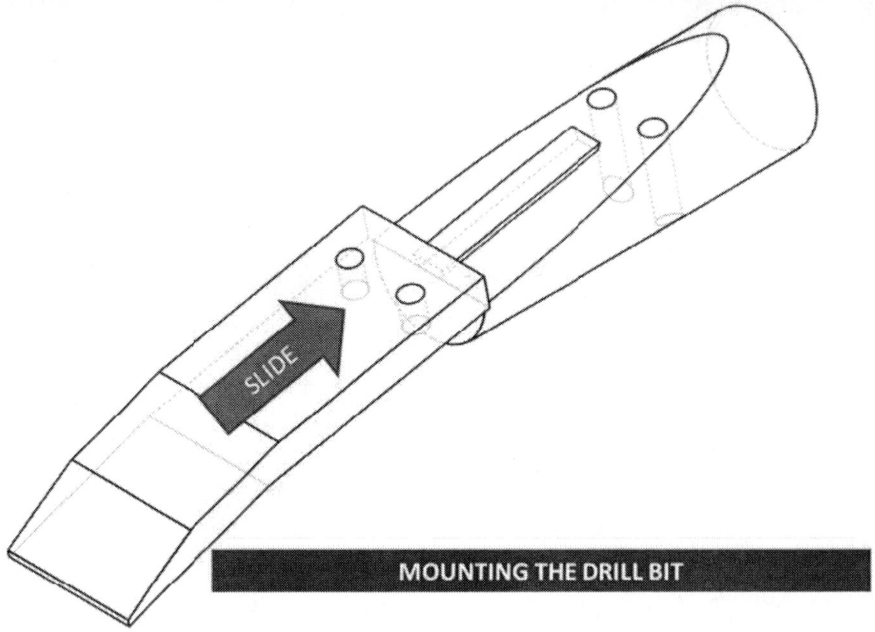

IMPROVED DESIGN

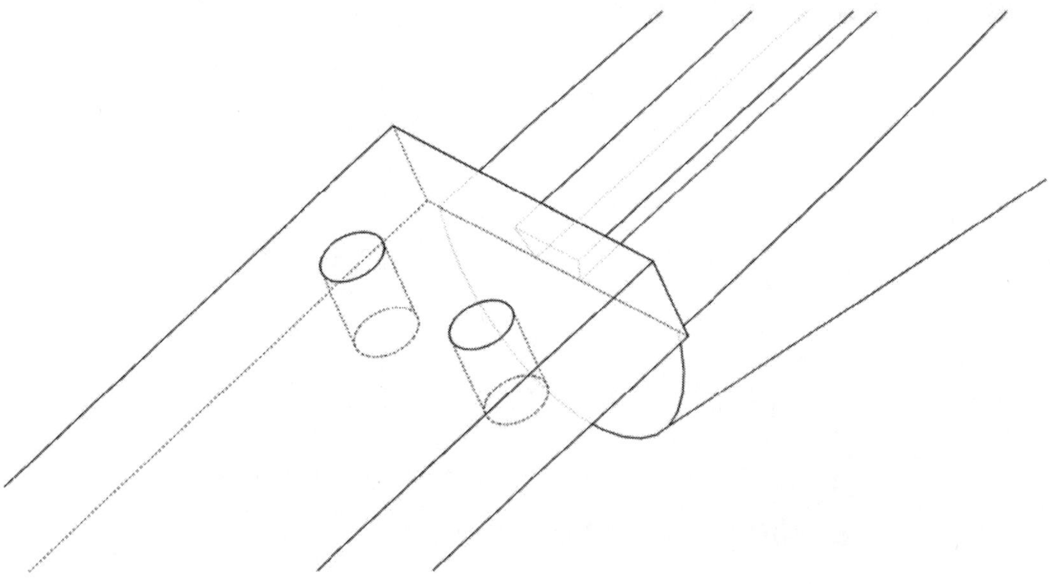

IMPROVED DESIGN

WEDGE-DOVETAIL HOLDER

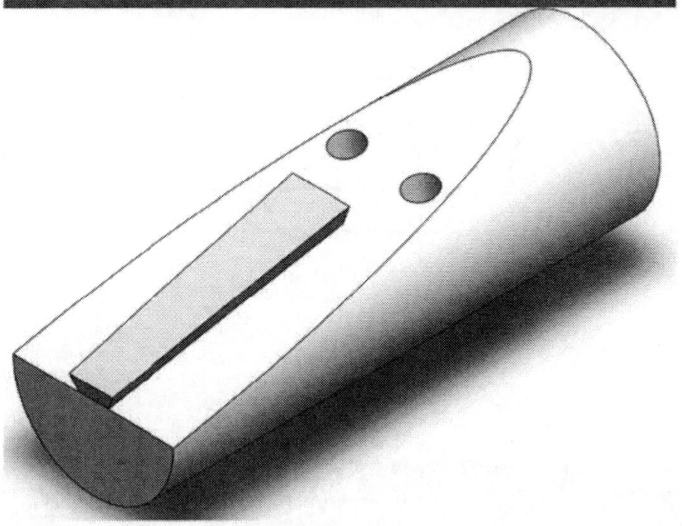

IMPROVED DESIGN

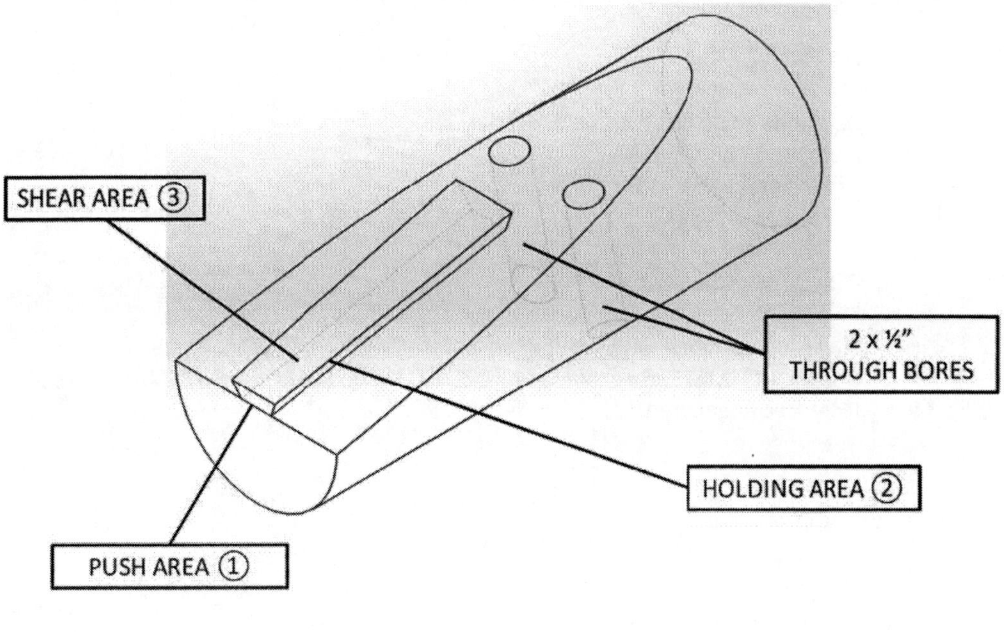

IMPROVED DESIGN

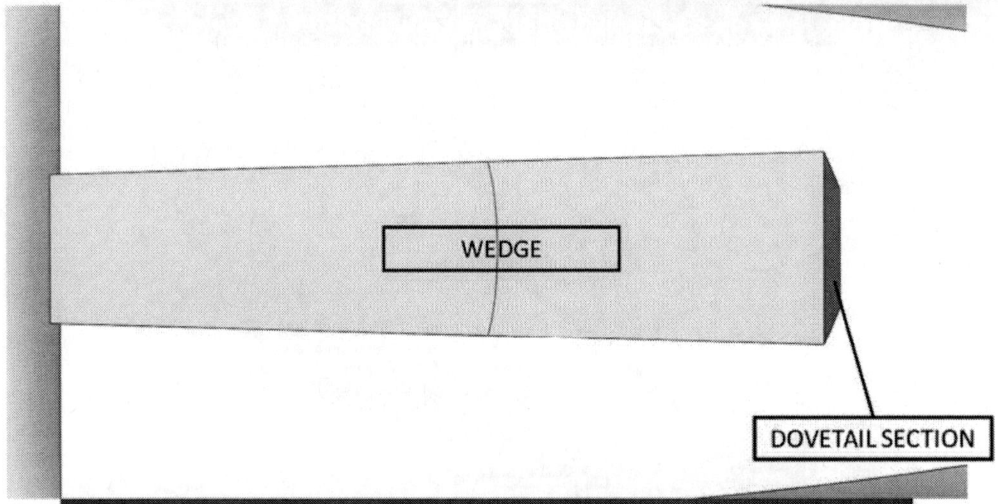

WEDGE-DOVETAIL DESIGN PROVIDES THE ULTIMATE GRIP FOR DRILL BIT, WHILE WITHSTANDING ALL SHEAR AND TENSILE STRESSES

IMPROVED DESIGN - ADVANTAGES

OPERATIVE

Less bolts to be fitted on field
Perfect drill bit alignment
Easy cleaning due to thru holes

PERFORMANCE

Shear forces acting EXCLUSIVELY on adaptor, NOT on BOLTS.
Drill bit ULTIMATE grip achieved by DOVETAIL and WEDGE-DOVETAIL holders
Increased shear area

THE TOTAL SOURCE

educational sessions

THANK YOU

Diego N. DELL'ERBA, PhD

PREMIER DRILL PIPE LTD.
A Tejas Tubular Products Company

THE TOTAL SOURCE
educational sessions

Compact, Powerful JT9

Gregg Wilder
Ditch Witch®
Product Sales Specialist III

Ditch Witch® JT9 Tier 4

JT9 Tier 4

- 66hp Deutz engine
- 16 gpm mud pump

Ditch Witch® JT9 Tier 4

Simple and Reliable

- High on board capacity of drill rod (Up to 350 feet)
- Compact HDD now consistently capable of 8" bores
  - Horsepower to provide both torque and mud volumes
- Direct hydraulic control
- No daily grease zerks

Ditch Witch® MR90 Recycling System

- 340 gal dirty tank
  110 gal clean tank
  300 gal fresh water
  Dry spoils hopper for onboard storage
- Can be pulled by one ton truck empty or full
- Can reduce the dumping of mud vac by 90% and vac costs by 60%
  - Based on 23 GPM for drilling and reaming on a single day
- Ideal for 5 GPM to 70 GPM jobsites

Ditch Witch® MR90 Recycling System

- Saves water and reduces transportation costs
- Simple hydraulic controls and pumps
- Reduces chemical costs & saves downtime
- Suitable for most soil conditions with proper mix

InSite™ Productivity Tools

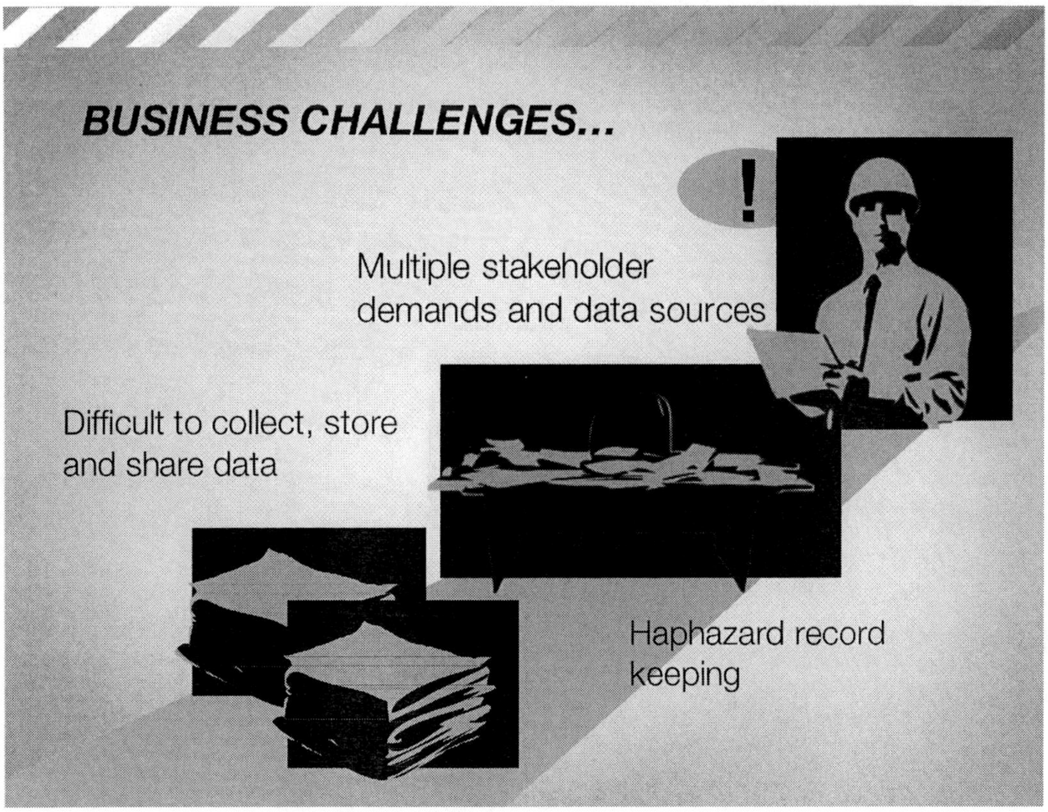
BUSINESS CHALLENGES...
!
Multiple stakeholder demands and data sources
Difficult to collect, store and share data
Haphazard record keeping

BUSINESS CHALLENGES...

The Impact

- Time wasted finding information
- Lost opportunities to improve crew performance
- Less time pursuing additional work

BUSINESS NEED...SET OF TOOLS

Set of tools to help contractors better manage projects from start to finish.

- Planning
- Mapping
- Producing
- Invoicing

InSite™ Productivity Tools

- Enables informed decisions at every job stage
- Eliminates mess and disorganization of manual paper processes
- Streamlines and helps job processes
- More time to pursue additional work

InSite™ Fleet

- InSite Fleet is a product that provides fleet/operations managers with GPS location to better locate and route equipment and crews to the jobsite
- Captures planned maintenance schedules and completed services
- Records both machine idle times and fuel consumed

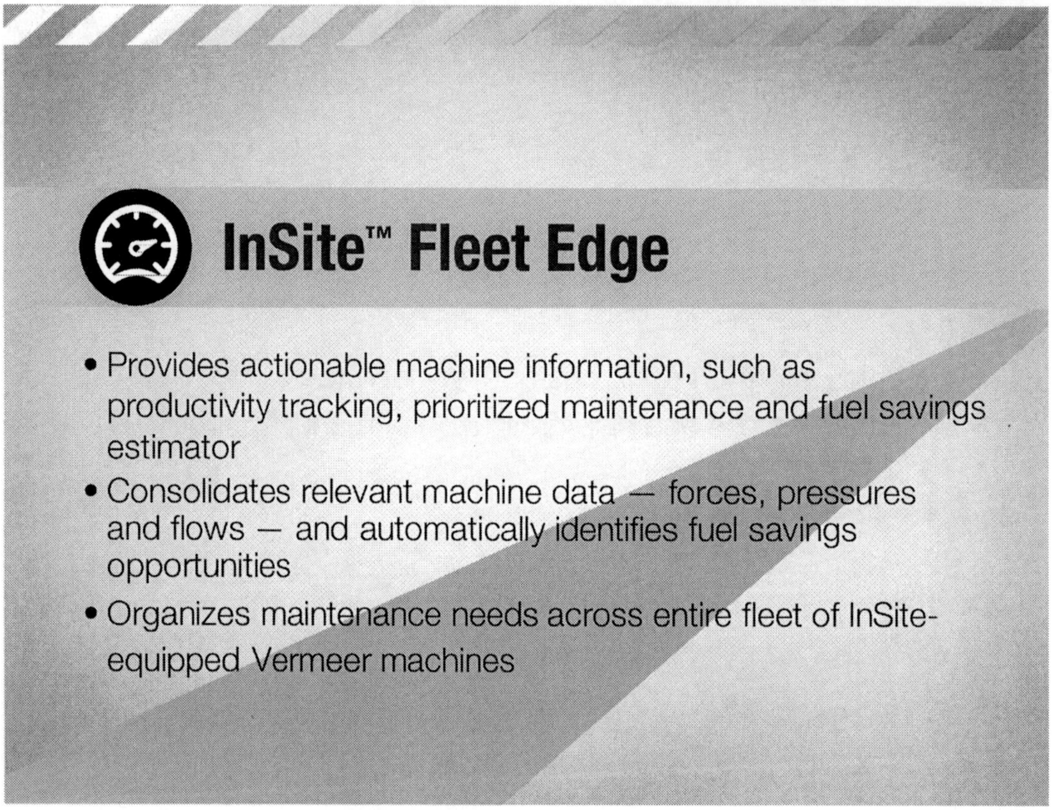
InSite™ Fleet Edge
• Provides actionable machine information, such as productivity tracking, prioritized maintenance and fuel savings estimator
• Consolidates relevant machine data — forces, pressures and flows — and automatically identifies fuel savings opportunities
• Organizes maintenance needs across entire fleet of InSite-equipped Vermeer machines

InSite™ Projects
• Planning tool maps jobsite and plans a bore
• Software application deploys easily to the field on a computer, tablet or phone
• Creates one repository for all information related to a project
• Offers ability to export a professional report on all aspects of the project

InSite™ BoreAid® Design Tool

- Takes user through the complete design and planning process of a bore in a sequential manner
- Considers soil conditions, product type, machine and other parameters
- Creates an electronic record to share with project owner or other stakeholders

InSite™ Mapping

- GPS-enabled function records white line and landmarks
- GPS-driven locator is used to white line a project site, mapping where a proposed excavation or bore will take place
- Uses GPS for better pinpointing of existing and proposed utility lines
- Records data into InSite Projects for planning purposes

InSite™ BoreAssist

- Load and review bore plan on the DCI® Aurora™ display
- Design simple bore plan on-rig
- Enables user to log drill data with a push of a button and compare results to plan

WORKING BETTER...

InSite™ Productivity Tools

Operations and Fleet Manager

InSite Fleet, InSite Projects and InSite BoreAid Design Tool (office function)

- Improves productivity and efficiency from start to finish
- Provides access to equipment data, configures jobsite plan, creates a job estimate, and later, expedites invoicing
- Consolidates data from a variety of tools

WORKING BETTER...
InSite™ Productivity Tools
Operations and Fleet Manager
Crew
InSite Mapping (jobsite function)
• Documents the position of in-process and existing utilities
• Helps build a more precise job plan

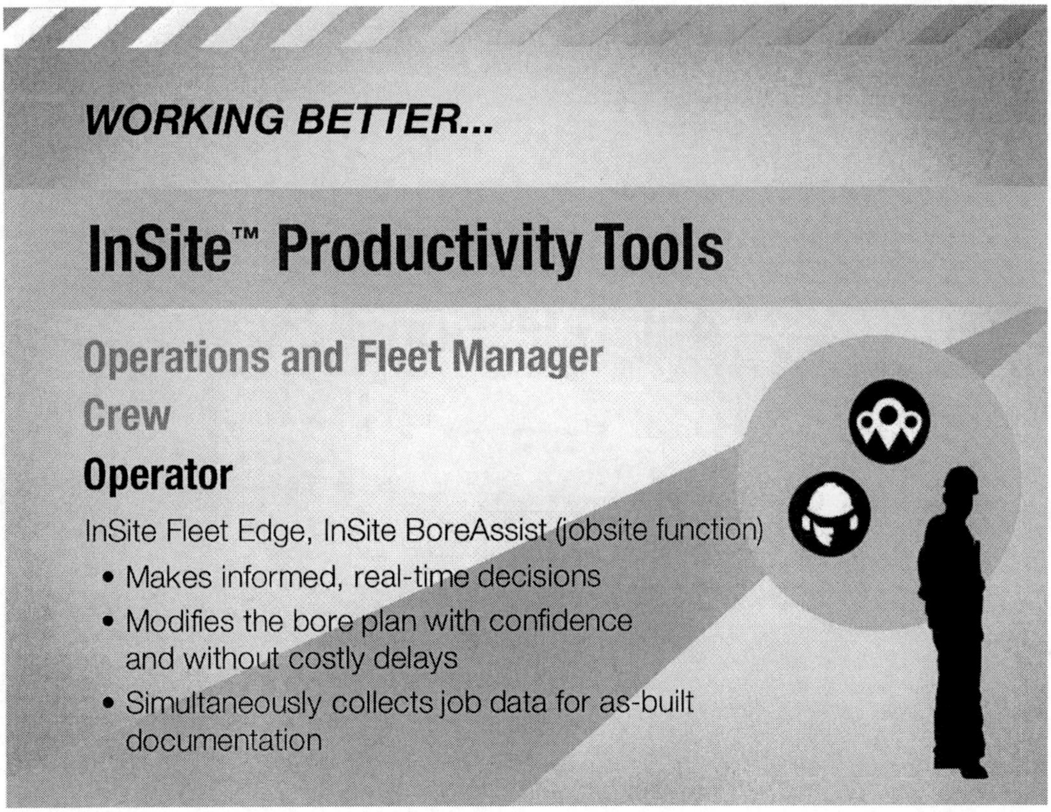
WORKING BETTER...
InSite™ Productivity Tools
Operations and Fleet Manager
Crew
Operator
InSite Fleet Edge, InSite BoreAssist (jobsite function)
• Makes informed, real-time decisions
• Modifies the bore plan with confidence and without costly delays
• Simultaneously collects job data for as-built documentation

WORKING TOGETHER. WORKING BETTER.
InSite™ Productivity Tools

Any Questions?
Vermeer

THE TOTAL SOURCE
educational sessions

PIPE JACKING DESIGN SOLUTIONS IN SIOUX CITY, IA

Patrick Minger, Minger Construction, Inc.
Robin Lorenzen, Akkerman

Minger Construction Co., Inc.

AKKERMAN
DRIVEN FOR CUSTOMER SUCCESS

OUTLINE

- INTRODUCTION
- PROJECT CHALLENGES
- PROJECT MAPS
- TRENCHLESS DESIGN CONSIDERATIONS
- PIPE JACKING CONSTRCUTION
- PROJECT SUCCESSES

INTRODUCTION

- Project Name: The I-29 Segment 2 Utility Replacement Project
- Location: Sioux City, IA
- General Contractor: Minger Construction, Inc.
- Engineer: HR Green, Inc.
- Purpose: To enhance the City's southern corridor near the Missouri River and Downtown Sioux City, improve safety, traffic operations and surrounding infrastructure.
- Timeline: October 2014 – October 2015

INTRODUCTION

- Located along Missouri River in Northwest, IA within Iowa Department of Transportation (DOT) right-of-way
- Collection system and treatment plant serves population of 150,000 along 69 square miles
- Existing collection system built in 1960's
- Three segments for renovation, Segment 2 is largest and includes the downtown corridor
- Segment 2 project costs: $14,497,818

PROJECT CHALLENGES

- Staging/scheduling
- Contaminated soils
- Buried utilities and debris
- Traffic control
- Temporary by-pass piping
- Geological conditions: clay, sand, fill material, rubble, high levels of groundwater

PROJECT MAPS: UPSTREAM

PROJECT MAPS: DOWNSTREAM

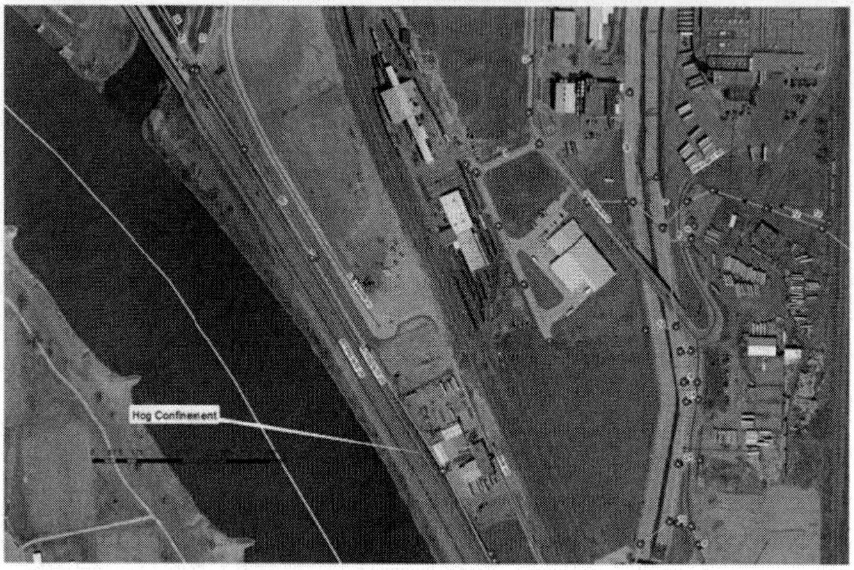

TRENCHLESS DESIGN CONSIDERATIONS

- Continuous access to I-29
- Deep sewers
- Minimal room for spoil during construction
- Tight staging limited/easements/restrictions
- Existing buried utilities
- Environmental review for crossing of river and railroad spurs
- Soil conditions including contaminated, high water content, unsuitable backfill

PIPE JACKING CONSTRUCTION

- 3,621-lf of 12-72-in VCP, RCP and Steel Casing installed with two Akkerman Guided Boring Machine (GBM) systems and an Akkerman Tunnel Boring Machine (TBM) system
- One 480-lf alignment using 72-in steel casing with a 54-in RCP carrier under a hog confinement building combined the use of GBM and TBM pipe jacking systems
- Range of Depth: 18-31-ft
- Range of Slope: 0.08-0.36%
- Additional trenchless work: Interceptor and inverted siphon replacement, completed by Gabes Construction

PIPE JACKING CONSTRUCTION

Bore #	Pipe	Length	Equip.	Start	End	Depth	Grade
11	12-in VCP	291-lf	GBM	4/6/15	4/15/15	30-ft	0.22%
12	12-in VCP	151-lf	GBM	4/16/15	4/22/15	18-ft	0.36%
1	24-in VCP	445-lf	GBM	10/23/14	11/6/14	18-ft	0.08%
2	24-in VCP	305-lf	GBM	11/11/14	11/21/14	22-ft	0.08%
3	24-in VCP	268-lf	GBM	11/24/14	12/5/14	24-ft	0.08%
4	24-in VCP	280-lf	GBM	12/12/14	12/19/14	24-ft	0.08%
5	24-in VCP	177-lf	GBM	1/7/15	1/14/15	19-ft	0.08%
6	36-in VCP	367-lf	GBM	1/19/15	1/27/15	22.5-ft	0.25%
10	54-in RCP	192-lf	TBM	4/5/15	4/10/15	30-ft	0.09%
7	54-in RCP	299-lf	TBM	2/9/15	2/22/15	30-ft	0.09%
8	54-in RCP/72-in SC	480-lf	GBM/TBM	3/1/15	3/13/15	30-ft	0.09%
9	54-in RCP	365-lf	TBM	3/23/15	3/28/15	31-ft	0.09%

PIPE JACKING CONSTRUCTION: GBM EQUIPMENT

- GBM 4812 & GBM 4815 Jacking Frames
- 200-265 ton jacking capacity, 100 tons pullback capacity, 20,000-26,000 ft-lbs rotational torque
- Powered Cutter Head (PCH) 28.5 & 36 – 10,500 and 20,000 ft-lbs of torque and two independent drive systems
- Guidance System
- P150Q and P250D Power Packs

PIPE JACKING CONSTRUCTION: TBM EQUIPMENT

- Series II TBM 600, 64,000-77,000 ft-lbs at 3,000 psi; cutting speed up to 11.9 rpm @120 gpm, closed-face cutter head
- Custom 800 ton jacking frame and thrust block
- 5000 Series Pump Unit

PIPE JACKING CONSTRUCTION: GBM EQUIPMENT

Step 1:

First, pilot tubes, with a soil appropriate steering head mounted on the front, establish line and grade along the bore path by using the guidance system. The dual-walled pilot tubes allow for lubrication from the steering head along the alignment. As pilot tubes are added, the steering head displaces soil and the operator makes steering corrections as necessary.

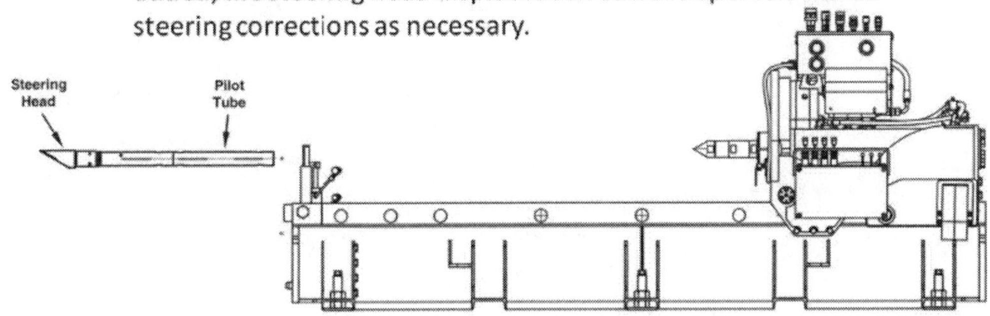

PIPE JACKING CONSTRUCTION: GBM EQUIPMENT

Step 1:

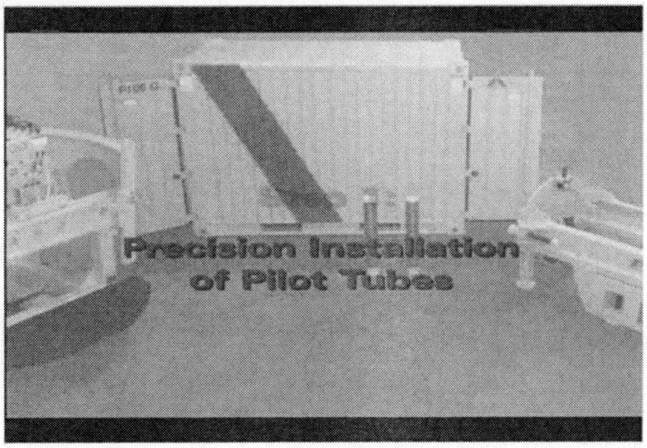

PIPE JACKING CONSTRUCTION: GBM EQUIPMENT

Step 2:

The jacking frame advances to install the pilot tubes across the full length of the drive. Casings and augers are connected to the last pilot tube using a reaming head. With the addition of each section of casing and auger in the launch shaft, a section of pilot tube is removed from the reception shaft.

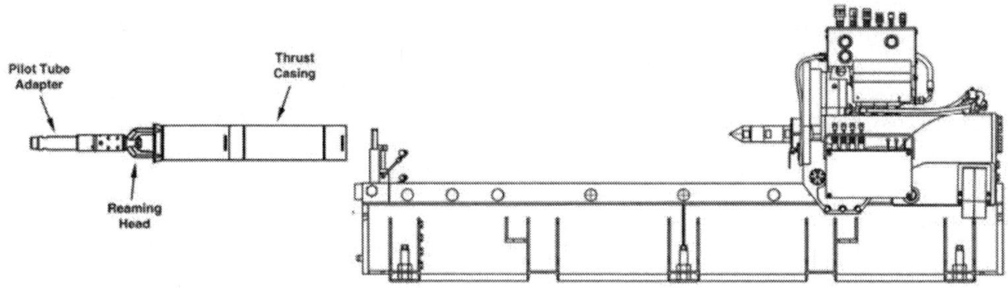

PIPE JACKING CONSTRUCTION: GBM EQUIPMENT

Step 2:

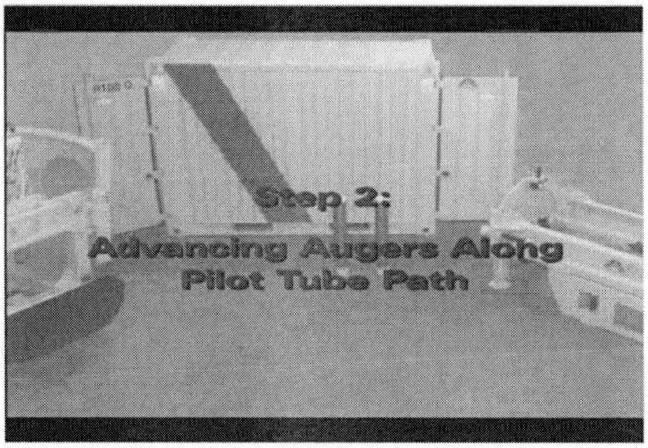

PIPE JACKING CONSTRUCTION: GBM EQUIPMENT

Step 3:

The final step is to connect the last casing and auger section to the PCH, and the PCH to the first section of final product pipe. The PCH tooling increases the bore diameter to the product pipe's OD and conveys the soil to the reception shaft.

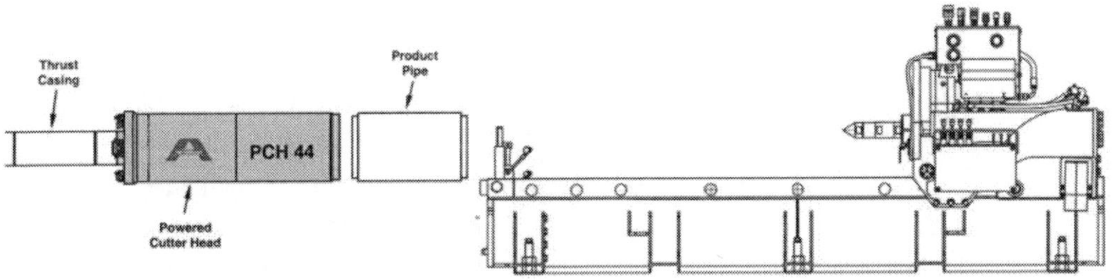

PIPE JACKING CONSTRUCTION: GBM EQUIPMENT

Step 3:

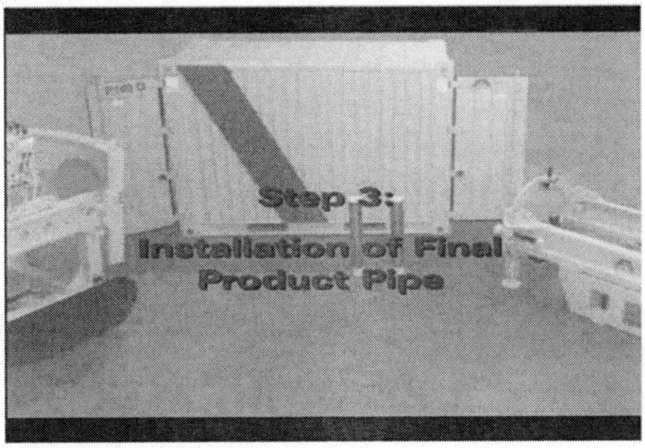

PIPE JACKING CONSTRUCTION : GBM

- The northwest region is comprised of commercial businesses, restaurants, car dealerships, hotels and embanked on the east edge by the Floyd River
- Contractor installed six runs of 24-in and one 36-in VCP drive along Dace Avenue and Floyd Boulevard, north of I-29 beginning in October 2014
- Temporary bypass sewer and storm sewer piping was required to avoid service disruptions
- 24-in alignments along the Dace Avenue connected to another project
- 36-in alignment along Floyd Boulevard connected to the existing sanitary sewer at the connection of I-29 and Leech Avenue
- All alignments in this area eventually connected to the new dual-siphon system

PIPE JACKING CONSTRUCTION: GBM

- The first 24-in VCP drive was originally designed as separate 269 and 176-lf and consolidated into one 465-lf reach at Minger Construction, Inc.'s recommendation
- Took place through unknown fill material in a sanitary landfill near Dace Avenue; bridge pilling and large boulders were discovered
- GBM installations were concluded with last two 12-in VCP runs that connected to siphon in April 2015

PIPE JACKING CONSTRUCTION: TBM

- Three 54-in RCP and one 72-in steel casing with a 54-in RCP carrier were installed beginning in February through April 2015 in the southwest region of the project zone, parallel to I-29, and the Missouri River
- The TBM system was used for a total of 1,348-lf and outfitted with an increase kit to match the diameter of the 72-in steel casing

PIPE JACKING CONSTRUCTION: GBM/TBM

- Most challenging crossing was under the hog processing facility in low blow count geology
- 480-lf of 72-in steel casing with a 54-in carrier RCP
- Originally designed for microtunneling and an 84-in casing
- Minger Construction, Inc. proposed solution to reduce casing diameter in order to re-use the onsite TBM with an increase kit, also provided contingency for TBM retrieval with an 84-in steel casing
- Two side-by-side pilot tubes passes installed three-feet apart, followed by 8-in casings which collected additional soil information, left in position 2-ft below anticipated steel casing to provide support for TBM in low blow counts
- Successfully completed in March 2015

PIPE JACKING CONSTRUCTION: PRODUCTION

- Best production with GBM 4815 system was 28, 2m sections or 180-ft, 24-in VCP in an 11-hour shift
- Best production with TBM system was 12, 8-ft section or 96-ft, of 54-in RCP in an 11-hour shift

PROJECT SUCCESSES

- Converted three open-cut to trenchless installations by combining drives
- Eliminated one manhole
- Eliminated need for contaminated soil removal and clean fill replacement
- Avoided disruption of billboard foundations and buried utilities
- Bid costs versus actual costs saved $535,355

THE TOTAL SOURCE

educational sessions

Questions?

Patrick Minger, Minger Construction, Inc.
patrickm@mingerconst.com

Robin Lorenzen, Akkerman
rlorenzen@akkerman.com

Minger Construction Co., Inc.

AKKERMAN
DRIVEN FOR CUSTOMER SUCCESS

THE TOTAL SOURCE

educational sessions

CIPP Rehabilitation of a 12-inch Ductile Iron Force Main in Palm Beach, FL

Andrew Costa, Business Development Manager

February 4, 2016

Insituform
an AEGION company

WHAT TO EXPECT

- **E-3 12-inch DIP Force Main Project – Town of Palm Beach**
  - Environmental concerns
  - Access/neighborhood
  - Job specifics and challenges
- **CIPP Pressure Liners**
  - Tube construction
  - Gravity vs. pressure

PRESSURE LINER CLASSIFICATIONS

Two Categories of Pressure Pipe Liners:

- **Interactive**
  Liner relies on radial support from host pipe to handle internal loads without failing

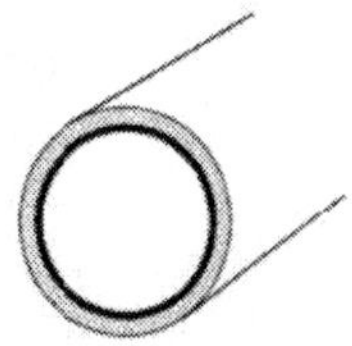

INTERACTIVE

- **Independent**
  Liner is designed for all applicable loads independent of the host pipe

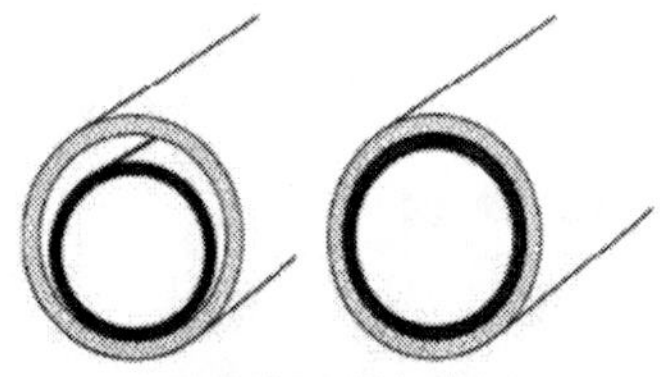

INDEPENDENT

Loose-fitting and close-fitting

AWWA LINER CLASSIFICATIONS

Industry Classification	Non-Structural	Semi-Structural		Fully Structural
AWWA Classification	**Class I**	**Class II**	**Class III**	**Class IV**
Trenchless Products	Cement Mortar Lining ——— Epoxy/ PU lining	CIPP ——— Modified Epoxy Lining	CIPP ——— Modified Sliplining	**CIPP** ——— Sliplining ——— Modified Sliplining ——— Polyester-Reinforced Polyethylene

CLASS IV LINER

Structurally <u>independent</u> of host pipe for internal loading

- **Addresses:**
  - Host pipe failure
  - Corrosion
  - Pin holes
  - Gaps
  - Joint leaks
- **Design life of 50+ years**

PRESSURE PIPELINE CONCERNS

Interior Corrosion and Sedimentation Buildup

- Poor water quality
- Decreased flow capacity

QUALITY/PERFORMANCE ISSUES

Structural Integrity

- Leakage
- Pinholes, pitting, thinning walls, etc.
- Severe deterioration

LIFESPAN ISSUES

TOWN OF PALM BEACH E-3 FORCE MAIN PROJECT

- 4,700 LF of 12-inc ductile iron force main
- Location: Bike Path corridor along Intracoastal
- High-end residential
- Operating pressure: 30 psi
- Lined with 7.5 mm composite CIPP liner
- Impregnated with vinyl ester resin
- Class IV fully-structural solution

TOWN OF PALM BEACH OVERVIEW

2003 "Best Place to Live"

9,000 Permanent Residents
20,000 Seasonal Residents

Highly Affluent Community

<u>Notables Residents:</u>

- **Howard Stern**
- **Donald Trump**
- **Jimmy Buffet**
- **Brad Pitt's Mom**

PROCESS OVERVIEW

- PIT CONSTRUCTION
- BYPASS / FLOW DIVERSION
- CLEANING / PIGGING
- CIPP LINING / CURING
- PRESSURE TESTING
- TERMINATION / CLOSURE
- BACKFILL / RESTORATION

CHALLENGES

- Access – Private Property
- Access – Proximity to Intracoastal
- Environmental Sensitivity
- Residents
- Water / Tides
- 8' Wide Bike Path
- Inaccurate As-Builts
- Utility Conflicts – Bends
- Access – Equipment

CHALLENGES

- Access – Private Property
- **Access – Proximity to Intracoastal**
- Environmental Sensitivity
- Residents
- Water / Tides
- 8' Wide Bike Path
- Inaccurate As-Builts
- Utility Conflicts – Bends
- Access – Equipment

CHALLENGES

- Access – Private Property
- Access – Proximity to Intracoastal
- **Environmental Sensitivity**
- Residents
- Water / Tides
- 8' Wide Bike Path
- Inaccurate As-Builts
- Utility Conflicts – Bends
- Access – Equipment

CHALLENGES

- Access – Private Property
- Access – Proximity to Intracoastal
- Environmental Sensitivity
- **Residents**
- Water / Tides
- 8' Wide Bike Path
- Inaccurate As-Builts
- Utility Conflicts – Bends
- Access – Equipment

CHALLENGES

- Access – Private Property
- Access – Proximity to Intracoastal
- Environmental Sensitivity
- Residents
- **Water / Tides**
- 8' Wide Bike Path
- Inaccurate As-Builts
- Utility Conflicts – Bends
- Access – Equipment

CHALLENGES

- Access – Private Property
- Access – Proximity to Intracoastal
- Environmental Sensitivity
- Residents
- Water / Tides
- **8' Wide Bike Path**
- Inaccurate As-Builts
- Utility Conflicts – Bends
- Access – Equipment

CHALLENGES

- Access – Private Property
- Access – Proximity to Intracoastal
- Environmental Sensitivity
- Residents
- Water / Tides
- 8' Wide Bike Path
- **Inaccurate As-Builts**
- **Utility Conflicts – Bends**
- Access – Equipment

CHALLENGES

- Access – Private Property
- Access – Proximity to Intracoastal
- Environmental Sensitivity
- Residents
- Water / Tides
- 8' Wide Bike Path
- Inaccurate As-Builts
- Utility Conflicts – Bends
- **Access – Equipment**

SOLUTIONS?......

MODIFIED EQUIPMENT

<u>Trailer Mounted Boiler</u>

- **Custom built for project**
- **Necessary for tight access**
- **Navigate width of path**
  - **Curves**
  - **Landscaping**
  - **Sprinklers**
  - **Homeowners**

SPECIALIZED EQUIPMENT

Torpedo Launcher CHIP Unit

(Controlled Head Inversion Process)

- Smaller footprint
- Provides increased hydrostatic head for inversion of specialized liner
- Increased safety factor
- Proprietary, patented process

Traditional inversion method uses tower to achieve head pressure for inversion.

CIPP LINERS – GRAVITY vs. PRESSURE

CIPP Liners are Comprised of:

1. Thermoset Resin
2. Felt Tube

Gravity Sewer Systems		Pressure Pipe Systems
• Primary Concern: Buckling from External Loading	• Primary Concerns: External Loading AND Internal (Tensile) Stresses	
• Polyester Resin / Standard Felt	• Vinyl Ester or Epoxy Resins	
• Felt has minimal contribution to cured physical properties	• Felt Tube Reinforced with Fibers	
Emphasis on Resin	Emphasis on Resin AND Tube	

REINFORCED TUBE

InsituMain® Reinforced Tube:

Greater resistance to hydrostatic loading and internal pressures (including surge)

Custom Engineered:

- Felt and fiberglass layers are project specific

Palm Beach E3 Project:

- 7.5 mm total liner thickness
- 2 layers of fiberglass – VE Resin
- Shot lengths 200 – 860 LF

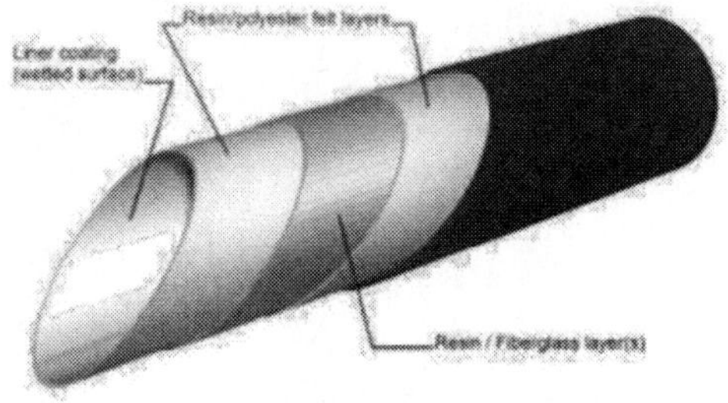

InsituMain® Glass Reinforced Composite Felt Liner

REINFORCED TUBE

In-house and third-party testing have independently concluded that standard felt tube with the use of epoxy or vinyl ester resin is insufficient for higher pressure applications.

Standard Felt (Gravity Systems)

VS.

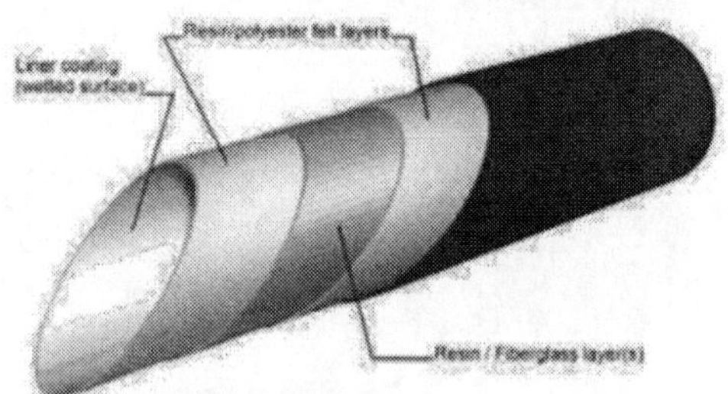

Glass Reinforced Composite Felt (Pressure Systems)

! The use of standard felt tube in pressure pipe applications above 10 psi is an outdated practice

PRESSURE TESTING

ASTM
F1216-09

> …recommended pressure and leakage test would be at twice the known working pressure or at the working pressure plus 50 psi, whichever is less.

30psi x 2 = 60psi

~~30psi + 50psi = 80psi~~

CLOSURE DETAILS

RELIANCE ON HOST PIPE IS DONE

CLOSURE DETAILS

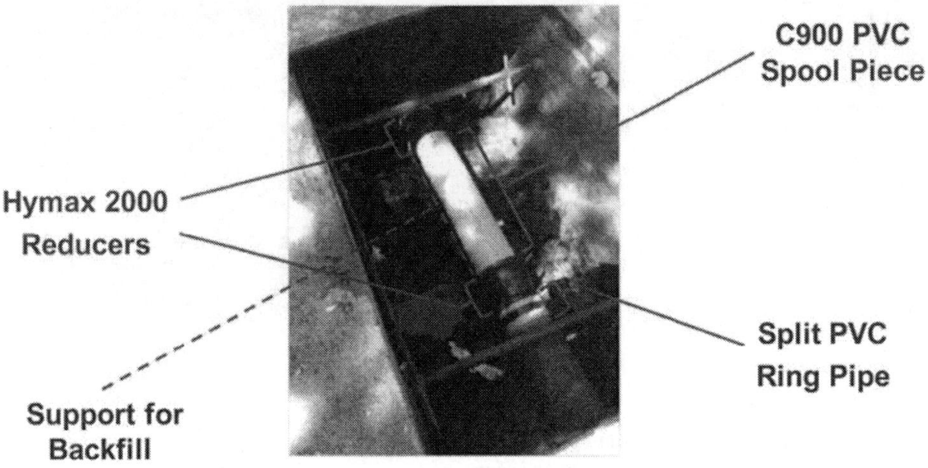

ADDITIONAL PALM BEACH FORCE MAIN PROJECTS

The E-3 Force Main Project – 2nd FM Project with the Town of Palm Beach

1st – E-6 Force Main

- Completed in Fall 2013
- ~5,000 LF of 12" DIP
- Composite Liner – 7.5mm

3rd – Royal Poinciana Force Main

- Completed in Fall 2015
- ~3,600 LF of 16" PCCP
- ~3,900 LF of 20" PCCP
- Composite Liner
  - 16" – 8.5 mm
  - 20" – 10.5 mm

16,000+ LF OF REHABILITATED FORCE MAIN

CIPP PRESSURE LINING CONSIDERATIONS

- Technical Envelope
  - 6-inch to 72-inch diameters
  - Operating pressures 10psi – 150psi
- Less invasive/smaller footprint than open-cut
- Cost-effective
- Efficient

QUESTIONS?

THANK YOU!

Contact: Andrew Costa

acosta@insituform.com
(813) 309-0385

www.insituform.com

THE TOTAL SOURCE
educational sessions

Fully Structural Renewal of 39-inch PCCP Water Transmission Main with Swagelining™ and HDPE

MPC Murphy Pipeline Contractors, Inc

GCWA
Gulf Coast Water Authority

Project Background

- Gulf Coast Water Authority and City of League City, Texas faced with need to renew critical 39-inch PCCP water transmission main along Calder Road
- PCCP main installed in 1971 which was susceptible to severe breaks due to internal wire corrosion
- Calder Road is scheduled for reconstruction and widening
- City of League City receives treated water from two sources
  - City of Houston's SE WTP to the north
  - Gulf Coast Water Authority's Thomas Mackey WTP to the south
- Full Structural Solution Required to meet:
  - All internal pressures (125 psi)
  - All external loading requirements
- No support from the host pipe

Design Considerations

- 39-inch PCCP provides critical interconnect between City of Houston WTP and Gulf Coast Water Authority WTP
- Pipe interconnects Thomas S. Mackey WTP, Texas City and SE Water Purification Plant, City of Houston
- Maintaining capacity and flow imperative
- Age (1971) and operating history (breaks and leaks) dictated a fully structural solution
- Thrust restraints at connections to lock in the existing pipe and prevent long-term longitudinal creep
- 1000 mm OD pipe (39.4”) inside a 38.5” ID host pipe

Design Phase & Selection Process

- Final ID Comparison between Swagelining & Slip Lining

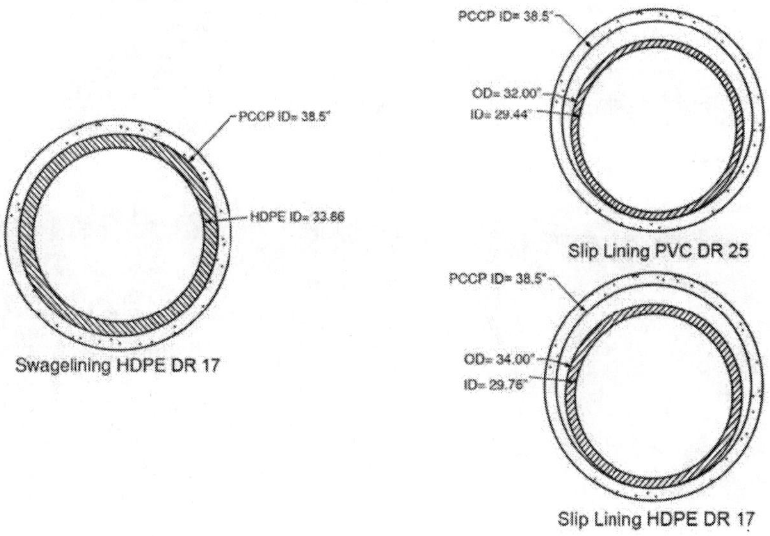

Design Phase & Selection Process

- Issued a Request for Competitive Sealed Proposal
- Received proposals from five contractors with alternatives
- 1000mm DR 17 HDPE 4710 was selected
- 1000mm (OD) pipe allowed maximum ID after installation
- Wall thickness was 2.32 inches
- Pipe design provided fully structural 125 psi operating pressure, 187 psi recurring surges and 250 psi occasional surges

Swagelining™ Process

SWAGELINING™
LIFE EXTENDING TECHNOLOGY

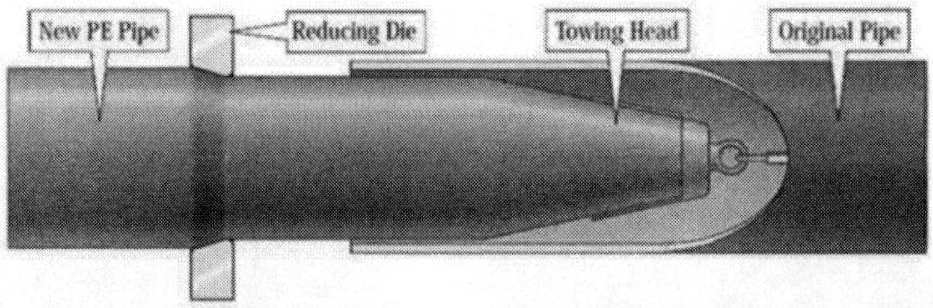

Swagelining™ Process

- History
  - Swagelining & Pipe Bursting developed by British Gas (1970's) and patented for gas distribution
  - Success led to potable water crossover
  - Swagelining™ over 28,200,000 feet installed in the past 32 years
- Today
  - Currently used to renew water, force main, gas, oil, mining, slurry, chemical, industrial, salt water and off-shore pipelines world wide

The Science behind Swagelining™

- Extensive physical testing program by British Gas
- Analysis of material behavior
- Customized Software Program
- Research & Development continued by Swagelining Limited
- Development Center Facility to support projects

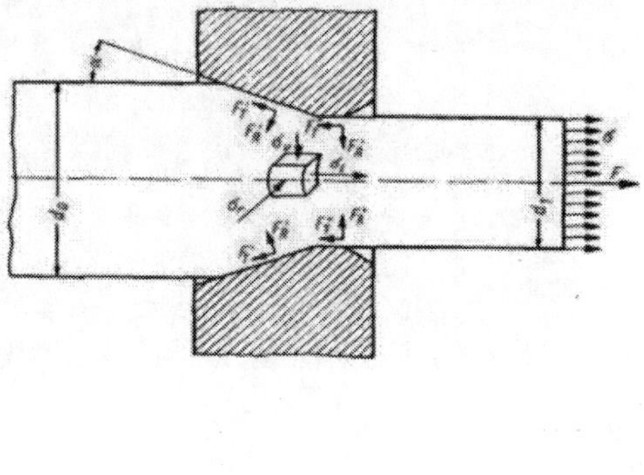

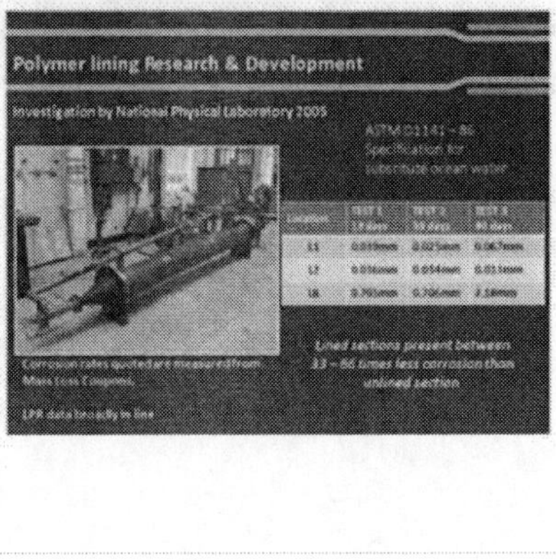

Construction Phase

- Project Details:
  - Renewal of 6,800 feet of existing 39" PCCP
  - Water Transmission Main
  - 12" by-pass
  - Four 36" butterfly valves
  - Three 36"x24" tees with gate valves
  - Two air release valves
  - Three blow-off valves
  - 39"x36" flange reducer

- 1000mm (39.37-inch) DR 17 HDPE PE 4710 selected through liner system design
- 4 pulls ranging from 1,250 feet to 2,100 feet in length
- Tight right-of-way with only one traffic lane shut down allowed

Swagelining™ Installation – Preparation of Host Pipe

- Camera/Man Entry inspection
- Cleaning
- Proving pig sent through host pipe to ensure free bore path

Swagelining™ Installation – Butt Fusion

- Sections of HDPE pipe are butt fused together
- External bead removed

Swagelining™ Installation – Shuttle Rods/Pull Back

Swagelining™ Installation – Reducing Die

- HDPE pipe has an OD slightly larger than the ID of the pipe to be renewed
- Pipe is pulled through reduction die which temporarily reduces its diameter from 39.37" down to 36.00"

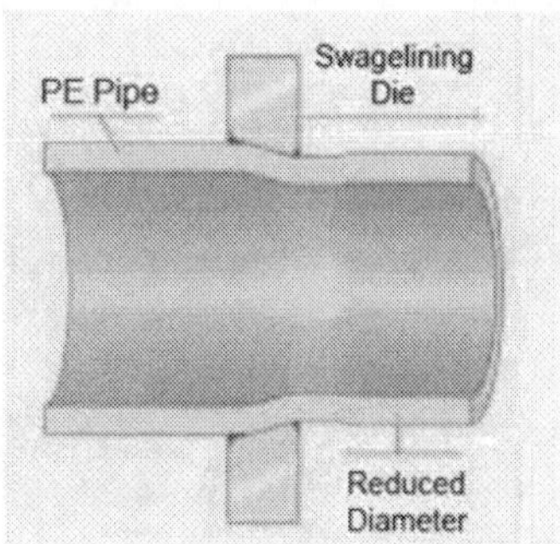

Swagelining™ Installation – HDPE enters receiving pit

- Pulling force removed
- Natural relaxation of HDPE
- 90% of reversion occurs in 2 hours
- Remaining reversion occurs overnight

Swagelining™ Process – Results

- No gaps - all annular space is eliminated
- Fully structural pipe
- Interactive pipe
- Thin walled lining

Swagelining™ Installation – Final Connections

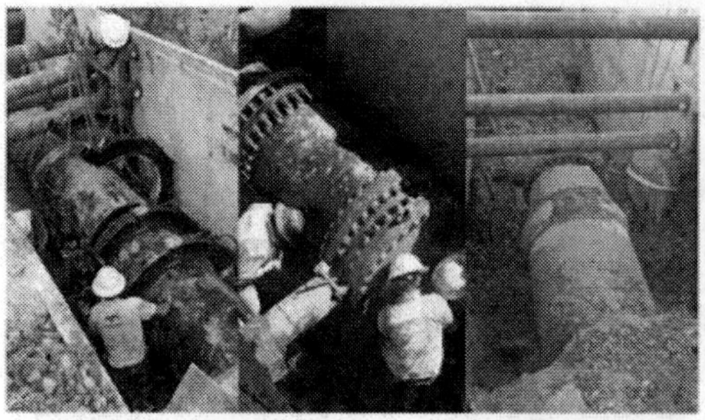

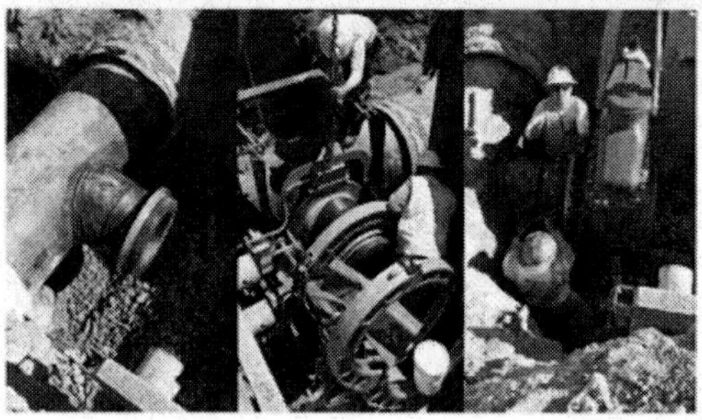

Challenges & Successes

- Installing fully structural HDPE (2.32-inch wall thickness through single swage die)
- Tight utility corridor
- Coordination of multiple agencies

- Technology meet all design parameters and increased flow and capacity
- Largest diameter, fully structural trenchless installation to date in North America
- Awarded 2014 Trenchless Project of the Year

THE TOTAL SOURCE

educational sessions

Fully Structural Renewal of 39-inch PCCP Water Transmission Main with Swagelining™ and HDPE

MPC Murphy Pipeline Contractors, Inc.

GCWA Gulf Coast Water Authority

Cured-In-Place Pressure Pipe Liner Design: Sanitary Force Mains and Water Mains

Norman E. "Ed" Kampbell, P.E.
Pipeline Rehabilitation Technologies Specialist, Rehabilitation Resource Solutions, LLC, 4862 Sarasota Ct., Hilliard, OH 43026-9096, ekampbell@sbcglobal.net

Introduction

At the 2014 UCT Conference this author presented a paper entitled "Design Considerations for Cured-In-Place Pressure Pipe Applications". Later that year at the No-Dig Conference Ian Moore presented a paper entitled "Design recommendations for performance limits for cured in place liner in cast iron pipe"; which built upon the No-Dig 2005 paper of Allouche and Moore entitled "Laboratory Examination of a Cured-In-Place Pipe Liner for Potable Water Distribution Pipes". These papers among many others have been demonstrating the need for a true design analysis that fits how close-fitting liners react when installed inside pressure pipes. Further, they also demonstrate that using the design equations in the often cited ASTM F-1216 design appendix can actually produce a wall thickness that is *under*-conservative. This paper summarizes the body of knowledge currently known so that those engineers who are designing CIPP for these applications (i.e. force mains, water transmission lines, etc.) have the references they need to design this new class of CIPP with enough understanding to properly err on the true side of conservatism.

Currently, an ASCE Task Group is finishing up their work on a new Manual of Practice for the Design of Close-Fit Flexible Liners for gravity pipe applications. At the heart of the new design methodology is the realization that existing dead loads and live loads acting on the host pipe at the time of lining will continue to be borne by the host pipe-soil structure interaction system after the lining has been installed. The only loads that come onto the liner for a gravity pipe application are the live loading and the external hydrostatic pressure exerted by the groundwater. There is no design condition requiring the soil load being borne by the liner (even a fractured one) because lining the pipe secures the state of equilibrium in the soil-structure interaction system required for lining to be a renewal solution (i.e. the condition state of fractured equilibrium). While the new Design MOP deals only with gravity pipes, all pressure pipe liner design must also look at the gravity pipe design condition to confirm that the lining installed will be structurally strong enough to accept the de-pressurized condition of the pipeline when undergoing maintenance or an internal negative pressure event (i.e. vacuum).

Pressure Lining Design Limit States

In the F1216 design appendix, the possibility of lining pressure pipe was addressed by the inclusion of equations X1.6 and X1.7 for the partially deteriorated (PD) and fully deteriorated (FD) design conditions, respectively. The equation X1.6 is based upon the liner spanning over a hole, or holes, in the host pipe and supporting the internal pressure in those gaps or potential gaps. Equation X1.7 treats the liner as the sole source of the resistance to the internal loads acting on the renewed pipe. In this equation is the term S_{TL} for the long-term tensile strength of the liner. The simplicity of employing these design equations is enticing while glossing over the very real need for additional testing to determine how the liner's performance is impacted under sustained versus cyclical pressure operating conditions, the condition assessment judgments that must be made regarding the host pipe's future degradation, surge pressure generation in the new "composite" structure, circumferential fracturing of the host pipe due to ground movement, design of the termination fittings for the liner, etc.

Figure 1 below is from the referenced 2005 paper by Allouche and Moore. Equation X1.6 (of F-1216) is represented by "e" and Equation X1.7 is represented by "a".

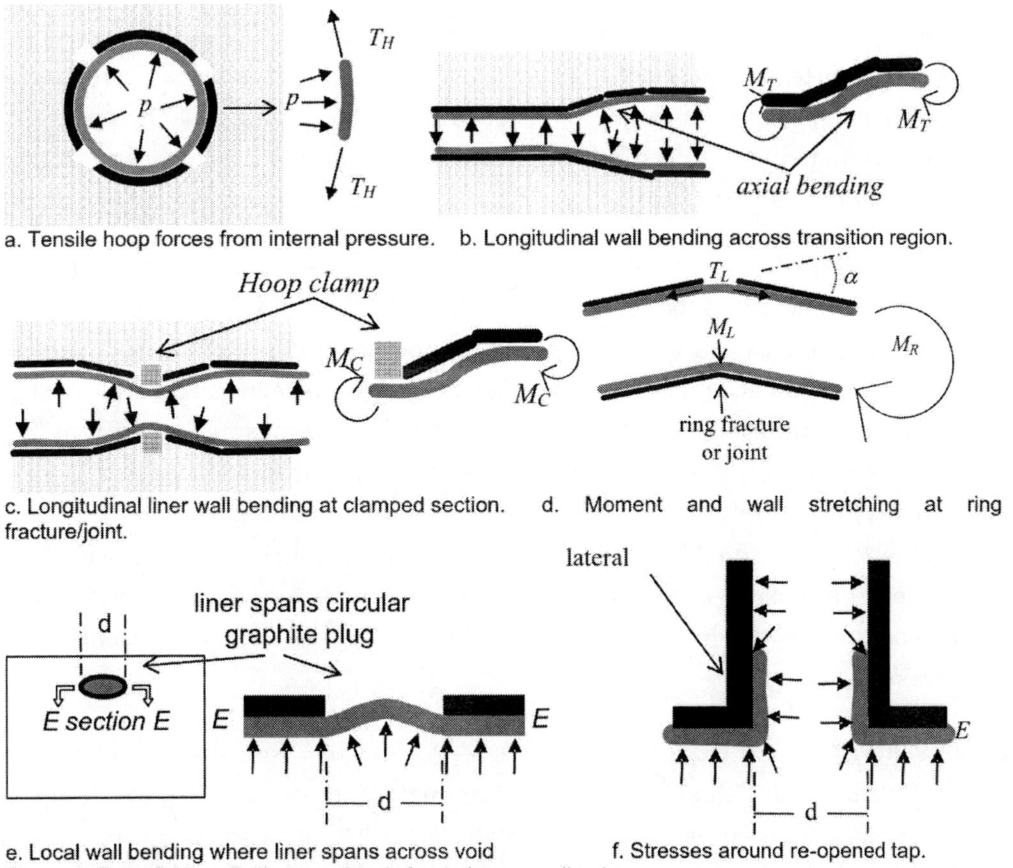

Figure 1. Potential liner limit states (black=host pipe, grey=liner).

Figure 1 illustrates six (6) potential design limit states. These limit states are defined as follows:

- LS1: Hoop tension induced by internal fluid pressure where the host pipe has fractured or split open, Figure 1a
- LS2: Axial bending in the liner wall where it spans from a section of host pipe that has fractured or split open, to one that is maintaining its integrity, Figure 1b

- LS3: Axial bending in the liner where it passes through a section fitted with a full circle clamp or similar restraining collar, Figure 1c
- LS4: Overall bending moment, M_G, or local bending moment, M_L, and stretching force T_L in the liner wall at ring fractures or joints, Figure 1d
- LS5: Local bending where the liner spans across circular "plugs" such as those formed by graphitization of cast iron pipe, Figure 1e
- LS6: Local stresses associated with the liner being in the vicinity of a service connection, Figure 1f

Allouche and Moore used Finite Element Analysis (FEA) analyzed to investigate bending in a 6-inch cast iron pipe for the condition states LS2 and LS3. Their analysis indicated that this longitudinal bending produces a small increase in circumferential tension relative to the tensile stress generated in the liner when it is in unconfined hoop tension (LS1), and an axial (longitudinal) bending stress about 20% higher than the hoop stress in unconfined tension. Neither the peak circumferential tension nor the peak axial tension was greatly influenced by the length of the restriction. Given the modest increases in tensile stress associated with bending across a restriction, Allouche and Moore stated that it appears to be reasonable to employ unconfined tension testing of the liner to characterize liner strength; relying on the factor of safety to support the added stress.

FEA was also used to investigate sections of liner spanning across circular perforations in the 6-inch cast iron pipe wall. That analysis indicated that perforations with a diameter of 0.39-inches and 0.0.79-inches generate tensile bending stresses less than the tension that occurs in this liner under unconfined hoop tension (LS1). However, with a perforation of 1.18-inches, the tensile bending stresses exceeded the unconfined hoop tension by around 30%; and a perforation of 1.57-inches generated tensile bending stresses of more than double (2X) the unconfined hoop tension. This lead Allouche and Moore to conclude that if the liner is being used in circumstances where perforations of 1.18-inches or greater are to be expected, the time to failure should be reduced relative to those obtained from testing of an unconfined liner under internal pressure. Further, if there are service connections having apertures of 1.0-inch or larger, any decision to leave the liner spanning the service connections that are no longer active may compromise liner service life.

Condition Assessment of the Host Pipe

While the works of Allouche and Moore have focused on metallic pressure pipes, the design engineer has been seeing an ever increasing number of non-metallic pipes (i.e. prestressed concrete cylinder pipe (PCCP) and PVC) that need to be renewed. Also, the size of the pipes being renewed is growing larger. In order to design a liner to perform in the candidate pipe the design engineer must first analyze the condition of the host pipe and the type(s) of failures being seen by the maintenance personnel.

The EPA report entitled Condition Assessment of Ferrous Water Transmission and Distribution Systems – State of Technology Review Report issued in 2009 placed metallic pipe failures into two categories; corrosion failures and mechanical failures. Although these causes are distinct, they are related in that mechanical failures may result from corrosion such that the pipe is no longer able to resist the applied internal and external forces acting it.

The criteria essential for corrosion by oxidation include

- Oxygen supply
- Moisture
- Soluble salts
- Cathodic and anodic sites

Conditions may vary widely along the length of a buried pipeline and can result in considerable variability found when investigating external corrosion. It is common to find a section of badly corroded pipe adjacent to a sound section. This situation is emphasized by the observation of corrosion "hot spots" in which accelerated corrosion can be focused on one location in the pipe exposed to conditions conducive to corrosion.

For iron pipes, graphitization is an important form of corrosion. Graphitization occurs where soil conditions including pH, dissolved salts, and organic content are favorable to anaerobic bacterial growth. For cast iron the metals goes into solution leaving behind a corrosion product that consists of a mass of residual graphite flakes interspersed with iron oxides. Over time the wall becomes a composite of gray cast iron and a graphite matrix with the corrosion product occupying the space previously occupied by the gray iron. It is difficult to identify graphitization during visual inspection because there is no apparent wall thickness loss. The graphitized material has some strength, although greatly reduced, which explains why graphitized gray cast iron pipes continue to function.

Often gray cast iron failures come from mechanical stress or a hydraulic shock (roadwork or ground movement) acting on the weakened structure. This is then classified as a mechanical failure rather than a corrosion failure.

General corrosion can occur in ductile iron in the form of graphitization, but the corrosion material that forms is weaker and more friable than that in gray iron. The graphite corrosion products are in the form of discrete nodules that are easily detached, and a structure is not retained.

Graphitization does not occur in steel pipes.

For all types of ferrous pipes, external corrosion in the form of isolated pitting, general corrosion, and graphitization is the most common form of deterioration, which can lead to leakage or failure.

Pitting corrosion is a concentration of corrosion in one particular area. The metal goes into solution preferentially at that spot rather than adjacent areas. Pitting corrosion is by far the most common deterioration mechanism for all ferrous pipes. It can occur internally, although it is more likely to be found externally; because of this, the design engineer must take advantage of the pipe's maintenance records and samples of removed pieces coming from any repair work that was done in the past. This first-hand knowledge of the piping's condition can be used to pinpoint where to employ more expensive condition assessment tools when warranted. As a group of pits in one area can weaken the wall sufficiently for the pipe to fail, utilities should maintain this database of information by cataloging forensic-type pictures and descriptions of the failure defects that prompted their maintenance actions.

Leakage through perforations in the pipe wall may lead to erosion of the surrounding soil and to structural failure. The interaction between a leaking pipe and its surrounding soil is complex. The relationship between head loss and flow is not linear, as a result of the interaction of the soil particles with the orifice (perforation), turbulent flow in the soil, changing geometry of the unconfined flow regime, hydraulic fracturing, and soil piping. Theoretical considerations suggest that small continuous leaks will drain away without a trace into granular soils. This cannot be expected of lower permeability clays and silts where hydraulic fracture is more likely, with leaks rapidly becoming visible as wet patches and water bursts at the ground surface.

Mechanical failures are more complex to investigate and interpret. The National Research Council of Canada (NRC) reported in 2003 that 23 out of 25 mechanical failures investigated were associated with corrosion pitting.

Exposed fractures and cracks indicate a mechanical failure. Six different mechanical failure modes have been identified for cast iron pipes (Makar, 2004). These modes are:

- **Circumferential failures**, typically found in small diameter pipes (<12-inches) and due to longitudinal stresses; pipe breaking in bending due to stresses from frost or swelling clay or by ground forces pulling the pipe apart (i.e. thermal contraction or longitudinal stresses near valves and fittings)

- **Longitudinal failures**, typically found in large diameter pipes (>20 inches) and caused by internal pressure which propagates either longitudinal splitting or cracking along its length; associated with water pressure surges, water hammer, and some type of weakening of the pipe wall...either through corrosion pitting or manufacturing defects.

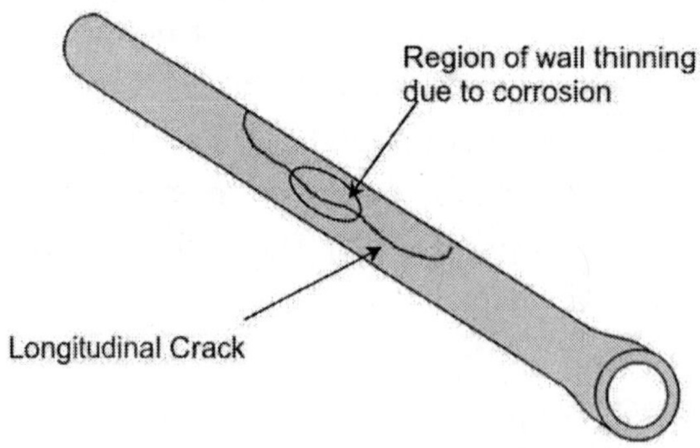

- **Wedge splitting failures**, typically found in large diameter grey cast iron pipes too thick to experience circumferential cracking when they experience high bending loads. The wedge of pipe splits off at the bell to relieve the bending stresses acting on the pipe.

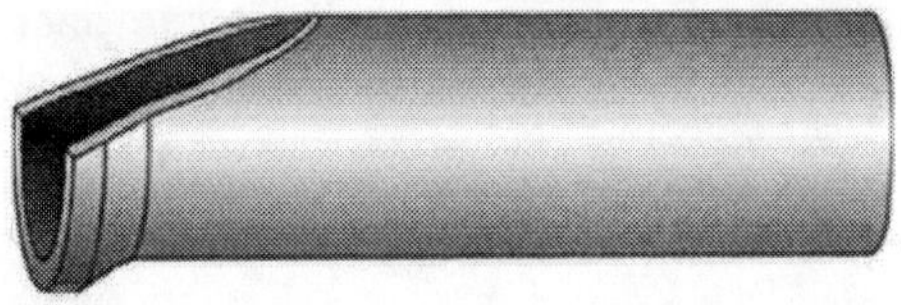

- **Spiral failures**, which have been found in medium diameters up to 20-inches typically start as a circumferential crack, but then a section of the pipe breaks and the crack propagates down the pipe barrel in a spiral fashion

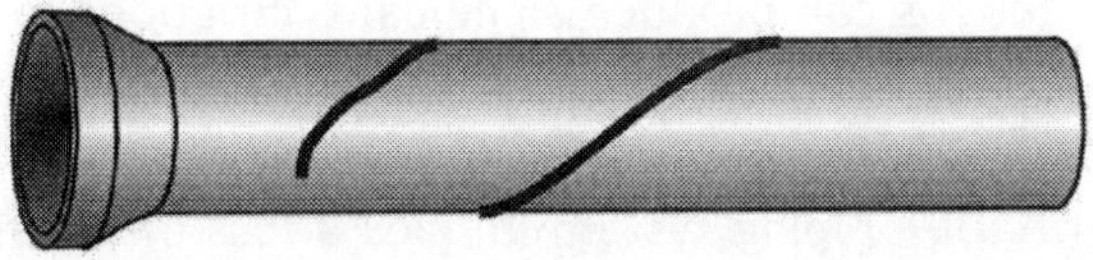

- **Bell splits,** normally occur in metallic pipes and take the form of a crack that starts at the bell of the pipe and travels a short way down its length before turning towards the pipe circumference and stopping. Normally this type of failure is associated with leadite-sulfur based joint sealant, and caused by the differences in the thermal expansion between metal and the non-metallic leadite

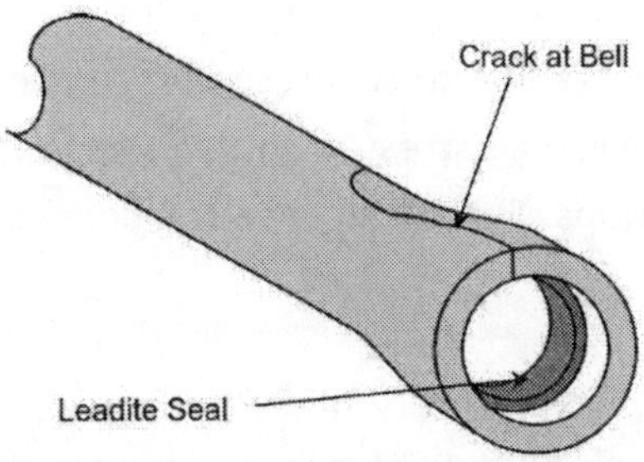

- **Perforations**, normally created by through-wall pitting.

The seventh failure mode of pressure piping – rupture/blow-out, mainly occurs in PCCP due to prestressed wire failure.

As one can clearly see from the information presented in this section and the previous section, the types of degradation present and the rate of this degradation determine the physical demands that the new liner will have to sustain. One cannot choose to ignore the host pipe's existence and its effects on the liner's structural performance as has been done in the past for the fully deteriorated design condition. As long as the existing pipe is in place it can have a significant impact on the loading exerted on the liner and its long-term durability.

Steady-State Pressure vs. Cyclical Pressure Pipe

From an infrastructure point of view, a water distribution system is an elaborate conveyance structure in which pumps move water through the system, control valves allow water pressure and flow direction to be regulated, and reservoirs smooth out the effects of fluctuating demands and provide reserve capacity for fire suppression and other emergencies. All of these distribution system components and their operations and complex interactions can produce significant variations in critical hydraulic components; of which operating pressure is one.

A loss of hydraulic integrity of a piping system implies a loss of positive line pressures, flow reversals, and rapid changes in velocity, among others. Rapid changes in pressure and flow caused by events such as rapid valve closures or pump stoppages and hydrant flushing can create pressure surges of excessive magnitude. These transient pressures, which are superimposed on the normal static pressures already present in the water line at the time the transient occurs, can strain the system leading to increased leakage and decreased system reliability, equipment failure, and even pipe rupture in extreme cases. Uncontrolled pump shutdown can lead to the undesirable occurrence of water-column separation, which can result in catastrophic failures due to severe pressure rises following the collapse of the vapor cavities. Vacuum conditions can create high stresses and strains that are much greater than those occurring during normal operating conditions. They can cause the collapse of thin-walled pipes if they are not designed to withstand such strains.

In a report jointly produced by the New Jersey Department of Environmental Protection and American Water (Voorhees, NJ) in December of 2005 on the *Susceptibility of Potable Water Distribution Systems to Negative Pressure Transients*, it was recommended that a holistic review of system conditions and utility procedures be undertaken to effectively minimize a system's susceptibility to pressure transients. The report stressed an operator's understanding of how his/her particular system behaves in response to a variety of events such as power outages, pump shutdowns, valve operations, flushing, firefighting, main breaks and other events that can create significant rapid drops in pressure and/or low pressure waves.

In the last few years there has been a significant rise in the use of liners for forcemain sewer renewal. This type of piping system is inherently cyclical in its operation which the design engineer must take into account when designing the fiber-reinforced CIPP. The magnitude of pressure surges and the number of repetitions will undoubtedly be driven almost exclusively by the host material's tensile modulus which must be accounted for in the new system's design. This will be discussed further in the next section on design considerations.

Pure Technologies in 2010 was asked to investigate the cause(s) of a prestressed concrete cylinder pipe force main that was operating between 32 and 35 psi according to the pipe owner's records, but yet had experienced multiple blow-outs; all while having a pressure capacity rating of 200 psi. The graph below was produced by Pure using their own pressure monitoring instruments set to a much higher polling frequency the results of which captured a much wider pressure variation and pressure excursions well above 200 psi. Given the length of service and the number of wire breaks subsequently found in the wall of the pipe gave Pure Technologies' engineers (and the owner) a clear indication of why the pipe was experiencing so many failures.

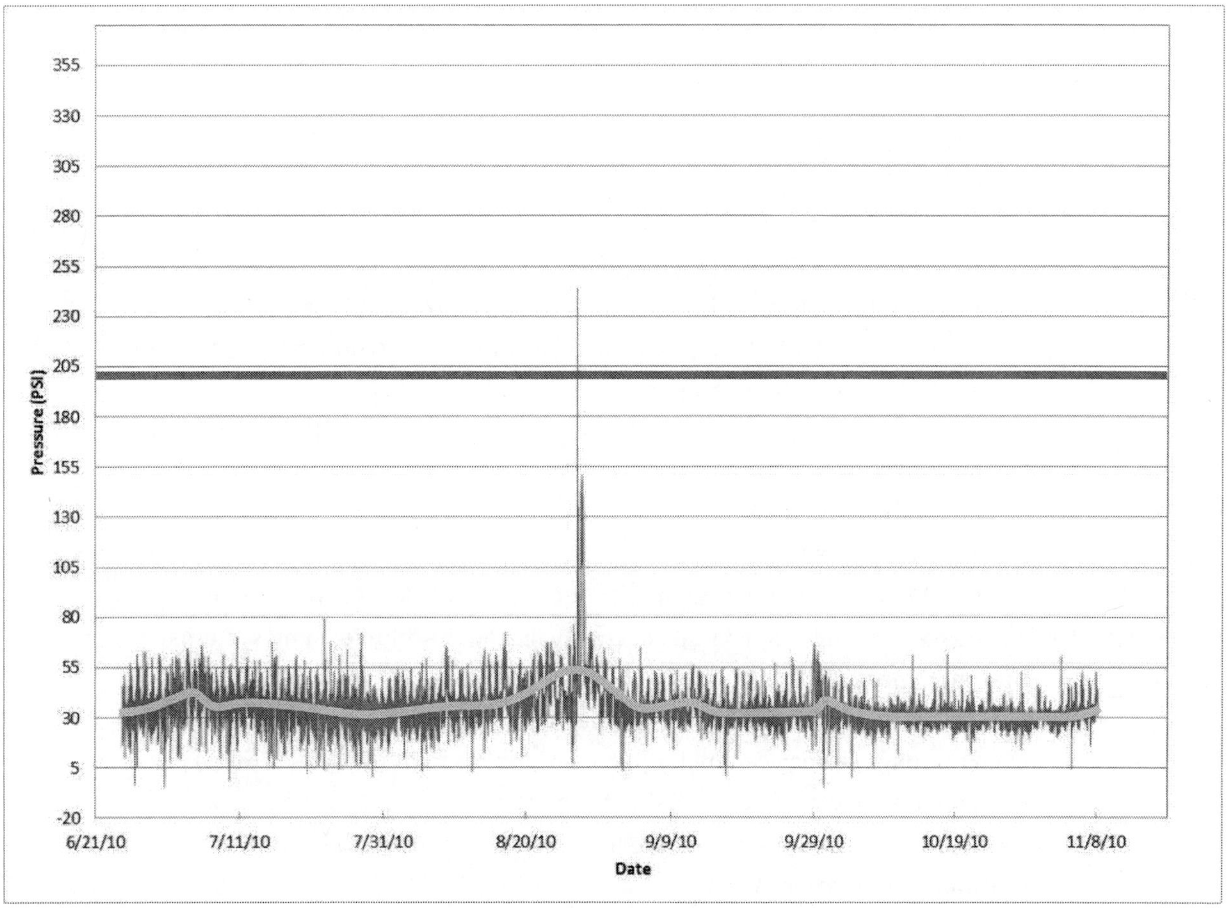

Design Considerations for CIPP in a Pressure Pipe Application

The design of a liner for a pressure pipe application, whether it is in a pressure distribution system or a forcemain system, must be based upon more than just the known normal operating pressure and the maximum operating pressure of the piping system to be renewed. It must account for the real world operating conditions which have the potential to occur and how the addition of a liner may make changes to that operational regime. In the design of gravity pipe liners the loads coming onto the liner are pretty straightforward to calculate; and some level of conservatism applied to make up the difference for the engineer's lacking in his/her diligence at accurately assessing the existing conditions will not typically lead to a failure of the liner. However, pressure piping systems have a greater number of variables in play which must be assessed sufficiently in order to produce the needed information for the design of the fiber-reinforced CIPP. Outlined below are the steps that this author feels must be undertaken when designing a liner for a pressure pipe application.

- Determine the current conditions for the pipe to be renewed
  - Host pipe material
  - Corrosion issues present (at the time of the lining)
    - Internal
    - External
    - Type
      - General area corrosion
      - Pit Corrosion
  - Appurtenances in the system
    - In-line valves
    - Branching fittings (e.g. tees)
    - Directional fittings (e.g. bends)
    - Air relief valving (with info on the sizing and capabilities)
    - Vacuum relief valving (with specific info on the sizing and capabilities)
    - Surge protection components installed
  - As-built plans showing horizontal and vertical alignments of the system
  - System operational information
    - Static pressure
    - Flows in reach to be lined
    - Pump(s) operating regime (especially important for forcemains)
- Determine the performance capabilities of the available alternative lining systems
  - Maximum pressure capabilities
  - Cyclical performance testing and validation
  - Lining through fittings capability (typically applies to directional fittings)
  - Position of structural fiber mat(s) in tube matrix
- Determine where the access points for the lining can and cannot be located
  - Maximum installation lengths
- Design Limit States
  - Gravity pipe design demands met
    - External hydrostatic pressure
    - Live load(s)
  - Pressure pipe performance demands met
    - Hydraulic performance analysis is performed to find the ...
      - NAOP (normal average operating pressure for reach to be lined)
      - MAOP (maximum allowable operating pressure for the reach to be lined)
      - Peak or Surge pressure due to operational changes
        - Planned

        - Un-planned
    - Max Vacuum or low pressure conditions
- Termination details design

The need to thoroughly gather information regarding the type of conditions present in the piping system at lining is irrefutable. One must understand the workings of the existing system and the associated demands that they will in turn place on the liner that is being designed. For example, pit corrosion versus general area corrosion requires consideration of the shape and aspect ratio of the perforations that may occur and the stresses that the liner may subsequently see over the design life of the proposed renewal.

The time dependent nature and the wide range of mechanical properties of the various fiber-reinforced CIPP systems necessitate that true qualification testing be performed in order to develop sufficient amounts of data to support the system manufacturer's wall thickness design and the engineer's subsequent wall thickness analysis. The number of samples tested must create a statistically significant representation of the product. Data from long-term testing (10,000 hours or more) using the methods like those contained within the ASTM standard D2992 can be used to statistically extrapolate to the needed design service life based on a particular pipe's internal pressure environment. We know from the engineering investigations done to date that the true tensile modulus retention factor for one currently available system is around 0.20; meaning that it really creeps under a pressurized loading condition.

Determining the potential surge pressure(s) that can occur in the piping can be quite challenging. Pressure surge, also commonly known as water hammer, results from an abrupt change of fluid velocity within the system. The magnitude of the pressure surge is a function of the fluid properties and velocity change, the modulus and wall thickness of the pipe material, the length of the line, and the speed at which the momentum of the fluid is changed. The relatively high compliance (low Young's modulus) of the fiber-reinforced CIPP by itself would produce a much lower surge pressure than a high modulus ductile iron pipe. The challenge facing the design engineer is how to predict the new composite structure's effective modulus as it pertains to the pressure transient's calculation. The Talbot equation which is shown below is used in the AWWA M45 Fiberglass Pipe Design. The values for *E*, *d*, and *t* are the modulus of elasticity, pipe inside diameter, and pipe wall thickness.

$$P_s = \left(\alpha/g\right)\left(SG/2.3\right)(\Delta V) \text{ ; where } \alpha = 12/\left[\left(\rho/g\right)\left(1/k + \left(d/E\right)(t)\right)\right]^{0.5}$$

Software programs such as KYPipes' Surge2000 can be used quite effectively to view the hydraulic pressures created by events such as a power loss at the pump station. The figure below is from a paper by Woods et al (2005), showing the calculated upper and lower bounds of the pressure envelope created with a sudden loss of power at a Canadian sewage lift station. The design flow rate was 951 GPM in the 10-inch diameter forcemain (720m of DIP at the PS; with the remaining 880m being PVC). The green dotted line represents the steady state HGL just prior to the station losing power. Note the area in light blue where cavitation occurs.

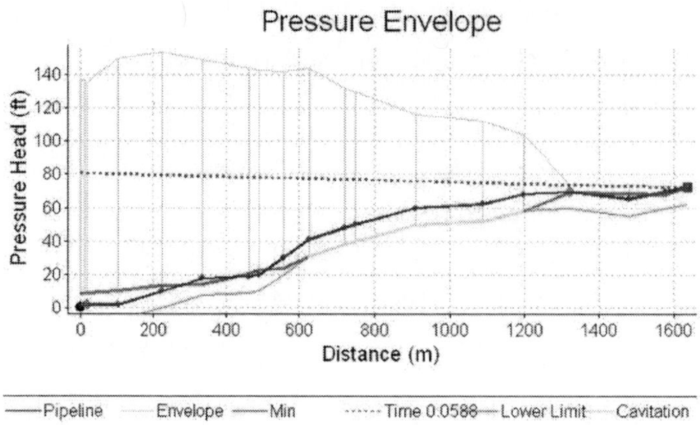

The wall thickness design which considers the positive and negative changes to the normal operating pressure must be carried forth into the field and the actual installation of the proposed liner. Fiber reinforcements don't stretch to size as do the needled polyester felt materials of the gravity pipe liners. This can pose a problem for the design engineer if not properly recognized during the project bid phase. Currently some liners have fiber reinforcing mats that are overlapped so that they can be expanded to the size of the host pipe while other liners use fiber reinforcing tubes or hoses that are manufactured to a specific diameter. In either type of construction there is the potential to form a fold in the reinforcement that downgrades the pressure carrying capabilities of the liner significantly if the fold crosses over an opening or perforation in the host pipe wall. Allouche et al (2005) in a paper presented at the annual ASCE Pipelines Conference on *The Design and Performance of Pressure Pipe Liners under Static and Cyclic Loading* reported that the burst pressure can be up to 280% lower than the value predicted by the design method in ASTM F-2207. Further, Allouche et al (2008) in a paper presented at the annual ASCE Pipelines Conference on *The Design and Performance of Pressure Pipe Liners under Static and Cyclic Loading* reported for host pipe pressures going between an average operating pressure of 60 psig and a maximum pressure of 120 psig that the outward deformation of a liner of the type used in their testing continued to increase until around 160,000 cycles (the cycle being defined as 20 seconds at 60 psig followed by 20 seconds at 120 psig). This lead to the following conclusions and recommendations:

1. Burst strength calculations in F2207 were consistent with their FEA for circular defects between 1.5 and 8 in. in diameter
2. Burst pressures with elongated shapes (in the axial direction) were predicted to be substantially lower than circular ones of the same cross-sectional area
3. A square shaped defect needs to be assumed for design purposes to reduce the over estimation of the burst strength for possible elongated defect shapes
4. Creep strain must be considered in the design of liners for pressure pipe applications (it was noted that the relative contribution of the creep strain to the deformation of the liner increased as the defect size was decreased)
5. The deformation of the liner during the primary creep phase appears to be directly related to the size of the defect, while the deformation during the secondary creep phase appears to reflect the mechanical properties of the liner's design.

Finally, the end terminations for the lining must be carefully developed to consider the tightness of fit that can be achieved between the liner and the host pipe. The slow build-up of pressure in the annular space due to an ill-conceived termination seal or the lack of the use of a termination seal can lead to an external buckling failure of the liner when a low pressure excursion suddenly occurs (this is much different than figuring a full vacuum head). Installation lengths must be factored into how the liner will move while in service. Below is a figure representing a possible termination detail that eliminates the opportunity for pressure build-up behind the liner while simultaneously protecting the liner from over-stressing as it exits the host pipe.

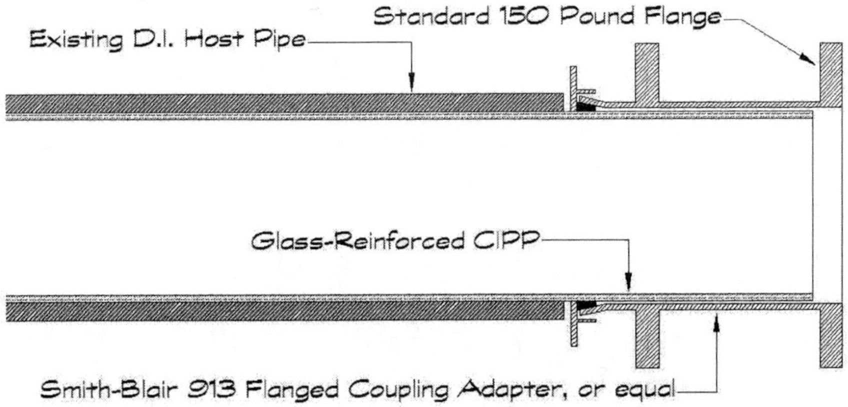

Conclusions

A well designed piping system should have controls and/or appurtenant works in place to prevent pressure surges, both positive and negative, from occurring. The use of a fiber-reinforced CIPP as a renewal system for existing pressure pipe applications is a viable alternative provided the design of the liner is tailored to meet the design limit states required in the proposed application. The presence of the host pipe cannot be ignored regarding its impact on the magnitude of the potential surge pressure(s) after lining and the CIPP's termination details must be derived from a careful consideration of the stress and strain levels that are likely to occur at these junctures if not well thought out.

More testing is needed by the fiber-reinforced CIPP systems' manufacturers to address the long-term performance of the bond that the reinforcing fibers must have with the resin systems used. The impact of pressure cycling and any limitations associated with the cumulative cycles that will be seen in forcemain applications must be assessed and communicated with the engineering community. The strain sensitivity of the liner system must be well documented to prevent over estimation of the finished CIPP performance and to assure delivery of the design service life desired.

Taking a traditionally "conservative approach" to choosing the defects that a particular pipe reach must be designed for as has been practiced in gravity pipe applications can actually produce the opposite engineering outcome; that is, an un-conservative design. The significant design differences between gravity pipe liners and pressure pipe liners must be brought forth in a consensus design practice that contains the necessary documentation to properly serve the design engineer.

THE TOTAL SOURCE

educational sessions

"Stop Cutting Corners !" Keyhole Coring is the Better Way

Marshall Pollock

Utilicor Technologies Inc.

Underground Construction Technology

Atlanta, Georgia

February 4th, 2016

Utilicor™ TECHNOLOGIES INC.

U.S. Utility Infrastructure

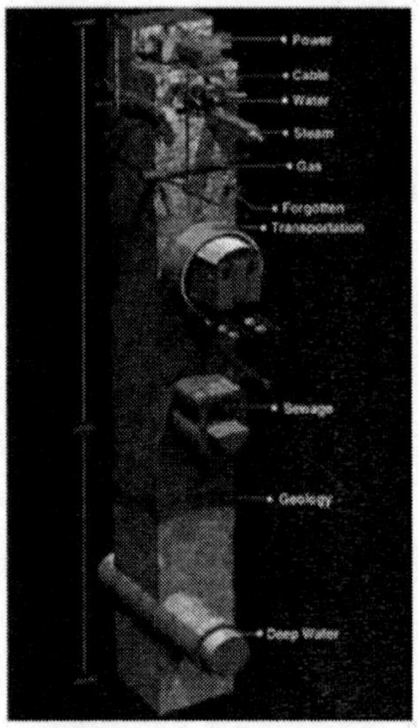

20 million miles of pipe, cable and wire - much of it buried in the ROW

- **Electrical cables**
  - High and low voltage
  - Signal cables
  - Cathodic protection cabinets and cables
  - Conduit systems, cable pits and chambers
- **Gas lines**
  - Distribution lines (high, medium and low pressure, cast iron, steel, Polyethylene)
  - Mains and Services (fittings, valves, service joints, casings, tracer wire, cathodic protection)
- **Petroleum Lines**
- **Communication cables**
  - Phone lines -- copper and fiber cptic
  - Conduit systems
  - Coaxial and data cables
- **Sewer/water mains and services**
  - Pits
  - Tunnels
  - Connections
  - Valves
- **Steam Lines**

Need to Repair or Replace

- Some of this infrastructure has been in the ground for **100 years** or more and **needs to be replaced**.
- New infrastructure **needs to be installed** to meet growing needs.
- Existing infrastructure needs to be **inspected and repaired**.
- **Utility cuts need to be made** to install or gain access to new or existing infrastructure buried under the roadways or sidewalks.
- This presentation discusses a ***Better Way*** to make and repair those utility cuts.

Our Road System is Deteriorating

- 28% major urban roads – interstates, freeways and other critical local routes – have pavements in poor condition, resulting in rough rides.

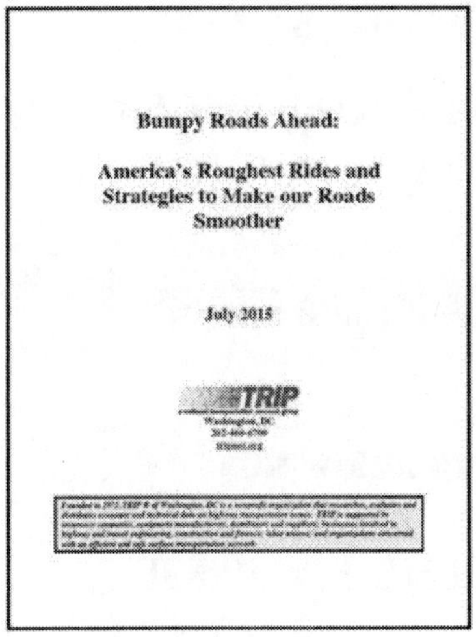
Bumpy Roads Ahead:

America's Roughest Rides and Strategies to Make our Roads Smoother

July 2015

TRIP

National Transportation Research Group
2015 TRIP Report

- Average of $516 annually in additional vehicle operating costs due to:
  - accelerated vehicle deterioration
  - additional maintenance needs
  - increased fuel consumption

Roads in Poor Condition

San Francisco-Oakland, CA	74%	Akron, OH	50%
Los Angeles, CA	73%	San Antonio, TX	49%
Concord, CA	62%	Milwaukee, WI	46%
Detroit, MI	56%	Riverside-San Bernardino, CA	46%
San Jose, CA	53%	El Paso, TX	46%
Cleveland, OH	52%	Oklahoma City, OK	45%
New York City-Newark, NY	51%	Tulsa, OK	45%
San Diego, CA	51%	New Haven, CT	45%
Grand Rapids, MI	51%	Bridgeport-Stamford, CT	44%
Honolulu , HI	51%	Birmingham, AL	43%

4

Outline

- Problems with Conventional Utility Cuts
- Keyhole Coring & Reinstatement – A Better Way
- Mechanics of Excavation:
  - Size, Shape, Cut and Repair Methods are Important
- Field Proven Repair Method
- The Coring and Reinstatement Process
- Environmental and Other Benefits
- Comparative Costs

5

News Item:

NEWS ITEM: August 22nd, 2014

Turner-Fairbanks Highway Research Center

- **FHWA** has commissioned a NEW **Best Practices Guide for Making and Restoring Utility Cuts in Pavement.** *Release Date: Early 2016.*
- **Keyhole Coring & Reinstatement is one of the featured technologies.**
- **DOTs and Municipalities are expected to follow this recommendation and encourage use of this technology in approving or permitting utility cuts and other small hole excavations in their streets.**

6

What's wrong with Conventional Utility Cuts?

- They settle and crack and result in a bumpy ride.
- They reduce performance life of the pavement.
- The sealant squeezes out allowing ground water to penetrate leading to premature pavement failure.
- They result in potholes that must be repaired over and over.
- They're ugly and don't match the rest of the pavement.
- Road no longer performs as a load bearing system.
- The road may need replacement sooner.

7

Conventional Excavation Methods

- **Damage:** Saw cuts, jackhammers, and backhoes can result in destruction of underground facilities, pavement damage or worker injury.

- **Delays:** Road work and utility cuts cause 400 million hours of traffic delay and disruption and costs drivers $6 Billion in wasted time.
- **Reduced Worker Safety:** Higher risk of injury from jackhammers and trench work.

- **Spoils Disposal:** Need to truck tons of pavement spoils through city streets to waste disposal dumpsite.
- **Reduced Pavement Life Expectancy:** Conventional utility cuts in the roadway can increase damage to adjacent pavement and potential for groundwater penetration which can significantly reduce the life expectancy of roads.

8

Conventional Utility Cut Repairs

Resulting in Potholes.

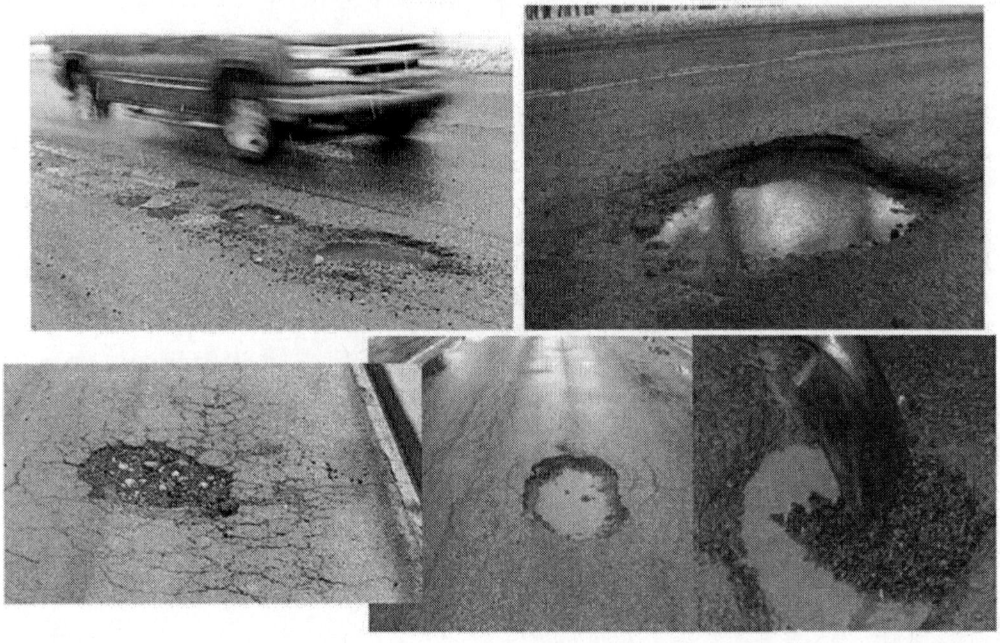

Leading to the Conclusion …

"There is no such thing as a Good Utility Cut Repair"

But there is a Better Way…

Keyhole Coring & Reinstatement

Keyhole Coring & Reinstatement (Video)

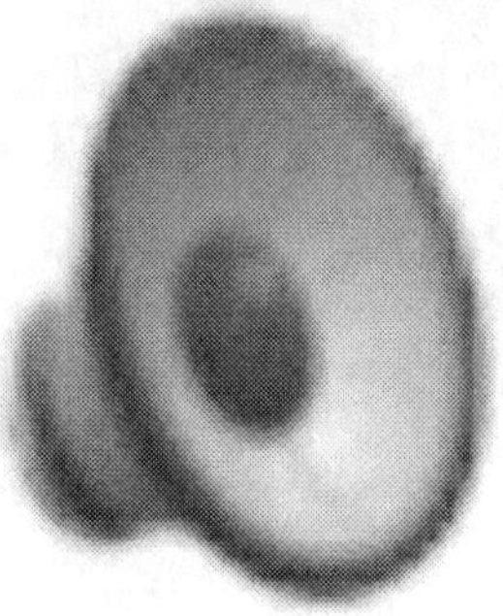

What is Keyhole Technology ?

- A **minimally invasive method** of accessing or viewing underground utilities through small holes or "keyholes" that are vacuum excavated through soft surfaces or through holes cut or cored through pavement.

14

Keyhole Pavement Cuts Mean…
Better Pavement Performance

Same Pavement Profile and **Appearance**

- **Perfect Match:** The reinstated core is a perfect match of profile, texture and color of the surrounding pavement … *it was cut from that pavement.*

Restores Load Transfer Capacity to Repaired Roadway

- **Mechanical Joint:** Utilibond bonding compound creates a mechanical joint with the remaining pavement to restore its load transfer capability to what it was prior to the excavation.

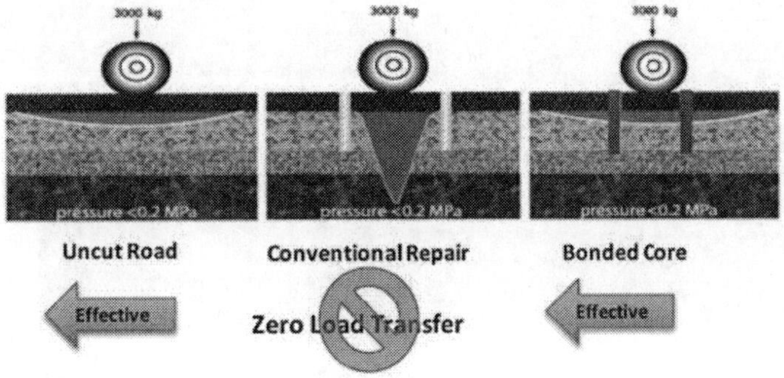

15

Reduced Traffic Delays and Disruption
Means Fewer Complaints

- **Shorter Road Closure:** Road can be safely reopened to traffic within 30 minutes of reinstatement as a **Permanent Repair.**
- **In and out the same day.** No need for further road closings for subsequent permanent pavement repair.

Single Road Closure

16

Greatly Improved Aesthetics

Final Repair

17

Basic Mechanics of Excavation & Repair

- **Size:** Smaller is better
- **Shape:** Rectangular cuts are a problem – Corner Cracks
- **Cut Method:** How you cut is important – Low Impact
- **Repair Method:** Waterproof joint, restores load transfer ability, aesthetically pleasing

18

Smaller is Better

- **Smaller is better and less intrusive**

Laparoscopic Surgery

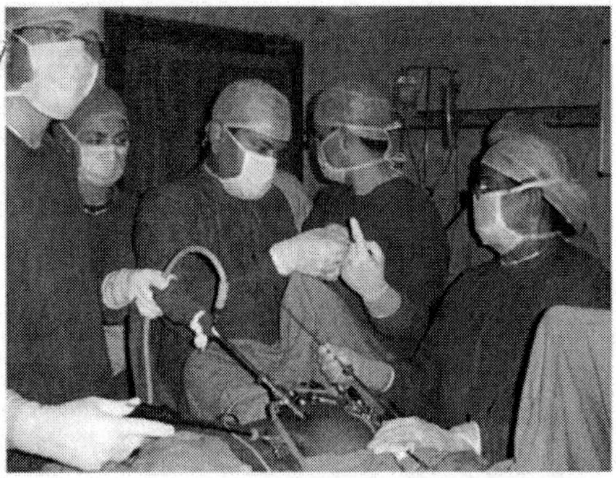

- Smaller Incision
- Short Recovery Period
- Faster Healing
- Smaller Scar
- Lower Cost

19

Smaller is Better

- **Smaller is better and less intrusive**

Laparoscopic Surgery

- Smaller Incision
- Short Recovery Period
- Faster Healing
- Smaller Scar
- Lower Cost

Keyhole Operation

- Smaller Opening (keyhole)
- Faster Restoration
- Less Damage to Road System
- Smaller Repair Footprint
- Lower Cost

✪ and safer -- no men in the hole

20

Circular Shape is Better

Shape Matters !

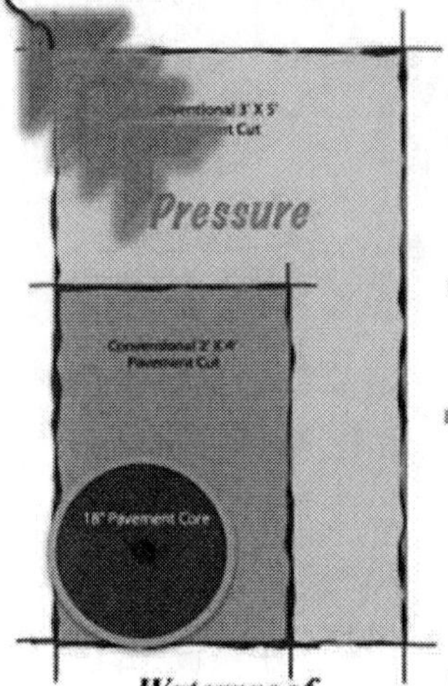

Waterproof
No Corner Cracks

- **Smaller Circular Shape – More Precise Excavation**
  - Waterproof: No saw over-cuts at the corners
  - Environmentally friendly, reuses materials, creates no spoils and no VOCs
  - Reduced surface scarring -- 10 to 20 times SMALLER than conventional restoration (1.75 ft^2 vs 24 ft^2 to $35ft^2$)
  - Circular geometry with no corner cracks
- **Proven Strength -- NO Failures**
  Reinstated core will support 50,000 lbs wheel load
- **Better Performance -- Reduced Delay**
  - Reduced pavement deterioration -- no potholes
  - Reduced traffic delay -- no additional road closing for repaving ... In and Out the same day.
  - Aesthetically pleasing – perfect surface match -- invisible
- **Reduces Pavement Restoration Cost by 87%**
- **Safer for Workers and Public**

21

Traffic Pressure = Corner Cracks

RECTANGULAR format has
FOUR TIMES MORE
Traffic Pressure in corners than CIRCULAR format, causing corner pressure cracks that allow ground water penetration.

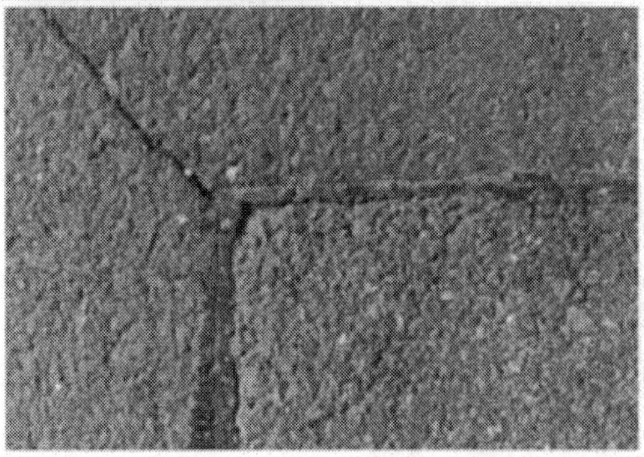

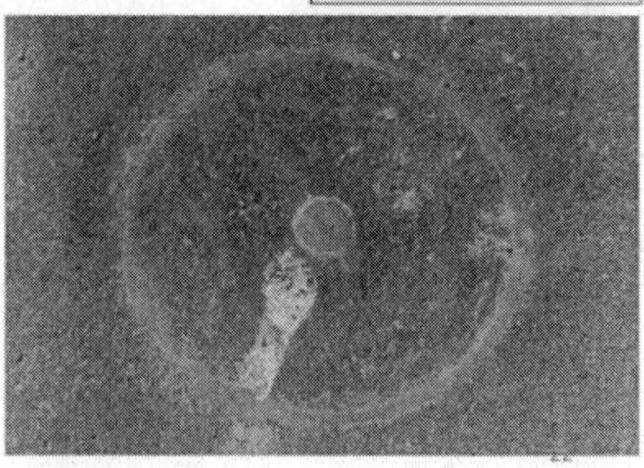

22

How You Cut is Important.

- Conventional excavation (Jackhammer, Backhoe, Pavement Breaker) weakens pavement well beyond the cut causing subsidence in "Zone of Influence"
- Precise coring operation eliminates Zone of Influence and reduces pavement damage

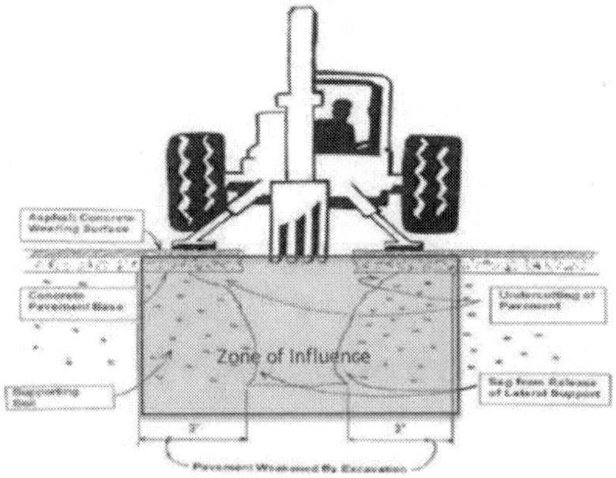

The precise coring process and vacuum excavation eliminates any "Zone of Influence".

23

Reduced Road Damage

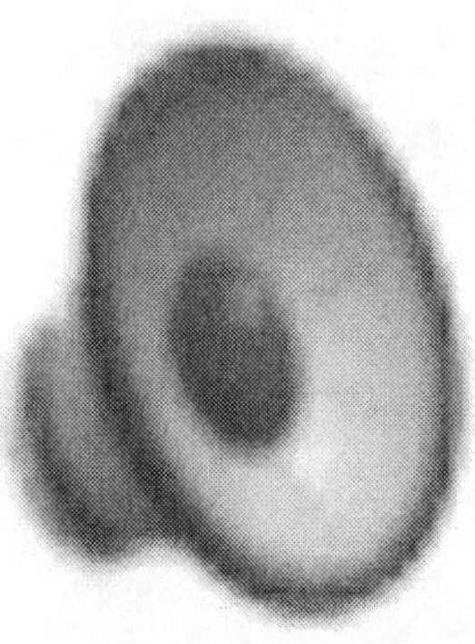

Proven Repair Method

Golder Associates 10 year Engineering Study (1992-2002):

Golder Associates Engineering Reports 1996 and 2003

- **Reinstated core reunites pavement sections with mechanical waterproof joint**
  - No sealants to squeeze out with action of traffic
  - No ground water penetration
- **Restores horizontal load transfer capability of pavement system**

- ***"Based on trials and testing we are satisfied that Process will ensure satisfactory long term performance of the pavement reinstatement."*** *Golder Associates 2003*

25

Golder Report Testing & Results

Comprehensive Laboratory Testing

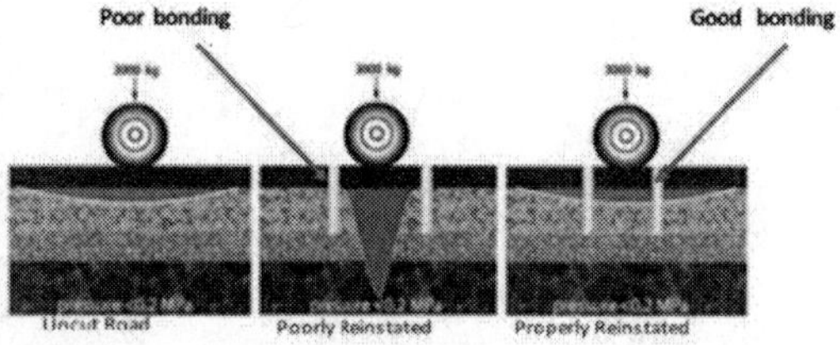

Effective Load Transfer

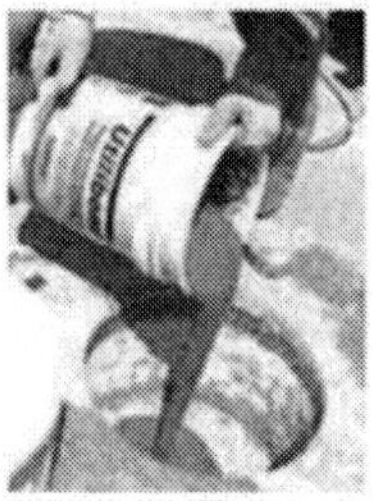

Fast Setting, Rapid Strength Gain, Convenient, Consistent, and Repeatable

Long-Term Performance through freeze-thaw cycles

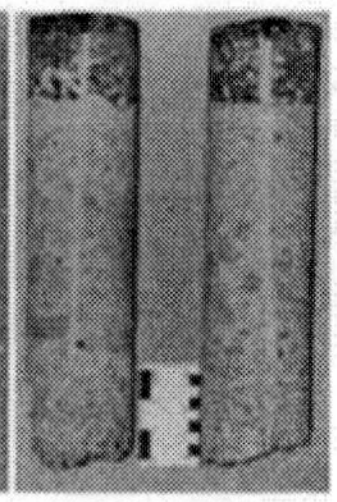

Mechanical Bond
No Voids

Restoration of Utility Cut Study (2000-2005)

Joint Project of 20 North American Cities and States, 7 utility companies, AGA and GTI to develop best restoration practice based on sound engineering principles

Toronto Field Experiment: Observations

CONVENTIONAL TRENCH	ROTARY CUT KEYHOLE
▪ **Noticeable failures** in conventional cut. ▪ Conventionally repaired **joint between road and the cut opened.** ▪ Visible **settlement in trench** along wheel path. ▪ **Joint seal material lost** under action of traffic. ▪ Sand cover at base of trench exposed to **higher than normal levels of moisture** (compared with keyhole).	▪ **No defects** in keyhole cut. ▪ Keyhole section established Oct. 2001 **continued to perform well** throughout life of project. ▪ **Surface remained level** with the road. ▪ The **bonding material remained intact** (no cracking or separation. ▪ **Created waterproof joint.**

US Army Corps of Engineers

National Research Council Canada 27

Findings & Conclusions: Minimally Invasive Keyhole Utility Cuts

- **No Distress:** The keyhole restored October 2001 showed no distress during the inspections performed during the May 2002 and April 2003 visits. The grout remained intact and the surface is level with the road.
- **Level -- No separation:** The surface of the restored keyhole remained at level with the road profile. The grout [Utilibond] surrounding the AC/PCC plug remained intact (no cracking or separation).
- **Lower Stress Levels:** The keyhole opening is quite small (18 in., 457 mm, in diameter) compared with the area of a tire print of heavy trucks in contact with the road surface. As a result, low stresses are transmitted to the underlying sections of the restored keyhole.
- **No Pressure Cracks:** Based on basic rules of mechanics, a circular cut shape in the AC is ideal for preventing propagation of cracks into the surrounding road area.

Conclusion: "The keyhole construction and restoration technique should be encouraged whenever feasible to minimize the need for opening large trenches in the future."

US Army Corps of Engineers

National Research Council Canada 28

APWA Supports Use of Keyhole Technology

(Report of Engineering and Technology Committee Utility and Public Right of Way Committee APWA (2007)

"When practical, APWA supports the use of minimally invasive keyhole technology **to facilitate utility exposure or daylighting.**"

"Keyhole technology has been **primarily used by the natural gas industry but this technology has the potential for being utilized on drinking water pipelines and service lines.**"

"Keyhole technology also has the potential to be used by government agencies for **subsurface utility engineering on urban reconstruction projects."**

29

Accepted and Approved by Municipalities

Keyhole coring and reinstatement has been ACCEPTED and APPROVED as the **"Better Way"** by dozens of municipalities across North America.

New York City DOT Officials, March 9th, 2006

- **"... no jack-hammering to disrupt or damage the roadbed** or the rest of the pavement and it is **much quieter** than other methods."
- "It **cuts the time on the road.** That means **less inconvenience to the public**, which to a government agency responsible to the public like us, ***is very important!"***
- "I like the smaller, circular hole and the fact that it **allows you to reopen the road to traffic within 30 minutes** of the repair."

30

Advantages for the Municipality & Public

- **Improved Appearance.** Almost invisible, matching circular core -- less than 1/10 the size of conventional road cut).
- **Reduced Damage** to the Pavement. Road restored to original design specification. No sunken patches or weakened or failed roads. *No Potholes.*
- **Cleaner, Safer, Less Intrusive** Worksite:
  - No jack-hammers or large excavation equipment
  - Less noise and mess and reduced disruption for neighbors
  - No spoil trucked through city.
- **Fewer Complaints** from Public.
- **Reduced Traffic** Disruption
- **Improved Logistics:** Restoration is immediate. Reopen road to traffic in 30 min.

31

Reduced Carbon Footprint

Reduced Carbon Footprint

- Coring and Reinstatement has an **OPERATING CARBON FOOTPRINT** ONE-SIXTH the size of conventional excavation and restoration methods (60 lbs vs 365 lbs)

  BUT if you add the CO_2 emitted in the production of the cement products used in each of the repairs

- **TOTAL CARBON FOOTPRINT** of the Keyhole Coring & Repair is **ONE-TWELFTH** the size of conventional method (69 lbs vs 845 lbs) 33

Carbon Footprint

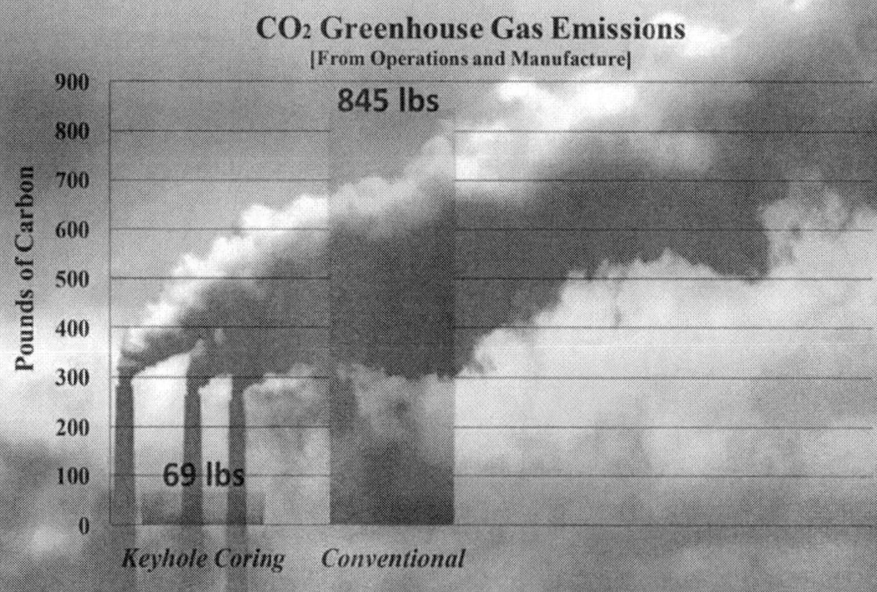

34

3.6 million Utility Cut Permits

(issued by municipalities every year)

35

Environmental Impact of Coring

- **20-25% of utility cuts can be small holes** (Gas Technology Institute)
- If 800,000 small hole utility cuts in NA were performed with keyhole coring and reinstatement:

Reduction in asphalt used:	**2 million tons** -- enough to resurface 650 miles of 4 lane highway
Reduction in spoil disposal:	**27 million cu. ft.** -- enough to fill 200,000 dump trucks
Reduction in work zone delay:	**2.8 million hours** **1.9 million gal fuel** **$520 million cost**
Restoration Cost Savings to Utilities:	**$340 to $900 million**
Reduction in GHG emissions:	**320,000 Tons of GHG Emissions** ≈ equal to CO_2 emissions from Average Power Plant (2.8 billion tons CO_2 ÷ 8000 power plants)

36

Comparative Costs

Cost to Cut and Repair 2 ft^2 Opening in Asphalt Pavement 6" Thick

Asphalt Pavement (6")	Coring	Conventional
Pavement Opening	Core: (20 minutes): $50 (2 men & truck = $150/hr.) **Total: $50**	Saw Cut (20 minutes) $50 Jack Hammer (30 minutes) $75 Haul & Dispose Spoil $85 **Total: $210**
Vacuum & Backfill	Equal	Equal
Restoration	**Reinstate Core** Utilibond (pail) $75 2 men & truck for 1 hr. = $150 (Includes waiting 30 minutes to reopen road to traffic) **Total: $225**	**Temporary Cold Patch:** $75 Maintenance: $75 **Sub Total: $150** **Permanent Repair** • Cut Back & Disposal: $195 • Repave 4x4 (@ $15 ft^2 = $240 **Sub Total: $425** **Total: $575**
Total Cut & Restore	**$275**	**$785**

Comparative Costs

Cost to Cut and Repair 2 ft^2 Opening in Composite Pavement 10"Thick

Composite Pavement (10")	Coring	Conventional
Pavement Opening	Core: (20 minutes): $50 (2 men & truck = $150/hr.) **Total: $50**	Saw Cut (20 minutes) $50 Jack Hammer (45 minutes) $115 Haul & Dispose Spoil $125 **Total: $290**
Vacuum & Backfill	Equal	Equal
Restoration	**Reinstate Core** Utilibond (pail) $75 2 men & truck for 1 hr. = $150 (Includes waiting 30 minutes to reopen road to traffic) **Total: $225**	**Temporary Cold Patch:** $75 Maintenance: $75 **Sub Total: $150** **Permanent Repair** • Cut Back & Disposal: $195 • Replace Concrete: $400 • Repave Asphalt 4x4: $240 **Sub Total: $835** **Total: $985**
Total Cut & Restore	**$275**	**$1,275**

Comparative Costs

Cost to Cut and Repair 2 ft^2 Opening in Concrete Sidewalk 4" Thick

Concrete Sidewalk Bay (4")	Coring	Conventional
Pavement Opening	Core: (15 minutes): $40 (2 men & truck = $150/hr.) **Total: $40**	Saw Cut (20 minutes) $50 Jack Hammer (30 minutes) $75 Haul & Dispose Spoil $85 **Total: $210**
Vacuum & Backfill	Equal	Equal
Restoration **NOTE:** Many municipalities require replacement of all sidewalk panels -- from expansion joint to expansion joint.	**Reinstate Core** Utilibond (pail) $75 2 men & truck for 1 hr. = $150 (Includes waiting 30 minutes to reopen road to traffic) **Total: $225**	**Temporary Cold Patch:** $75 Maintenance: $75 **Sub Total: $150** **Permanent Repair** Replace 1 Panel: $500 Replace 3 Panels: $1,500 **Sub Total: $500 or $1,500** **Total: $650 or $1,650**
Total Cut & Restore	**$265**	**$860 or $1,860**

39

Summary – Keyhole Coring Advantages

- Precise Coring – No Heavy Equipment
  - No Zone of Influence
  - Easier on the back of workers
- No Pressure Corner Cracks or Saw Over-Cuts
  - No ground water penetration
- Restores Load Transfer Capability of Pavement System
- Reunites Pavement Sections with Mechanical Waterproof Joint
  - No sealants that can squeeze out with action of traffic needed
  - No ground water penetration
- Environmentally Friendly – Consumes Fewer Resources
  - Reuses original pavement core – no spoil to truck away – no new paving materials required.
  - No VOCs to escape -- 1/12 Carbon Footprint of conventional methods
- Reduced Public Inconvenience – Aesthetically Pleasing – Cost Effective
  - Shorter and fewer road closings -- Reduced traffic delays
  - Saves money for Taxpayer/Ratepayer/Contractor/Utility.

40

Questions and Answers

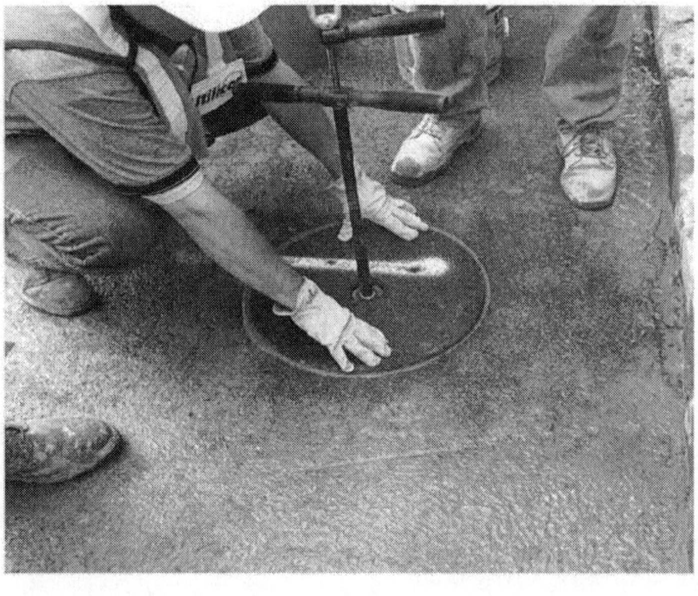

Marshall Pollock
Utilicor Technologies
888-572-6666
mpollock@utilicor.ca

www.utilicor.ca
for **more information**

Utilicor™
TECHNOLOGIES INC.

41

THE TOTAL SOURCE

educational sessions

Thermal Grouting

PSE&G (NJ) Southern and Northeastern Reinforcement Projects

Constellation Group LLC

www.cgllc.us

Thermal Grouting-PSE&G

UCT 2016

Public Service Electric and Gas (PSE&G)-New Jersey

Schedule of 14 Upgrade and Improvement Projects in New Jersey

$4 Billion from 2012-2017 Budget

$1.5 Billion to Date

Total includes Gas and Maintenance Costs

Southern Reinforcement

Two 230Kv Overhead Circuits
Three 230Kv Underground Circuits-16 miles
$435 Million
Should be online now

Thermal Grouting-PSE&G
UCT 2016

Northeast Grid Reinforcement

Upgrading 138 Kv to 230 Kv
15 miles of Underground Transmission
3000 new jobs in construction
Scheduled online Summer 2016

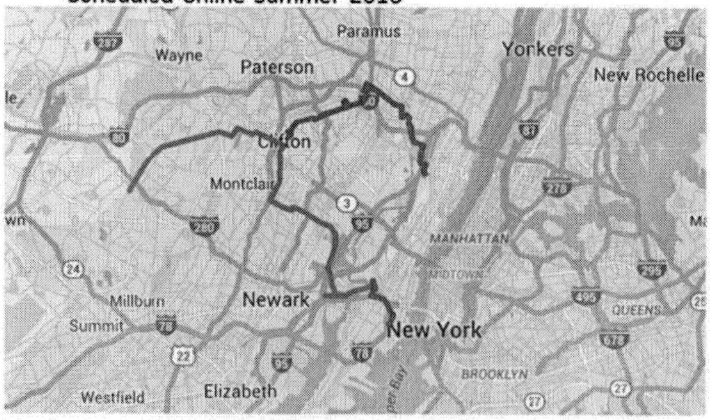

Thermal Grouting-PSE&G

UCT 2016

Trenchless Work
Jack and Bore

Thermal Grouting-PSE&G

UCT 2016

Trenchless Work
Microtunneling
Crossing 10A

Photo Courtesy of
Northeast Remsco

Thermal Grouting-PSE&G

UCT 2016

Horizontally Directionally Drilled (HDD)

Thermal Grouting-PSE&G

UCT 2016

Installing the Bundle

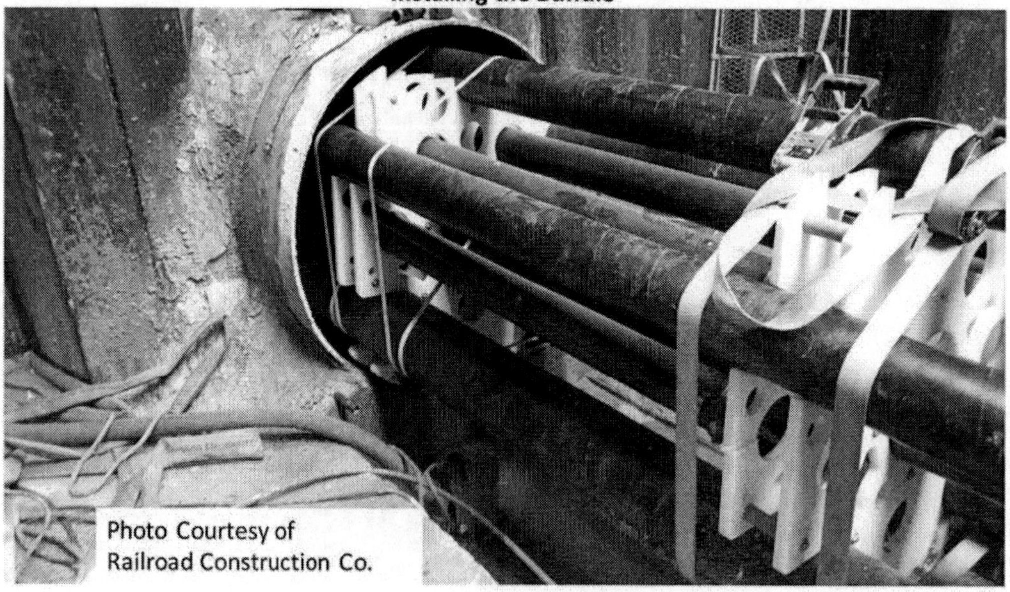

Thermal Grouting-PSE&G UCT 2016

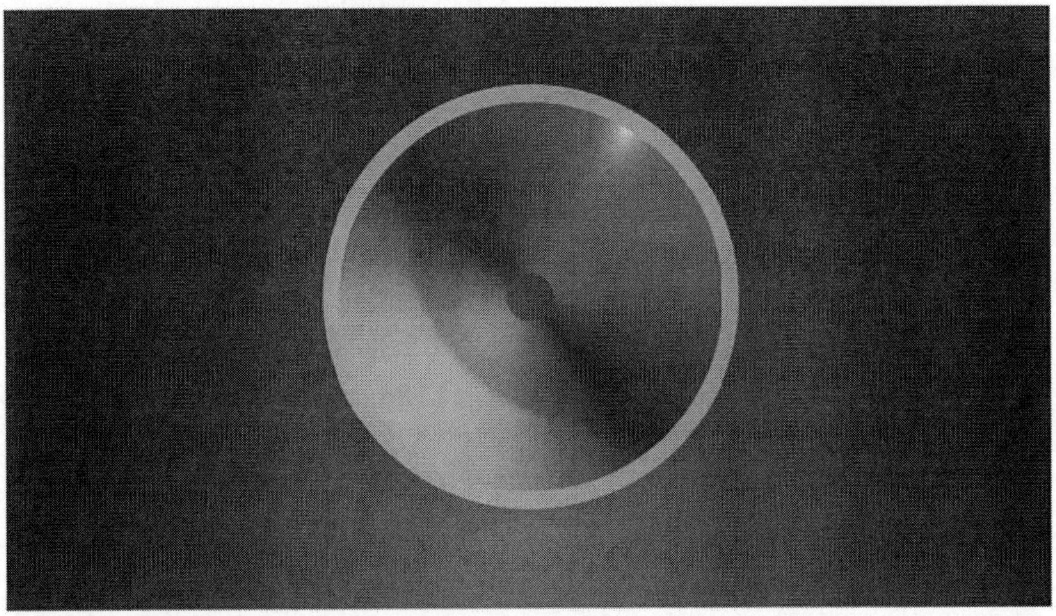

Animation Courtesy of Carson Corporation

Thermal Grouting-PSE&G UCT 2016

FAQ

What about Heat of Hydration?

Typical Thermal Grouts have High Cement Content
Often 500-1000 pounds per CY

Laboratory tests:	6X12 Cylinder- Pennoni	33^0 F Rise at 18h 45m
	6X12 Cylinder- CGLLC	32^0 F Rise at 18h+/-

Mock-up and Production
Vancouver VCCT

Large Volume Insulated Container
RCC/PSE&G Southern Reinforcement

Real World Monitoring of Production Bores
VCCT and RCC/PSE&G

Thermal Grouting-PSE&G

UCT 2016

I Cubic Meter HoH Test

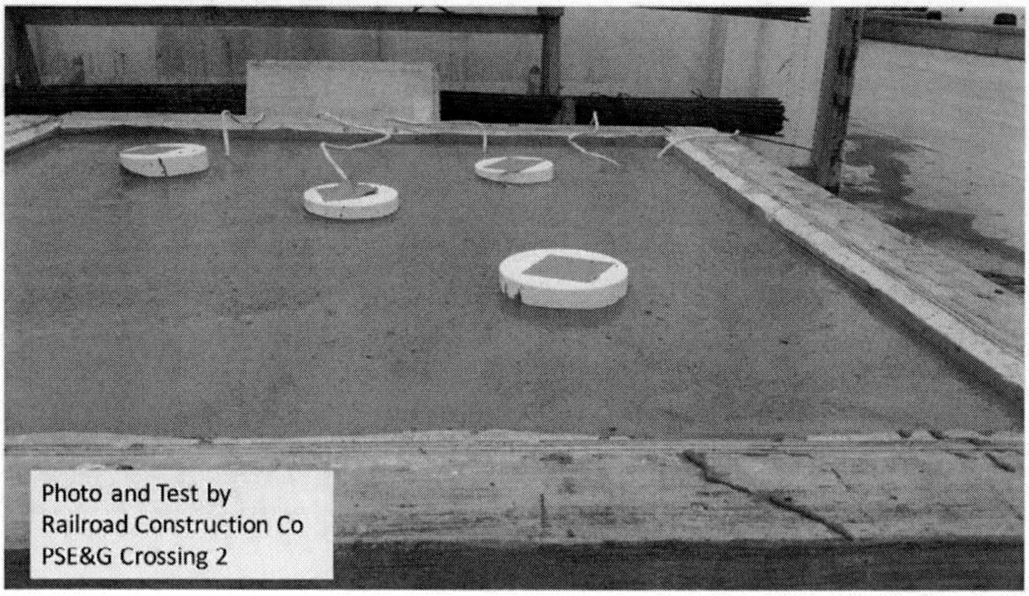

Photo and Test by
Railroad Construction Co
PSE&G Crossing 2

Thermal Grouting-PSE&G

UCT 2016

Heat of Hydration
1 CM Insulated Box

Probe in Grout

Events

Time (hrs)	Date	Event Description	Temperature (° F)	Maturity (°C-Hrs)
1.02	Tue, 03/3/2015 01:19 PM	MIN TEMPERATURE	39.2	6
21.18	Wed, 03/4/2015 09:29 AM	MAX TEMPERATURE	190.4	1148
44.62	Thu, 03/5/2015 08:55 AM	LAST READING	181.4	3177

Probe in PVC filled with Water

Events

Time (hrs)	Date	Event Description	Temperature (° F)	Maturity (°C-Hrs)
0.45	Tue, 03/3/2015 12:42 PM	MIN TEMPERATURE	41	3
26.8	Wed, 03/4/2015 03:03 PM	MAX TEMPERATURE	188.6	1519
44.73	Thu, 03/5/2015 08:59 AM	LAST READING	179.6	3047

Thermal Grouting-PSE&G

UCT 2016

Heat of Hydration In-situ

Crossing Two Probes in about 50 feet

Probe in Grout

Events

Time (hrs)	Date	Event Description	Temperature (° F)	Maturity (°C-Hrs)
0	Wed, 04/8/2015 10:02 AM	MIN TEMPERATURE	53.6	0
47.78	Fri, 04/10/2015 09:49 AM	MAX TEMPERATURE	109.4	842
126.47	Mon, 04/13/2015 04:30 PM	LAST READING	87.8	3769

Probe in PVC Conduit filled with Water

Events

Time (hrs)	Date	Event Description	Temperature (° F)	Maturity (°C-Hrs)
21.63	Thu, 04/9/2015 07:42 AM	MIN TEMPERATURE	46.4	245
49.22	Fri, 04/10/2015 11:17 AM	MAX TEMPERATURE	120.2	852
126.42	Mon, 04/13/2015 04:29 PM	LAST READING	80.6	3645

Thermal Grouting-PSE&G

UCT 2016

Vancouver City Central Transmission

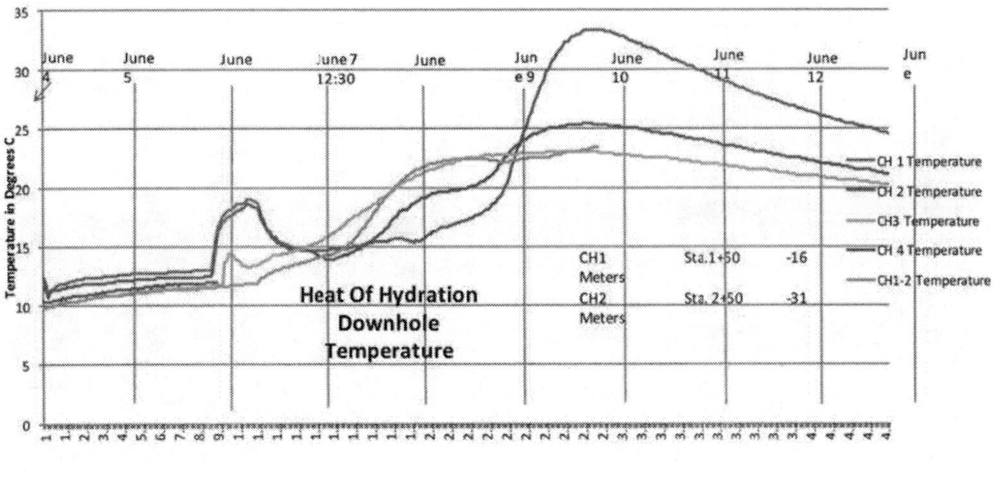

FAQ
What are the pumping pressures?

Pumping Pressures are dependent on:

- Viscosity of grout
- Density of grout
- Certain physical properties
- Length and diameter of grout lines
- Rate of pumping
- Other material characteristics

FAQ
How do we know the casing is full?

Theoretical volume calculations

C18					
	ID inches	Length	Number	Total Volume	
Casing	46.75	156	1	1858.6	CF
	OD Inches				CF
Conduits	4.5	156	2	-34.44	CF
	10.75	156	2	-196.55	CF
	6.625	156	1	-37.33	CF
	2.375	156	0	0.00	CF
		156	0	0.00	CF
		156	0	0.00	CF
		156	0	0.00	CF
Grout Tubes OD	4.5	225	1	-24.84	CF
Grout Tubes ID	3.875	225	1	18.42	CF
Spacers	0.9	156	31.2	-28.08	CF
				1555.8	CF
				57.6	CY

Thermal Grouting-PSE&G UCT 2016

All PSE&G Crossings Required Thermal Grout

Various Methods/Materials:

<u>Cementitious Grout</u>:
- Ready Mixed
- Mixed on Site
- Hybrid Mixing

<u>Bentonite Grout</u>:
- Mixed on Site

Thermal Grouting-PSE&G UCT 2016

Ready Mix- Delivered to Site

Photo- Railroad Construction Co.

Auger Mixed- on site manufacturing

Hybrid:

Ready Mix
- Cement
- Water
- Admixtures

Added on Site
- Specialty Sand

Thermal Grouting-PSE&G

UCT 2016

Bentonite Mixing- Added to Empty Ready Mix Truck, Then Sand Added

Thermal Grouting-PSE&G

UCT 2016

Summary:

Thousands of Cubic Yards of Thermal Grout Installed
Smallest Crossing: 12 cubic yards
Largest Crossing: 380 cubic yards
Upcoming- multiple HDD Crossings: 900 cubic yards each

Unofficial Record:

Largest Thermal Grouting Project Total to Date

Thermal Grouting-PSE&G

UCT 2016

Additional Information:

Guy Dickes, Constellation Group LLC
GUY@CGLLC.US 410-484-0672
WWW.CGLLC.US

Thank you:

UNDERGROUND CONSTRUCTION TECHNOLOGY

SPONSORS

EFFICIENCY PRODUCTION, INC.

America's Trench Box Builder™

Pre-Engineered Shoring Systems for Cross-Trench Utility Challenges

Presented by Mike Ross,
Shoring Specialist

SPECIAL OPERATIONS

SHORING DIVISION

Underground Contractors and utility owners are faced with a variety of challenges during everyday operations

Limited easements, congested work area, soil and water conditions are part of a daily battle fought by the Competent person

One problem that can bring production to a slow crawl is the presence of a crossing utility

1926.651 (b) Underground installations

(3) When excavation operations approach the estimated location of underground installations, the exact location of the installation shall be determined by safe and acceptable means

1926.651 (b) Underground installations

(4) While the excavation is open, underground installations shall be protected, supported or removed as necessary to safeguard employees.

Creating a safe excavation that protects employees from the hazards of both the utility and possibility of a cave-in can be a difficult and expensive task

Manufacturers dedicate a good share of R & D to the development of equipment that can safely accommodate crossing utilities at reasonable cost

Resulting designs give the Competent Person reliable and modular equipment with pre-engineered Manufacturer's Tabulated Data (MTD)

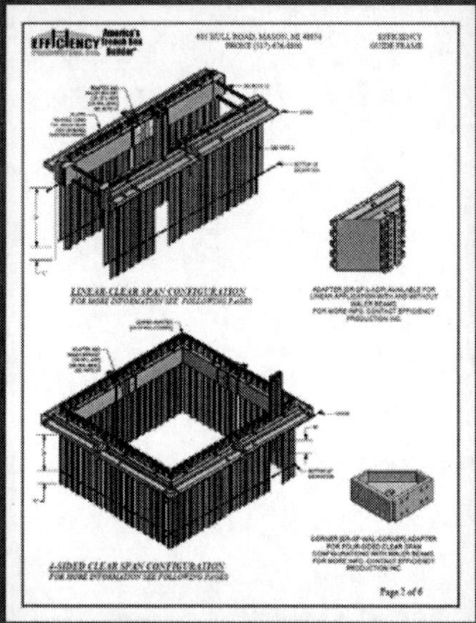

Manufactured Systems for Crossing Utilities

Hydraulic Shoring

Hydraulic Shoring

Hydraulic shoring relies on pressure arches created by cylinder thrust to shore banks. This makes the system a perfect solution for crossing utilities

Hydraulic Shoring

If raveling or sloughing is not a concern – Hydraulic shoring can be used without plywood sheeting

Sheeting Frames

Sheeting frames utilize a reverse cantilever design. Frame acts as waler system that *cages* sheeting from above allowing excavation ahead of sheeting.

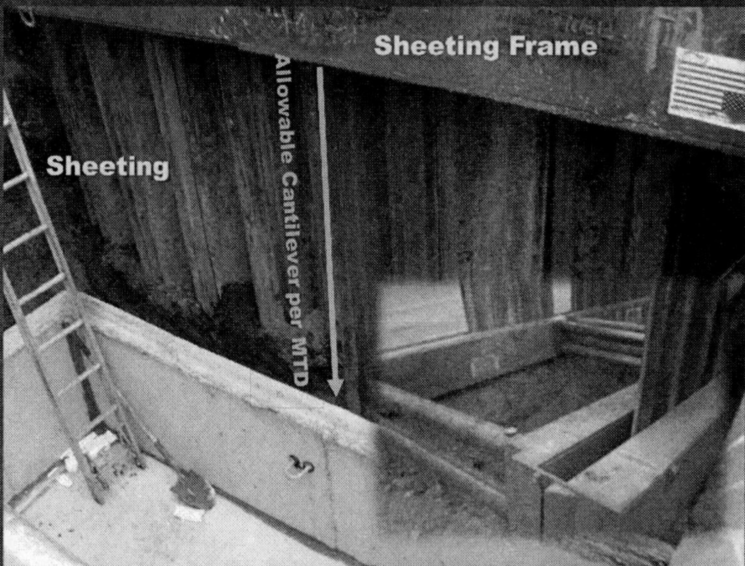

Sheeting Frames

Sheeting frames utilize a reverse cantilever design. Frame acts as waler system that *cages* sheeting from above allowing excavation ahead of sheeting.

Sheeting Frames

The greater the design cantilever – the larger and heavier the frame

Sheeting Frames

The greater the design cantilever – the larger and heavier the frame

Lightweight Aluminum Sheeting Frames

A modular lightweight aluminum sheeting frame for use with smaller excavation equipment is now available

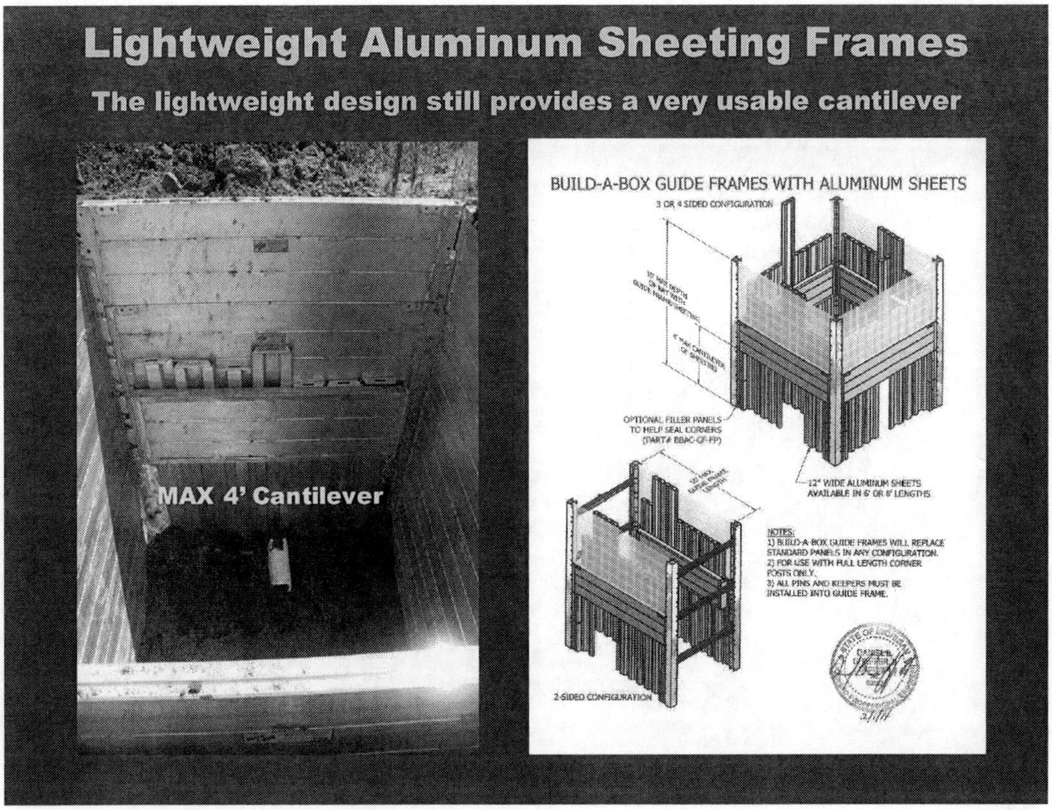
Lightweight Aluminum Sheeting Frames
The lightweight design still provides a very usable cantilever
MAX 4' Cantilever
BUILD-A-BOX GUIDE FRAMES WITH ALUMINUM SHEETS
3 OR 4 SIDED CONFIGURATION
OPTIONAL FILLER PANELS TO HELP SEAL CORNERS (PART# BBAC-GF-FP)
12" WIDE ALUMINUM SHEETS AVAILABLE IN 6' OR 8' LENGTHS
NOTES:
1) BUILD-A-BOX GUIDE FRAMES WILL REPLACE STANDARD PANELS IN ANY CONFIGURATION.
2) FOR USE WITH FULL LENGTH CORNER POSTS ONLY.
3) ALL PINS AND KEEPERS MUST BE INSTALLED INTO GUIDE FRAME.
2-SIDED CONFIGURATION

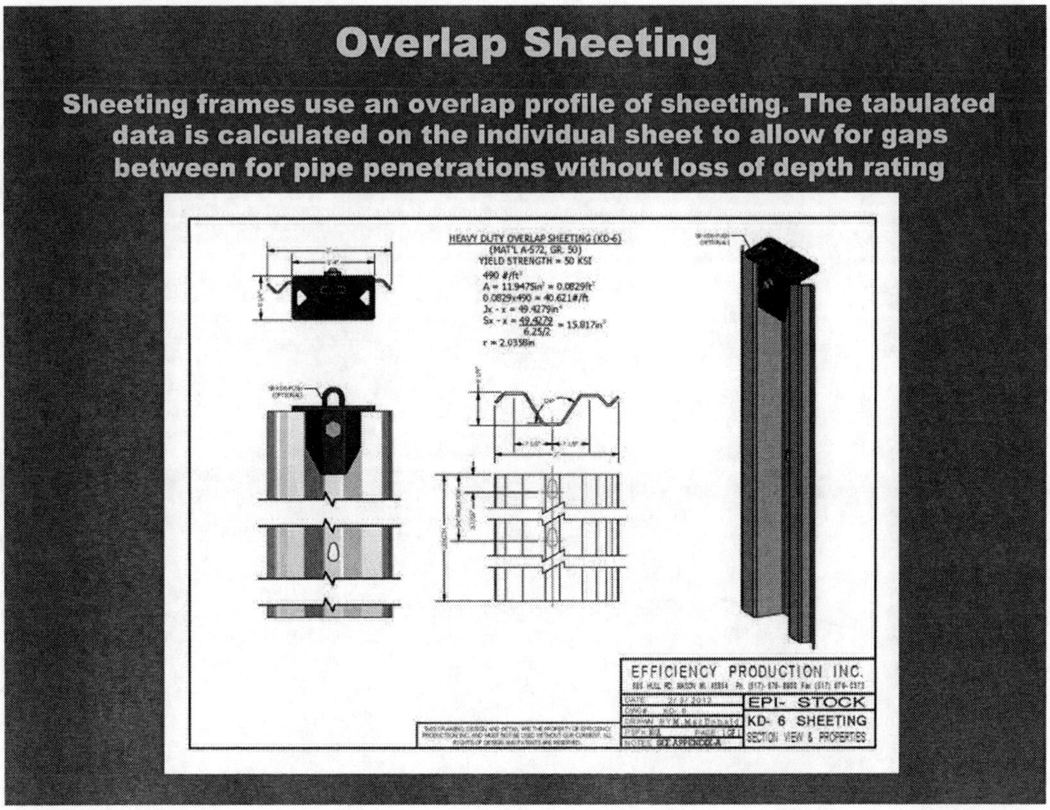
Overlap Sheeting
Sheeting frames use an overlap profile of sheeting. The tabulated data is calculated on the individual sheet to allow for gaps between for pipe penetrations without loss of depth rating
HEAVY DUTY OVERLAP SHEETING (KD-6)
(MAT'L A-572, GR. 50)
YIELD STRENGTH = 50 KSI
EFFICIENCY PRODUCTION INC.
EPI- STOCK
KD- 6 SHEETING
SECTION VIEW & PROPERTIES

Manufactured systems provide the Competent Person with readily available solutions to a frustrating problem

Distributors throughout the U.S. and Canada can match the proper equipment to make the crossing both efficient and productive.

The manufacturer's engineering and development produce a quality system with easily applied tabulated data.

This ensures that the problem is solved safely with minimal equipment & labor

EFFICIENCY PRODUCTION, INC.

America's Trench Box Builder™

Pre-Engineered Shoring Systems for Cross-Trench Utility Challenges

Presented by Mike Ross,
Shoring Specialist

AUTHOR INDEX